Mechanical Reliability: Theory, Models and Applications

B. S. Dhillon
Professor, Department of Mechanical Engineering
University of Ottawa, Ottawa, Canada

AIAA EDUCATION SERIES
J. S. Przemieniecki
Series Editor-in-Chief
Air Force Institute of Technology
Wright-Patterson Air Force Base, Ohio

Published by
American Institute of Aeronautics and Astronautics, Inc.
370 L'Enfant Promenade, SW, Washington, DC 20024

Text Published in the AIAA Education Series

Re-Entry Vehicle Dynamics
Frank J. Regan, 1984
Aerothermodynamics of Gas Turbine and Rocket Propulsion
Gordon C. Oates, 1984
Aerothermodynamics of Aircraft Engine Components
Gordon C. Oates, Editor, 1985
Fundamentals of Aircraft Combat Survivability Analysis and Design
Robert E. Ball, 1985
Intake Aerodynamics
J. Seddon and E. L. Goldsmith, 1985
Composite Materials for Aircraft Structures
Brian C. Hoskin and Alan A. Baker, Editors, 1986
Gasdynamics: Theory and Applications
George Emanuel, 1986
Aircraft Engine Design
Jack D. Mattingly, William Heiser, and Daniel H. Daley, 1987
An Introduction to The Mathematics and Methods of Astrodynamics
Richard H. Battin, 1987
Radar Electronic Warfare
August Golden Jr., 1988
Advanced Classical Thermodynamics
George Emanuel, 1988
Aerothermodynamics of Gas Turbine and Rocket Propulsion, Revised and Enlarged
Gordon C. Oates, 1988
Re-Entry Aerodynamics
Wilbur L. Hankey, 1988

American Institute of Aeronautics and Astronautics, Inc.
Washington, DC

Library of Congress Cataloging in Publication Data

Dhillon, B. S.
Mechanical reliability.

(AIAA education series)
Bibliography: p.
Includes index.
1. Reliability (Engineering) I. Title. II. Series.
TA169.D45 1988 620'.00452 88-3431
ISBN 0-930403-38-X

FOREWORD

In designing engineering systems, the consideration of their reliability has become of paramount importance. Not only is reliability important for safety but also for product competitiveness in the world markets. In the early days of the aviation and automobile industries, the reliability consideration received low priority as attention was focused mainly on improvements in performance, weight reduction, and lower manufacturing costs. But this has changed; reliability in engineering design has now become an essential component of any engineering design or manufacturing activity.

Thus, *Mechanical Reliability: Theory, Models and Applications* by B.S. Dhillon is a particularly welcome addition to the AIAA Education Series. In the past, most textbooks on engineering reliability concentrated heavily on the theoretical aspects — basically quantitative analytical methods and modeling, and did not directly address design concepts for improved reliability. Professor Dhillon's text, in addition to the necessary introductory material on the basic mathematical concepts for analysis and modeling, discusses at some length various approaches to engineering design for reliability. It is this feature of the book that will make it particularly useful in any engineering design activity, either for practicing engineers or for upper-level undergraduate or graduate-level courses in engineering design.

Dhillon is an authority on reliability and quality control in engineering, having published over 180 papers as well as several books on various aspects of engineering reliability and related areas. Some of his books have been translated into Russian, German, and Chinese. Dhillon is a recipient of prestigious reliability awards from the American Society for Quality Control and the Society of Reliability Engineers.

J.S. PRZEMIENIECKI
Editor-in-Chief
AIAA Education Series

TABLE OF CONTENTS

PREFACE

Nowadays greater emphasis than ever before is being placed on reliability during the planning, design, and operation of systems for use in defense, aerospace, and nuclear power generation. Over the years of the fairly recent past, many advances in mechanical reliability have resulted in a large amount of published literature – witness the appendix of this book – and concepts, models, and applications have been developed. This book covers these developments and recent related progress in the field

Some sections give background on general reliability. Throughout the book emphasis is made on the structure of concepts rather than on minute details and mathematical rigor. Topics are treated in such a manner that the reader needs no previous knowledge to understand the contents. The text cites over 1400 references, most listed in the Appendix, as well as some at the end of each chapter. In addition, the book contains several examples, along with corresponding solutions, to help in understanding its contents.

Twelve chapters and the appendix comprise *Mechanical Reliability: Theory, Models, and Applications*. Chapter 1 briefly discusses history, terms and definitions, and scope of the text. It lists related periodicals and conference proceedings.

Chapter 2 reviews basic reliability mathematics and concepts useful in understanding subsequent chapters.

Chapter 3 presents general expressions for reliability, hazard rate, failure density function, mean time to failure, and cumulative distribution function. It also covers general and time-dependent reliability models, mean time to failure of reliability networks, reliability analysis techniques, and failure rate evaluation models for mechanical items.

Chapter 4 is devoted to modes of mechanical failures, failure causes, casting and weld defects in pressure vessels and piping, and safety factors and margins.

The next two chapters, Chapters 5 and 6, deal with reliability evaluation of mechanical components and systems and mechanical-reliability modeling, respectively. Basically, Chapter 5 presents several interference theory models, and Chapter 6 covers optimization in structural reliability, modeling of an earthquate protection system and fire-related failures in buildings, and models related to mechanical reliability.

Chapter 7 is completely devoted to human errors in mechanical systems. Several different aspects of human errors are discussed.

Chapter 8 covers maintenance and models for mechanical-equipment replacement or maintenance.

Chapter 9 concerns failure data-collection and analysis. Several important aspects of failure data collection and analysis are addressed in this chapter

Chapter 10 treats transit-system reliability modeling, setting forth several mathematical models dealing with transit vehicles.

Chapter 11, in two major parts, covers automotive-vehicle reliability and robot safety and reliability.

Chapter 12 treats such topics as fatigue reliability, pipeline reliability, oil-tank, steam-turbine-blade, shaft and bearing failures, and reliability evaluation of electromechanical devices. The appendix lists over 1200 references on mechanical reliability and related areas.

This text should be useful to mechanical, aeronautical, structural, and reliability engineers, as well as students of aeronautical, mechanical, reliability, and structural engineering. Researchers, administrators, and others should find it a valuable reference source.

The author wishes to thank referees, colleagues, and editors for their invisible inputs. In particular, thanks to J.S. Przemieniecki, Series Editor-in-Chief, for his intererst in this project and suggestions. Thanks also to Jeanne Godette, AIAA Managing Editor for Books, for her invaluable assistance and interest in this text. I am grateful to S.N. Rayapati for preparing diagrams for the book. The author is indebted to his relatives and friends for their interest and encouragement at moments of need. Last but not least, I thank my wife Rosy for typing the original manuscript and for her help in proofreading.

B.S. DHILLON
University of Ottawa
Ottawa, Ontario, Canada

1
INTRODUCTION

This chapter briefly presents the historical background of mechanical reliability. A number of selected periodicals and conference proceedings publishing articles on mechanical reliability and related areas are listed. Selective terms and definitions concerned with mechanical reliability are given and the scope of the book is briefly described.

1.1 BRIEF BACKGROUND OF MECHANICAL RELIABILITY

Usually, World War II is regarded as the starting point of the reliability discipline. It was the Germans who made use of basic reliability concepts to improve the reliability of their rockets during that war. Therefore, it may be said that the beginning of the mechanical reliability field was also World War II because of the reliability problems with these rockets.

In 1951, Weibull of the Swedish Royal Institute of Technology published a statistical distribution for material strength and life length.[1] This distribution (Weibull distribution) has played an important role in the development of mechanical reliability, as well as of reliability in general. The contributions of Freudenthal in the 1950's and thereafter advanced structural reliability.[2–4]

The United States Armed Forces and the National Aeronautics and Space Administration (NASA) also played a pivotal role in the development of the mechanical reliability field. Basically, their efforts led to the in-depth emphasis on mechanical reliability in the 1960's. This serious effort was the result of several factors. For example, in the case of the space program, three incidents can be cited[5]: 1) the loss of Syncom I in space, in 1963, due to a bursting high-pressure gas tank; 2) the failure of Mariner III in 1964 (believed to be the result of a mechanical failure); and 3) the frequent failure of components such as valves, pyrotechnics, and regulators in the Gemini spacecraft systems.

During the 1960's, NASA initiated and completed several projects concerned with mechanical reliability. Some of these were designing specified reliabilities into mechanical components with time-dependent stress and strength distributions, making cumulative fatigue damage surveys, and investigating the reliability of structures and components subjected to random dynamic loading. Many people have contributed to the field of mechanical reliability, some of whom are listed in Refs. 1–10. A comprehensive list of publications on mechanical reliability and related areas is presented in the Appendix.

1.2 PERIODICALS AND CONFERENCE PROCEEDINGS

There are a number of journals and conference proceedings publishing articles on mechanical reliability and related areas. Some of those are given below.

Journals

Structual Safety

IEEE Transactions on Reliability (Institute of Electrical and Electronics Engineers)

Transactions of ASME, Journal of Vibration, Acoustics, Stress, and Reliability in Design (American Society of Mechanical Engineers)

Reliability Engineering

Journal of Spacecraft and Rockets (American Institute of Aeronautics and Astronautics)

Microelectronics and Reliability

Journal of the Structural Division (American Society of Civil Engineers)

Journal of the Engineering Mechanics Division (American Society of Civil Engineers)

Automotive Engineering — formerly *SAE Journal* (Society of Automotive Engineers)

Astronautics and Aeronautics (American Institute of Aeronautics and Astronautics)

Conference Proceedings

Proceedings of ASME Design Engineering Division's Conferences on Reliability

Proceedings of ASCE Speciality Conferences on Probabilistic Mechanics and Structural Reliability

Proceedings of North Atlantic Treaty Organization (NATO) Advanced Study Institute on Reliability Theory and Its Application in Structural and Soil Mechanics, Martinus Nijhoff Publishers, the Hague, the Netherlands, 1982.

Proceedings of the Japan Society of Mechanical Engineers International Symposium on Hazard-Free Operation against Potential Emergencies (HOPE), 1977.

Proceedings of Annual Reliability and Maintainability Symposium (published by various engineering societies).

1.3 TERMS AND DEFINITIONS

This section presents selected terms and definitions related to mechanical reliability.[10–12]

Availability: the probability that an item is available for operation at the moment of need.

Constant failure period: that period in which failures occur at an approximately uniform rate.

Cracks: linear ruptures of metal under stress.

Durability: the ability of an item to withstand use and exposure to the environment.

Duty cycle: a given operating time of an item, followed by a given time of shutdown.

Failure: the termination of the ability of a system, a subsystem, or a part to carry out its required mission.

Failure mechanism: the process that results in a failure.

Failure mode: the physical process or processes that occur or combine their effects to generate failure.

Failure rate: the number of failures of an item (i.e., a system, a subsystem, or a part) per unit measure of life. Examples of the units used to measure life are time, events, miles, and cycles.

Forced defect: the failure caused by stress testing.

Mean time to failure (exponential distribution): the sum of the operating time of a number of items over the total number of failures.

Mechanical failure: any change in the shape, size, or material properties of a structure, piece of equipment, or equipment part that renders it unfit of carrying out its specified function adequately.

Mission time: the time during which the item (i.e., the system, the subsystem, or the part) is carrying out its assigned mission.

Misuse failure: the failure attributable to the application of stresses beyond the specified designed capacities of the item (i.e., the system, the subsystem, or the part).

Model: the mathematical representation of a process in question.

Random failure: any failure whose occurrence cannot be predicted.

Reliability: the probability that an item (a system, a subsystem, or a component) will carry out its assigned mission under specified conditions for a stated time period.

Steady-state condition (statistical): that condition where the probability of being in a specific state is independent of time.

Strength: the ability of an item to resist failure due to external environment and loading.

Stress: any factor (i.e., temperature, environment, electric current) that tends to produce a failure.

Useful life: the length of time an item functions with an acceptable level of failure rate.

1.4 SCOPE OF THE TEXT

Even though its history can be traced back to World War II, the field of mechanical reliability is still considered to be relatively new when compared with electronic reliability. This might be due to the fact that the degree of advances made in electronic reliability outweighs the advances in mechanical reliability. The recent well-publicized failures such as Space Shuttle Challenger and the Chernobyl (USSR) and Three Mile Island (USA) nuclear accidents have forced greater emphasis on mechanical reliability.

Mechanical reliability has certain concepts, models, and applications of its own. This book emphasizes these specifics as well as recent progress made in the mechanical reliability field. Engineers and others needing information on such specifics generally encounter a great deal of difficulty in finding it. There are only a small number of books concerned with mechanical reliability and none covers such topics as failure data banks for mechanical equipment, robot reliability, transit system reliability, mechanical equipment replacement, and human errors. This book is an attempt to fulfill this vital need. The topics covered in the book are of current interest. Previous knowledge is not necessary to digest its contents, since two chapters on basic reliability principles and the necessary mathematics are provided to give sufficient background to any reader. Many solved examples are included which makes the text more self-explanatory. This book should be useful to reliability, mechanical, aeronautical, and structural engineers, as well as to students of reliability, aeronautical, mechanical, and structural engineering. In addition, researchers, administrators, and others will find it a valuable reference source.

References

[1]Weibull, W., "A Statistical Distribution Function of Wide Applicability," *Journal of Applied Mechanics*, Vol. 18, 1951, pp. 293–297.

[2]Freudenthal, A. M. and Gumbel, E. J., "Failure and Survival in Fatigue," *Journal of Applied Physics*, Vol. 25, 1954, pp. 110–120.

[3]Freudenthal, A. M., "Safety and the Probability of Structural Failure," *Transactions of American Society of Civil Engineers*, Vol. 121, 1956, pp. 1337–1397.

[4]Freudenthal, A. M., *Fatigue in Aircraft Structures*, Academic Press, New York, 1957, p. 308.

[5]Redler, W. M., "Mechanical Reliability Research in the National Aeronautics and Space Administration," *Proceedings of the Reliability and Maintainability Conference*, IEEE, New York, 1966, pp. 763–768.

[6]Kececioglu, D., "Reliability Analysis of Mechanical Components and Systems," *Nuclear Engineering and Design*, Vol. 19, 1972, pp. 259–290.

[7]Kececioglu, D. and Cormier, D., "Designing a Specified Reliability Directly Into a Component," *Proceedings of the Third Annual Aerospace Reliability and Maintainability Conference*, SAE, Warrendale, PA, 1964, pp. 546–565.

[8]Lipson, C., Sheth, N. J., and Disney, R. L., "Reliability Prediction-Mechanical Stress/strength Interference," Rome Air Development Center, Griffiss Air Force Base, NY, Rept. RADC-TR-66-710, 1967.

[9]Kececioglu, D. and Haugen, E. B., "A Unified Look at Design Safety Factors, Safety Margins and Measures of Reliability, *Proceedings of the Annual Reliability and Maintainability Conference*, ASME, New York, 1968, pp. 522–530.

[10]Sadlon, R. J., "Analysis Techniques for Mechanical Reliability," Reliability Analysis Center, Rome Air Development Center, Griffiss Air Force Base, NY, Rept. NPS-1, 1985, pp. 159–166.

[11]Naresky, J. J., "Reliability Definitions," *IEEE Transactions on Reliability*, Vol. 19, 1970, pp. 198–200.

[12]*Maintainability Engineering Handbook*, Direction of Commander, U.S. Naval Ordnance Systems Command, Rept. NAVORD OD 39223, 1970, pp. G1–G9 (available from Superintendent of Documents, U.S. Government Printing Office, Washington, DC 20402).

Exercises

1.1 Discuss the history of reliability.

1.2 Define the following terms: item, failure mode, redundancy, burn-in.

2
THE MATHEMATICS OF RELIABILITY

This chapter presents the mathematics related to mechanical reliability. Several selected definitions concerned with differentiation, Laplace transform, probability, probability density function, mean time to failure, variance, mean time to repair, etc., are presented. Six probability distributions for use in mechanical reliability analyses are discussed: exponential, gamma, Erlangian, Weibull, lognormal, and normal. Their probability density and cumulative distribution functions are given. The reliability, mean, hazard rate, and variance expressions are given in a table for Weibull, normal, lognormal, gamma, and exponential distributions. The Markov technique is described. A Markov model representing a repairable mechanical equipment is presented.

2.1 INTRODUCTION

In general, mathematics has played an instrumental role in the development of mechanical reliability studies. In particular, probability and statistics are the backbone of mechanical reliability.

The history of probability starts in the 17th century. Men like Chevalier de Mere, Blaise Pascal (1623–1662), Pierre Fermat (1601–1665), Karl Gauss (1777–1855), and Pierre Laplace (1749–1827) have played important roles in its development. Probability concepts have been applied to reliability-related engineering problems since the early 1930's.

In subsequent sections, this chapter presents some mathematical concepts considered useful for mechanical reliability studies. Because it is highly mathematical and is applicable in mechanical reliability work, the Markov technique is also covered in this chapter. One Markov model representing mechanical equipment is presented.

2.2 SELECTED DEFINITIONS

This section presents a number of definitions.

Definition I

This definition is concerned with differentiation. If $z = f(t)$, then

$$\frac{dz}{dt} = \lim_{\Delta t \to 0} \frac{f(t + \Delta t) - f(t)}{\Delta t} \tag{2.1}$$

where Δt is the finite interval and $f(t)$ a function of t.

Two of the rules of differentiation are

$$\frac{d(\lambda t^{\beta})}{dt} = \lambda \beta t^{\beta - 1} \tag{2.2}$$

$$\frac{d(e^{\lambda t})}{dt} = e^{\lambda t} \frac{d(\lambda t)}{dt} = \lambda e^{\lambda t} \tag{2.3}$$

where λ is a constant, β a positive integer, and t a variable.

Example 2.1

Assume that the failure probability of a hydraulic control system is given by

$$F(t) = 1 - e^{-\lambda t} \tag{2.4}$$

where t is a time variable, $F(t)$ the failure probability of the hydraulic control system at time t, and λ the constant failure rate of the hydraulic control system.

Obtain an expression for the failure density function of the control system by differentiating Eq. (2.4) with respect to time t.

Differentiating Eq. (2.4) with respect to time t, results in

$$\frac{dF(t)}{dt} = f(t) = -e^{-\lambda t} \frac{d(-\lambda t)}{dt} = \lambda e^{-\lambda t} \tag{2.5}$$

where $f(t)$ is the failure density function of the hydraulic control system.

Definition II

This definition is concerned with the Laplace transform. The Laplace transform of a time function is defined as

$$\mathscr{L}[f(t)] = F(s) = \int_0^{\infty} e^{-st} f(t)\, dt \tag{2.6}$$

where t is a time variable, $f(t)$ the time function whose Laplace transform has to be found, s the Laplace transform variable, $\mathscr{L}$ the Laplace transform operator, and $F(s)$ the Laplace transform of $f(t)$.

Using Eq. (2.6), one can obtain the Laplace transform of a given time function. The Laplace transform of $\mathrm{d}f(t)/\mathrm{d}t$ is

$$\mathscr{L}\left[\frac{\mathrm{d}f(t)}{\mathrm{d}t}\right] = sF(s) - f(0) \tag{2.7}$$

Similarly, we get

$$\lim_{t\to\infty} f(t) = \lim_{s\to 0} sF(s) \tag{2.8}$$

Example 2.2

Prove Eq. (2.8).
From Eqs. (2.6) and (2.7), we get[1]

$$\mathscr{L}\left[\frac{\mathrm{d}f(t)}{\mathrm{d}t}\right] = \int_0^\infty e^{-st}\frac{\mathrm{d}f(t)}{\mathrm{d}t}\cdot \mathrm{d}t = sF(s) - f(0) \tag{2.9}$$

The limit of

$$\left[\int_0^\infty e^{-st}\frac{\mathrm{d}f(t)}{\mathrm{d}t}\cdot \mathrm{d}t\right]$$

as $s \to 0$ is

$$\begin{aligned}\lim_{s\to 0}\int_0^\infty e^{-st}\frac{\mathrm{d}f(t)}{\mathrm{d}t}\,\mathrm{d}t &= \int_0^\infty \frac{\mathrm{d}f(t)}{\mathrm{d}t}\,\mathrm{d}t = \lim_{Q\to\infty}\int_0^Q \frac{\mathrm{d}f(t)}{\mathrm{d}t}\,\mathrm{d}t \\ &= \lim_{Q\to\infty}\,[f(Q) - f(0)] = \lim_{t\to\infty} f(t) - f(0)\end{aligned} \tag{2.10}$$

Similarly, the limit of $[sF(s) - f(0)]$ as $s \to 0$ is

$$\lim_{s\to 0} sF(s) - f(0) \tag{2.11}$$

Thus, from Eqs. (2.10) and (2.11), we get

$$\lim_{t\to\infty} f(t) - f(0) = \lim_{s\to 0} sF(s) - f(0) \tag{2.12}$$

Equation (2.12) results in

$$\lim_{t\to\infty} f(t) = \lim_{s\to 0} sF(s) \tag{2.13}$$

One should note here that Eqs. (2.8) and (2.13) are exactly the same.

Example 2.3

Assume that the reliability of a fluid flow valve is given by

$$R(t) = f(t) = e^{-(\lambda_0 + \lambda_c)t} \tag{2.14}$$

where $R(t)$ is the reliability of the fluid flow valve, t the time, λ_0 the constant open-mode failure rate of the valve, and λ_c the constant closed-mode failure rate of the valve.

Obtain the final value of Eq. (2.14) using the both sides of Eq. (2.13).

Using Eq. (2.14) in the left-hand side of Eq. (2.13) yields

$$\lim_{t \to \infty} e^{-(\lambda_0 + \lambda_c)t} = 0 \tag{2.15}$$

Finding the Laplace transform of Eq. (2.14) by using Eq. (2.6) yields

$$\begin{aligned} F(s) &= \int_0^\infty e^{-st} \cdot e^{-(\lambda_0 + \lambda_c)t} \, dt \\ &= \int_0^\infty e^{-(s + \lambda_0 + \lambda_c)t} \, dt \\ &= \left[-\frac{e^{-(s + \lambda_0 + \lambda_c)t}}{(s + \lambda_0 + \lambda_c)} \right]_0^\infty \\ &= \frac{1}{s + \lambda_0 + \lambda_c} \end{aligned} \tag{2.16a}$$

Substituting the above result into the right-hand side of Eq. (2.13) results in

$$\lim_{s \to 0} \frac{s}{s + \lambda_0 + \lambda_c} = 0 \tag{2.16b}$$

Thus, the end results of Eqs. (2.15) and (2.16b) are the same.

Definition III

This is concerned with probability. The probability of the occurrence of event, say Y, is defined by

$$P(Y) = \lim_{m \to \infty} (N/m) \tag{2.17}$$

where N is the number of times event Y occurs and m the number of times an experiment is repeated.

Normally $P(Y)$ is simply approximated by (N/m) because it is practically impossible to perform an infinite number of repeated experiments.

Definition IV

This definition deals with the cumulative distribution function. For our purposes, the cumulative distribution function of a continuous random variable is defined as

$$F(t) = \int_0^t f(x)\,\mathrm{d}x \tag{2.18}$$

where t is the time, $f(t)$ the failure (probability) density function, and $F(t)$ the cumulative distribution function.

At $t = 0$ and ∞, we get

$$F(0) = 0 \tag{2.19}$$

and

$$F(\infty) = 1 \tag{2.20}$$

Example 2.4

Assume that the failure times of a military tank are represented by the following probability density function:

$$f(t) = \frac{2t}{\theta}\exp\left(-\frac{t^2}{\theta}\right) \tag{2.21}$$

where t is the time and θ the scale parameter.

Prove that the total area under the probability density function curve is equal to unity as given by Eq. (2.20).

In this case, in order to find the total area under the curve, let $t = \infty$ in Eq. (2.18). Thus, from Eqs. (2.18) and (2.21), we get

$$F(\infty) = \int_0^\infty \frac{2t}{\theta}\exp\left[-\frac{t^2}{\theta}\right]\mathrm{d}t = \frac{2}{\theta}\left[\frac{\theta}{2}\Gamma(1)\right]$$

$$= \frac{2}{\theta}\left[\frac{\theta}{2}(0!)\right] = 1$$

Definition V

This definition describes the reliability function. The reliability function is given by

$$R(t) = 1 - F(t) = 1 - \int_0^t f(x)\,\mathrm{d}x \tag{2.22}$$

where $R(t)$ is the reliability function.

The above equation can be used to obtain a reliability function from a given failure density function $f(t)$.

Example 2.5

Using the given failure density function [Eq. (2.21)] of Example 2.4, obtain an expression for the reliability function.

Substituting Eq. (2.21) into Eq. (2.22) results in

$$R(t) = 1 - \int_0^t \frac{2x}{\theta} \exp\left(-\frac{x^2}{\theta}\right) dx = \exp\left(-\frac{t^2}{\theta}\right) \tag{2.23}$$

Definition VI

This definition deals with the hazard rate. The hazard rate $\lambda(t)$ is defined as

$$\begin{aligned}\lambda(t) &= \frac{f(t)}{R(t)} = \frac{f(t)}{1 - F(t)} \\ &= \frac{-d\ln[1 - F(t)]}{dt} = \frac{-d\ln R(t)}{dt}\end{aligned} \tag{2.24}$$

Example 2.6

Using the resulting Eq. (2.23) of Example 2.5, obtain an expression for the hazard rate.

Substituting Eq. (2.23) into Eq. (2.24), we get

$$\begin{aligned}\lambda(t) &= \frac{-d\ln[\exp(-t^2/\theta)]}{dt} \\ &= -\left[\frac{1}{\exp(-t^2/\theta)}\right] \cdot \left[\exp\left(-\frac{t^2}{\theta}\right)\right]\left(-\frac{2t}{\theta}\right) \\ &= \frac{2t}{\theta}\end{aligned} \tag{2.25}$$

Definition VII

Definition of the failure (probability) density function is given by

$$f(t) = \frac{dF(t)}{dt} = -\frac{dR(t)}{dt} \tag{2.26}$$

Example 2.7

Using Eqs. (2.23) and (2.26), prove that the failure density function is

$$f(t) = \frac{2t}{\theta} \exp\left(-\frac{t^2}{\theta}\right) \tag{2.27}$$

The above equation is the same as Eq. (2.21).

Inserting Eq. (2.23) into Eq. (2.26) yields

$$\begin{aligned} f(t) &= -\frac{\mathrm{d}[\exp(-t^2/\theta)]}{\mathrm{d}t} \\ &= -\left[\exp\left(-\frac{t^2}{\theta}\right)\right]\left(-\frac{2t}{\theta}\right) \\ &= \frac{2t}{\theta} \exp\left(-\frac{t^2}{\theta}\right) \end{aligned} \tag{2.28}$$

Equations (2.27) and (2.28) are identical.

Definition VIII

This definition deals with the mean value (mean time to failure) of a failure distribution. The mean value or mean time to failure (MTTF) is defined by

$$\mathrm{MTTF} = \int_0^\infty t f(t)\,\mathrm{d}t = \int_0^\infty R(t)\,\mathrm{d}t \tag{2.29}$$

where t is the time, $f(t)$ the failure density function, and $R(t)$ the reliability function.

Alternatively, the mean time to failure is expressed as

$$\mathrm{MTTF} = \lim_{s \to 0} R(s) \tag{2.30}$$

where s is the Laplace transform variable and $R(s)$ the Laplace transform of the reliability function.

Example 2.8

With the aid of Eq. (2.29), prove that the mean time to failure is given by Eq. (2.30). Rewrite Eq. (2.29) as

$$\mathrm{MTTF} = \lim_{t \to \infty} \int_0^t R(x)\,\mathrm{d}x \tag{2.31}$$

The Laplace transform of $[\int_0^t R(x)\,dx]$ is

$$\mathscr{L}\left[\int_0^t R(x)\,dx\right] = \frac{R(s)}{s} \tag{2.32}$$

Using Eq. (2.13), Eq. (2.31) can be written as

$$\text{MTTF} = \lim_{s \to 0} s\,\mathscr{L}\left[\int_0^t R(x)\,dx\right] \tag{2.33}$$

Substituting Eq. (2.32) into Eq. (2.33) results in

$$\text{MTTF} = \lim_{s \to 0}\left[\frac{sR(s)}{s}\right] = \lim_{s \to 0} R(s) \tag{2.34}$$

This result is the same as given by Eq. (2.30).

Example 2.9

The reliability of a pump seal is given by

$$R(t) = e^{-\lambda t} \tag{2.35}$$

where t is the time and λ the constant failure rate of the pump seal.

Using Eqs. (2.29) and (2.30), calculate the mean time to failure of the seal if $\lambda = 0.0005$ failures/h.

Using Eqs. (2.29) and (2.35), we get

$$\text{MTTF} = \int_0^\infty e^{-\lambda t}\,dt = \left[-\frac{e^{-\lambda t}}{\lambda}\right]_0^\infty = \frac{1}{\lambda} \tag{2.36}$$

Finding the Laplace transform of Eq. (2.35) yields

$$R(s) = 1/(s + \lambda) \tag{2.37}$$

Substituting for $R(s)$ in Eq. (2.30) leads to

$$\text{MTTF} = \lim_{s \to 0} \frac{1}{s + \lambda} = \frac{1}{\lambda} \tag{2.38}$$

We note here that the mean time to failure solutions obtained through two different means are the same. Thus, by substituting the given data into Eq. (2.36) or (2.38) results in

$$\text{MTTF} = \frac{1}{0.0005} = 2000 \text{ h}$$

The mean time to failure of the pump seal is 2000 h.

Definition IX

This definition is concerned with variance. For our purposes, we describe variance as a measure of the dispersion of values about the mean (mean time to failure). For continuous distributions, the variance is defined as

$$\sigma^2 = \int_0^\infty (t - \text{MTTF})^2 f(t)\, \mathrm{d}t$$
$$= \int_0^\infty t^2 f(t)\, \mathrm{d}t - (\text{MTTF})^2 \tag{2.39}$$

where σ^2 is the variance, t the time, $f(t)$ the failure density function, and MTTF the mean time to failure.

Alternatively, the variance is given by

$$\sigma^2 = -2 \lim_{s \to 0} R'(s) - \left[\lim_{s \to 0} R(s) \right]^2 \tag{2.40}$$

where $R(s)$ is the Laplace transform of the reliability function and s the Laplace transform variable. The prime on $R(s)$ denotes differentiation with respect to s.

Example 2.10

Find the variance of the time to failure associated with Eq. (2.35) given in Example 2.9 by using Eq. (2.40).

The Laplace transform of Eq. (2.35) is obtained from Eq. (2.37) as follows:

$$R(s) = 1/(s + \lambda) \tag{2.41}$$

Differentiating the above equation with respect to s leads to

$$\frac{\mathrm{d}R(s)}{\mathrm{d}s} = -\frac{1}{(s + \lambda)^2} \tag{2.42}$$

Substitution of Eqs. (2.41) and (2.42) into Eq. (2.40) yields

$$\sigma^2 = -2 \lim_{s \to 0} \left[\frac{-1}{(s + \lambda)^2} \right] - \left[\lim_{s \to 0} \left(\frac{1}{s + \lambda} \right) \right]^2$$
$$= \frac{2}{\lambda^2} - \frac{1}{\lambda^2} = \frac{1}{\lambda^2} \tag{2.43}$$

Definition X

This definition is concerned with mean time to repair (MTTR), which is defined by

$$\text{MTTR} = \int_0^{\infty} [1 - F_r(t)]\, \mathrm{d}t \tag{2.44}$$

where $F_r(t)$ is the cumulative distribution function of the repair time.

The function $F_r(t)$ is also known as the maintainability function and is given by

$$F_r(t) = \int_0^{t} f_r(x)\, \mathrm{d}x \tag{2.45}$$

where t is the time and $f_r(t)$ the probability density function of the repair time (of an item).

Example 2.11

The repair times of a fan are described by an exponential distribution as

$$f_r(t) = \mu e^{-\mu t} \tag{2.46}$$

where $f_r(t)$ is the probability density function of the fan repair time, t the time, and μ the distribution parameter.

Obtain an expression for mean time to repair of the fan. Substituting Eq. (2.46) in Eq. (2.45) yields

$$F_r(t) = \int_0^{t} \mu e^{-\mu x}\, \mathrm{d}x = 1 - e^{-\mu t} \tag{2.47}$$

Using the above result in Eq. (2.44), we obtain

$$\text{MTTR} = \int_0^{\infty} 1 - [1 - e^{-\mu t}]\, \mathrm{d}t = \frac{1}{\mu} \tag{2.48}$$

The mean time to repair of the fan is given by $(1/\mu)$.

Definition XI

This definition deals with the (time-dependent) repair rate, which is expressed as

$$\mu_r(t) = \frac{\mathrm{d}F_r(t)/\mathrm{d}t}{1 - F_r(t)} \tag{2.49}$$

where $\mu_r(t)$ is the time-dependent repair rate.

Example 2.12

For the fan of Example 2.11, obtain an expression for repair rate. Substitution of Eq. (2.47) into Eq. (2.49) yields

$$\mu_r(t) = \frac{d(1 - e^{-\mu t})/dt}{e^{-\mu t}}$$

$$= \mu e^{-\mu t}/e^{-\mu t}$$

$$= \mu \tag{2.50}$$

The repair rate of the fan is constant and is given by μ.

2.3 PROBABILITY DISTRIBUTIONS FOR USE IN MECHANICAL RELIABILITY

This section presents selected probability distributions useful for performing mechanical reliability analysis.[2,3] Table 2.1 presents the probability density, reliability, and hazard rate functions, and the mean time to failure and variance formulas associated with five different distributions. The failure (probability) density and cumulative distribution functions of six distributions are presented below.

Exponential Distribution

This is probably the most commonly used distribution in performing reliability analyses. Failure times of some mechanical items may be described by this distribution. In 1952, Davis[4] discussed the application of exponential distribution to life studies. The probability density and cumulative distribution functions are

$$f(t) = (1/\theta)e^{-t/\theta}, \qquad 1/\theta > 0 \text{ and } t \geqslant 0 \tag{2.51}$$

and

$$F(t) = 1 - e^{-t/\theta} \tag{2.52}$$

where $f(t)$ is the failure density function, t the time, $F(t)$ the cumulative distribution function, and θ the scale parameter (mean time to failure).

For $\theta = 1/\lambda$, the above functions become

$$f(t) = \lambda \exp(-\lambda t), \qquad \lambda > 0 \text{ and } t \geqslant 0 \tag{2.53}$$

and

$$F(t) = 1 - \exp(-\lambda t) \tag{2.54}$$

The parameter λ is known as the constant failure rate.

Table 2.1 Selected Probability Distributions

Distribution	Failure (probability) density function $f(t)$	Reliability function $R(t)$	Mean (mean time to failure)	Hazard rate $\lambda(t)$	Variance σ^2
Exponential	$\lambda \exp(-\lambda t),\ \lambda > 0,\ t \geqslant 0$	$\exp(-\lambda t)$	$1/\lambda$	λ	$1/\lambda^2$
Gamma	$\dfrac{\lambda(\lambda t)^{m-1}\exp(-\lambda t)}{\Gamma(m)}$ $\lambda > 0,\ m > 0,\ t \geqslant 0$	$1 - \dfrac{\Gamma(m,\lambda t)}{\Gamma(m)}$	m/λ	$\dfrac{1}{\lambda(t)} = \int_0^\infty \left(1 + \dfrac{x}{t}\right)^{m-1} e^{-\lambda x}\,dx$	m/λ^2
Lognormal	$\dfrac{(2\pi)^{-1/2}}{t}\exp\left[-\dfrac{1}{2}\left(\ln t/\theta - \dfrac{m}{\theta}\right)^2\right]$ θ and m are constants with dimensions of time.	$1 - \int_0^t f(x)\,dx$	$\theta \exp\left(\dfrac{1}{2} + \dfrac{m}{\theta}\right)$	$\dfrac{f(t)}{R(t)}$	$4.670775\theta^2\exp(2m/\theta)$
Truncated normal	$\dfrac{c(2\pi)^{-1/2}}{\sigma}\exp\left[-\dfrac{(t-m)^2}{2\sigma^2}\right]$ $t \geqslant 0,\ \sigma > 0,\ m \geqslant 0$ and $c = 2/\left[1 + \mathrm{erf}\left(\dfrac{m}{\sqrt{2}\sigma}\right)\right]$ erf(.) is the error function	$1 - \int_0^t f(x)\,dx$	$\dfrac{\sigma c}{\sqrt{2\pi}}\exp\left(-\dfrac{m^2}{2\sigma^2}\right) + m\dfrac{f(t)}{R(t)}$		$\int_0^\infty (t - \mathrm{MTTF})^2 f(t)\,dt$ MTTF is the mean time to failure
Weibull	$\dfrac{b}{\theta}\left(\dfrac{t}{\theta}\right)^{b-1}\exp\left[-\left(\dfrac{t}{\theta}\right)^b\right]$ $b, \theta > 0,\ t \geqslant 0$	$\exp\left[-\left(\dfrac{t}{\theta}\right)^b\right]$	$\theta\Gamma\left(\dfrac{1}{b} + 1\right)$	$\dfrac{b}{\theta}\left(\dfrac{t}{\theta}\right)^{b-1}$	$\theta^2\left\{\Gamma\left(1 + \dfrac{2}{b}\right) - \left[\Gamma\left(1 + \dfrac{1}{b}\right)\right]^2\right\}$

Example 2.13

The failure times of a pump are exponentially distributed with a mean value of 400 h. Calculate the failure probability of the pump for an 8 h mission.

Using the given data in Eq. (2.52), we obtain

$$F(8) = 1 - \exp(-8/400) = 0.0198$$

The failure probability of the pump is 0.0198.

Gamma Distribution

This is a two-parameter distribution and was considered, in 1961, as a model in life test problems.[5] The failure density and cumulative distribution functions are

$$f(t) = \frac{\lambda(\lambda t)^{m-1}}{\Gamma(m)} \exp(-\lambda t), \qquad \lambda > 0,\ m > 0,\ t \geqslant 0 \tag{2.55}$$

and

$$F(t) = 1 - \frac{\Gamma(m,\lambda t)}{\Gamma(m)} \tag{2.56}$$

where m is the shape parameter, t the time, $\Gamma(m)$ the gamma function, $\Gamma(m,\lambda t)$ the incomplete gamma function, and $\lambda = 1/\theta$ (θ is the scale parameter).

The following are the special cases of the gamma distribution:

1) For integer values of m, the gamma distribution becomes the special Erlangian distribution. It means $\Gamma(m) = (m - 1)!$

2) For $m = 1$, the gamma distribution becomes the exponential distribution.

3) For $\lambda = 0.5$ and $m = 0.5n$, the gamma distribution becomes a chi-square distribution, where n is an integer (the number of degrees of freedom).

Erlangian Distribution

This distribution is named after A. K. Erlang, a Danish telephone engineer, who applied it to study congestion in telephone systems. The distribution is a special case of the gamma distribution. The failure density and cumulative distribution functions are given by

$$f(t) = \frac{\lambda(\lambda t)^{m-1} \exp(-\lambda t)}{(m-1)!}, \qquad \lambda > 0,\ m \geqslant 1,\ t \geqslant 0 \tag{2.57}$$

and

$$F(t) = 1 - \exp(-\lambda t) \sum_{i=0}^{m-1} \frac{(\lambda t)^i}{i!} \tag{2.58}$$

where m is an integer (shape parameter). The other symbols are the same as for the gamma distribution.

Weibull Distribution

This distribution is named after its originator[6] and is used often in mechanical and general reliability studies.[7] The failure density and cumulative distribution functions are

$$f(t) = \frac{b}{\theta}\left(\frac{t}{\theta}\right)^{b-1} \exp\left[-\left(\frac{t}{\theta}\right)^{b}\right], \qquad b, \theta > 0,\ t \geqslant 0 \tag{2.59}$$

and

$$F(t) = 1 - \exp[-(t/\theta)^b] \tag{2.60}$$

where b is the shape parameter and θ the scale parameter.

The two particular and special cases of the Weibull distribution are as follows:

1) For $b = 1$, the Weibull distribution becomes the exponential distribution. Thus, from Eqs. (2.59) and (2.60), we get

$$f(t) = \frac{1}{\theta} \exp\left(-\frac{t}{\theta}\right) \tag{2.61}$$

and

$$F(t) = 1 - \exp\left(-\frac{t}{\theta}\right) \tag{2.62}$$

These two equations are the same as Eqs. (2.51) and (2.52).

2) For $b = 2$, the Weibull distribution becomes the Rayleigh distribution. Thus, setting $b = 2$ in Eqs. (2.59) and (2.60) results in the following Rayleigh probability density and cumulative distribution functions:

$$f(t) = \frac{2t}{\theta^2} \exp\left[-\left(\frac{t}{\theta}\right)^2\right] \tag{2.63}$$

and

$$F(t) = 1 - \exp\left[-\left(\frac{t}{\theta}\right)^2\right] \tag{2.64}$$

Example 2.14

Assume that the failure times of a ball bearing are Weibull distributed with shape and scale parameter values of 4 and 100 h, respectively. Calculate the failure probability of the ball bearing for a 50 h mission.

Substituting the specified data into Eq. (2.60), we obtain

$$F(50) = 1 - \exp\left[-\left(\frac{50}{100}\right)^4\right] = 0.0606$$

The failure probability of the ball bearing is 0.0606.

Lognormal Distribution

This is a two-parameter distribution and is quite useful to represent the distribution of equipment repair times. The probability density[8] and cumulative distribution functions are

$$f(t) = \frac{(2\pi)^{-\frac{1}{2}}}{t\theta} \exp\left[-\frac{(\ell n t - m)^2}{2\theta^2}\right], \qquad t \geqslant 0 \tag{2.65}$$

and

$$F(t) = \frac{(2\pi)^{-\frac{1}{2}}}{\theta} \int_0^t \frac{1}{x} \exp\left[-\frac{(\ell n x - m)^2}{2\theta^2}\right] dx \tag{2.66}$$

where m and θ are the parameters.

Letting $y = (\ell n x - m)/\theta$, we get

$$\frac{dy}{dx} = \frac{1}{x\theta} \tag{2.67}$$

$$\therefore \quad dy = \frac{dx}{\theta x} \tag{2.68}$$

Using the above results in Eq. (2.66) yields

$$F(t) = (2\pi)^{-\frac{1}{2}} \int_{-\infty}^{(\ell n t - m)/\theta} \exp\left(-\frac{y^2}{2}\right) dy \tag{2.69}$$

We should note here that this distribution implies that the logarithms of the lifetimes are distributed normally.

Normal Distribution

This is a widely used statistical distribution also known as the Gaussian distribution. The probability density and cumulative distribution functions are defined as

$$f(t) = \frac{(2\pi)^{-\frac{1}{2}}}{\sigma} \exp\left[-\frac{(t-m)^2}{2\sigma^2}\right], \qquad -\infty < t < \infty \tag{2.70}$$

and

$$F(t) = \frac{(2\pi)^{-\frac{1}{2}}}{\sigma} \int_{-\infty}^{t} \exp\left[-\frac{(y-m)^2}{2\sigma^2}\right] dy \tag{2.71}$$

where m is the mean and σ the standard deviation.

Letting $W = (y - m)/\sigma$, we obtain

$$\frac{dW}{dy} = \frac{1}{\sigma} \tag{2.72}$$

$$\therefore \quad dy = \sigma\, dW \tag{2.73}$$

Thus, we can rewrite Eq. (2.71) as

$$F(t) = (2\pi)^{-\frac{1}{2}} \int_{-\infty}^{(t-m)/\sigma} \exp\left(-\frac{W^2}{2}\right) dW \tag{2.74}$$

We should note here that m and σ are also the parameters of the distribution in question. If $m/\sigma \geqslant 3$, then $f(t)$ of this distribution function is very close to the density function of the truncated normal distribution given in Table 2.1. Because life must be positive, the fraction of the distribution below zero must be small for a satisfactory approximation. For practical purposes, it may be a satisfactory failure model if m is at least 2 or 3 times greater than σ.

2.4 MARKOV TECHNIQUE

This method is very powerful and is quite useful in performing reliability and availability analyses in situations where the failure and repair rates are constant. It can probably handle more reliability problems than any other technique. However, the method has some shortcomings, one of which is solving a set of differential equations, particularly in the case of large, complex systems. Under such conditions, finding analytical solutions can be a very difficult task. The following assumptions are associated with this method in developing equations for reliability systems:

1) The probability of a transition occurring in a small time interval Δt from system state i to state j is $\theta\, \Delta t$, where θ is a constant. This constant takes the dimensions of the occurrences per unit time, e.g., failures per unit time or repairs per unit time.

2) All occurrences are independent.

3) The transition probability of more than one transition occurrences in a small time Δt is slight and is neglected.

The application of this approach is demonstrated in the following description of a Markov model used for a repairable piece of mechanical equipment. The model represents a system having two states: the normal operating

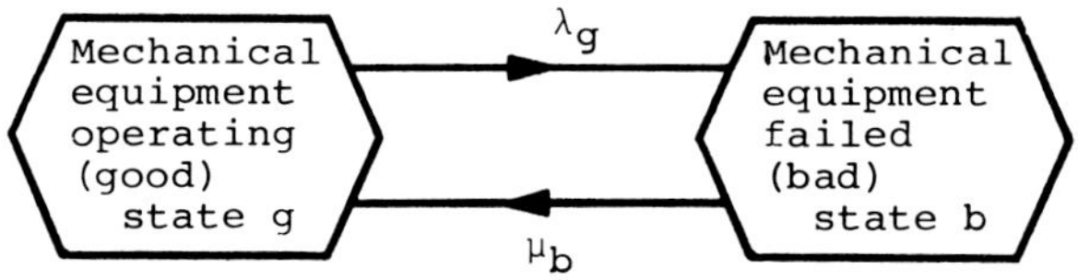

Fig. 2.1 Mechanical system transition diagram.

state and the failed state. More specifically, the mechanical equipment can be in either an operating (up) or a failed (down) state. Examples of this type of equipment are a diesel generator, pulverizer, or pump. If the failure and repair rates of such a piece of equipment are known, one can obtain values for equipment availability and unavailability. The system transition diagram is shown in Fig. 2.1.

In this model, it is assumed that failure and repair times are exponentially distributed and that the repaired equipment is as good as new. The following symbols are associated with this mathematical model:

- i = state of the mechanical equipment; $i = g$ (operating normally), $i = b$ (failed)
- $P_b(t + \Delta t)$ = probability of the mechanical equipment being in bad (failed) state b at time $t + \Delta t$
- $P_g(t + \Delta t)$ = probability of the mechanical equipment being in good state g at time $t + \Delta t$
- $P_i(t)$ = probability that the mechanical equipment is in state i at time t; $i = g,b$
- s = Laplace transform variable
- t = time
- Δt = small time interval
- λ_g = constant failure rate of the mechanical equipment
- $\lambda_g \Delta t$ = probability of mechanical equipment failure in time interval Δt
- μ_b = constant repair rate of the mechanical equipment
- $\mu_b \Delta t$ = probability of mechanical equipment repaired in time interval Δt
- $1 - \lambda_g \Delta t$ = probability of no failure in time interval Δt
- $1 - \mu_b \Delta t$ = probability of no repair in time interval Δt

Using the above assumptions and symbols, we can write the following equations:

$$P_g(t + \Delta t) = P_g(t)(1 - \lambda_g \Delta t) + P_b(t)\mu_b \Delta t \tag{2.75}$$

$$P_b(t + \Delta t) = P_b(t)(1 - \mu_b \Delta t) + P_g(t)\lambda_g \Delta t \tag{2.76}$$

In the limiting case, Eqs. (2.75) and (2.76) become

$$\lim_{\Delta t \to 0} \frac{P_g(t + \Delta t) - P_g(t)}{\Delta t} = \frac{dP_g(t)}{dt} = -\lambda_g P_g(t) + P_b(t)\mu_b \tag{2.77}$$

and

$$\lim_{\Delta t \to 0} \frac{P_b(t + \Delta t) - P_b(t)}{\Delta t} = \frac{dP_b(t)}{dt} = -\mu_b P_b(t) + P_g(t)\lambda_g \tag{2.78}$$

At time $t = 0$, $P_g(0) = 1$ and $P_b(0) = 0$.

Finding the Laplace transforms of the above two differential equations, we get

$$sP_g(s) - P_g(0) = -\lambda_g P_g(s) + P_b(s)\mu_b \tag{2.79}$$

$$sP_b(s) - P_b(0) = -\mu_b P_b(s) + P_g(s)\lambda_g \tag{2.80}$$

Rearranging Eqs. (2.72) and (2.73) leads to

$$P_g(s)(s + \lambda_g) = 1 + P_b(s)\mu_b \tag{2.81}$$

$$P_b(s)(s + \mu_b) = P_g(s)\lambda_g \tag{2.82}$$

From Eq. (2.82), we get

$$P_b(s) = \frac{P_g(s)\lambda_g}{(s + \mu_b)} \tag{2.83}$$

Substitution of Eq. (2.83) into Eq. (2.81) yields

$$P_g(s) = \frac{(s + \mu_b)}{s(s + \mu_b + \lambda_g)} \tag{2.84}$$

Substituting Eq. (2.84) into Eq. (2.83) results in

$$P_b(s) = \frac{\lambda_g}{s(s + \mu_b + \lambda_g)} \tag{2.85}$$

Finding the inverse Laplace transforms of Eqs. (2.84) and (2.85) results in

$$P_g(t) = \frac{\mu_b}{\lambda_g + \mu_b} + \frac{\lambda_g}{\lambda_g + \mu_b} \exp[-(\lambda_g + \mu_b)t] \tag{2.86}$$

and

$$P_b(t) = \frac{\lambda_g}{\lambda_g + \mu_b} - \frac{\lambda_g}{\lambda_g + \mu_b} \exp[-(\lambda_g + \mu_b)t] \tag{2.87}$$

Thus, from Eqs. (2.86) and (2.87), the equipment availability and unavailability are

$$A_{me}(t) = P_g(t) = \frac{\mu_b}{\lambda_g + \mu_b} + \frac{\lambda_g}{\lambda_g + \mu_b} \exp[-(\lambda_g + \mu_b)t] \tag{2.88}$$

and

$$UA_{me}(t) = P_b(t) = \frac{\lambda_s}{\lambda_g + \mu_b} - \frac{\lambda_g}{\lambda_g + \mu_b} \exp[-(\lambda_g + \mu_b)t] \tag{2.89}$$

where $A_{me}(t)$ is the availability of the equipment and $UA_{me}(t)$ is the unavailability of the equipment.

Substituting Eqs. (2.84) and (2.85) into the right-hand side of Eq. (2.13) leads to

$$A_{me} = \lim_{s \to 0} \left[\frac{s(s + \mu_b)}{s(s + \mu_b + \lambda_g)} \right] = \frac{\mu_b}{\mu_b + \lambda_g} \tag{2.90}$$

and

$$UA_{me} = \lim_{s \to 0} \left[\frac{s\lambda_g}{s(s + \mu_b + \lambda_g)} \right] = \frac{\lambda_g}{\mu_b + \lambda_g} \tag{2.91}$$

where A_{me} is the steady-state availability of the equipment and UA_{me} the steady-state unavailability of the equipment.

One should note here that the sum of the equipment availability and unavailability is equal to unity, i.e.,

$$\begin{aligned} A_{me}(t) + UA_{me}(t) &= \frac{\mu_b}{\lambda_g + \mu_b} + \left\{ \frac{\lambda_g}{\lambda_g + \mu_b} \exp[-(\lambda_g + \mu_b)t] \right\} \\ &\quad + \frac{\lambda_g}{\lambda_g + \mu_b} - \frac{\lambda_g}{\lambda_g + \mu_b} \exp[-(\lambda_g + \mu_b)t] \\ &= \frac{(\lambda_g + \mu_b)}{(\lambda_g + \mu_b)} = 1 \end{aligned} \tag{2.92}$$

and

$$\boxed{A_{me} + UA_{me} = \frac{\mu_b}{\mu_b + \lambda_g} + \frac{\lambda_g}{\mu_b + \lambda_g} = 1} \tag{2.93}$$

Example 2.15

Assume that the constant failure and repair rates of a feed water pump are 0.005 failures/h and 0.006 repairs/h, respectively. The pump has two states: operating and failed. In addition, the repaired pump is as good as new. Compute the availability of the pump at time $t = 50$ h.

Using this given data in Eq. (2.88) yields

$$A_{me}(50) = \frac{(0.006)}{(0.005) + (0.006)} + \frac{(0.005)}{(0.005) + (0.006)} e^{-(0.005 + 0.006)(50)}$$

$$= 0.8077$$

The availability of the pump is 0.8077.

References

[1]Spiegel, M. R., *Theory and Problems of Laplace Transforms*, McGraw-Hill, New York, 1965, pp. 20–21.

[2]Kao, J. H. K., "Statistical Models in Mechanical Reliability," *Proceedings of the National Symposium on Reliability*, IEEE, New York, 1965, pp. 240–247.

[3]Singpurwalla, N. D., "Statistical Fatigue Models: A Survey," *IEEE Transactions on Reliability*, Vol. 20, 1971, pp. 185–186.

[4]Davis, D. J., "An Analysis of Some Failure Data," *Journal of American Statistical Association*, June 1952, pp. 113–150.

[5]Gupta, S. and Groll, P., "Gamma Distribution in Acceptance Sampling Based on Life Tests," *Journal of American Statistical Association*, Dec. 1961, pp. 942–970.

[6]Weibull, W., "A Statistical Distribution Function of Wide Applicability," *Journal of Applied Mechanics*, Vol. 18, 1951, pp. 293–297.

[7]Lieblein, J. and Zelen, M., "Statistical Investigation of the Fatigue Life of Deep-Groove Ball Bearings," *Journal of Research of the National Bureau of Standards*, Vol. 5, 1956, pp. 273–316.

[8]Aitchison, J. and Brown, J. A. C., *The Lognormal Distribution*, Cambridge University Press, Cambridge, England, 1957.

Exercises

2.1 Discuss in detail the history of probability.

2.2 Differentiate the following function:

$$f(t) = \exp\left[-\left(\frac{t}{\theta}\right)^b\right] \tag{2.94}$$

where t is the time variable and b and θ the constants.

2.3 Find the Laplace transform of the following function:

$$f(t) = \frac{\lambda(\lambda t)^{m-1} \exp(-\lambda t)}{(m-1)!} \tag{2.95}$$

where t is the time variable and m and λ the constants.

2.4 Find the inverse Laplace transform of

$$F(s) = \frac{(s + A)}{s(s + A + B)} \tag{2.96}$$

where s is the Laplace transform variable and A and B the constants.

2.5 What probability distributions are the special cases of the following two distributions: Weibull and gamma?

2.6 Prove that the total area under the curve of the following probability density function is equal to unity:

$$f(t) = \lambda e^{-\lambda t}, \qquad \lambda > 0,\ t \geqslant 0 \tag{2.97}$$

where t is time and λ the distribution parameter.

2.7 Prove that the mean value m of the Rayleigh distribution is given by

$$m = \theta \Gamma(3/2) \tag{2.98}$$

where θ is the scale parameter.

2.8 Assume that the failure times of a power generator are Rayleigh distributed with the scale parameter value of 150 h. Calculate the mean time to failure of the generator using Eq. (2.98).

3
RELIABILITY EVALUATION MODELS AND TECHNIQUES

This chapter presents a number of reliability evaluation models and techniques. General expressions for reliability, cumulative distribution function, hazard rate, failure density function, and mean time to failure are presented. Reliability equations for series, parallel, standby, and r-out-of-m systems are developed. In addition, the mean time to failure and hazard rate expressions for series and parallel networks are given. Mean-time-to-failure equations for series, parallel, r-out-of-m, standby, and bridge networks are presented. Three techniques concerned with the reliability analysis of engineering systems are given; the network reduction method, fault tree technique, and failure mode and effects analysis (FMEA) approach. A method for allocating failure rates to system components is discussed and four failure rate evaluation models for mechanical items presented. These mathematical models are concerned with pumps, pump seals, pump shafts, and filters.

3.1 INTRODUCTION

The field of reliability has been receiving significant attention for over three decades and is thus sufficiently established so that studies are now beginning to branch out into various specialized areas. Mechanical reliability is one of those areas. Many people have contributed in the development of the reliability discipline. A number of general reliability evaluation models and techniques have been developed. Some of these techniques and models can also be applied in specialized areas. This chapter presents selected general reliability evaluation models and techniques that can be useful in mechanical reliability analysis. In addition, a number of failure rate evaluation models for mechanical items are presented. Some of the topics covered are redundant networks, network reduction method, fault trees, failure mode and effects analysis, and failure rate allocation.

3.2 GENERAL EXPRESSIONS FOR RELIABILITY, HAZARD RATE, FAILURE DENSITY FUNCTION, MEAN TIME TO FAILURE, AND CUMULATIVE DISTRIBUTION FUNCTION

General Equation for Reliability

This equation is expressed as follows[1]:

$$R(t) = \exp\left[-\left(\int_0^t \lambda(x)\,dx\right)\right] \tag{3.1}$$

where t is the time, $R(t)$ the reliability of an item at time t, and $\lambda(t)$ the hazard rate (time-dependent failure rate) of an item.

Equation (3.1) can be used to calculate reliability of a component, subsystem, or system whose failure times are described by statistical distributions such as the exponential or Weibull.

Example 3.1

The failure times of a diesel generator are described by the Weibull distribution. The generator's hazard rate is given by

$$\lambda_D(t) = 3t^2/600 \tag{3.2}$$

where $\lambda_D(t)$ is the diesel generator's hazard rate and t the time.

Calculate the generator's reliability for a 6 h mission.

Substituting Eq. (3.2) and the specified data into Eq. (3.1) yields

$$R(t) = \exp\left[-\int_0^6 \left(\frac{3x^2}{600}\right) dx\right] = \exp\left[-\frac{(6)^3}{600}\right] = 0.6977$$

Thus, the reliability of the diesel generator for the specified mission time is 0.6977.

General Expression for Cumulative Distribution Function

This is given by

$$\begin{aligned} F_c(t) &= 1 - R(t) \\ &= 1 - \exp\left[-\left(\int_0^t \lambda(x)\,dx\right)\right] \end{aligned} \tag{3.3}$$

where $F_c(t)$ is the cumulative distribution function.

For a given hazard rate function, the cumulative distribution function can be obtained from Eq. (3.3).

Example 3.2

Assume that the failure times of a vehicle are represented by the exponential distribution. The hazard rate (failure rate) of the vehicle is 0.005 failures/h. Calculate the failure probability of the vehicle for a 5 h mission.

Substituting the specified data into Eq. (3.3) leads to

$$F_c(5) = 1 - \exp\left[-\int_0^5 (0.005)\, \mathrm{d}x\right]$$
$$= 1 - \exp[-(0.005)(5)] = 0.0247$$

Thus, the failure probability of the vehicle is 0.0247.

General Expression for Hazard Rate

The hazard rate of an item is given by

$$\lambda(t) = -\frac{1}{R(t)} \frac{\mathrm{d}R(t)}{\mathrm{d}t} \tag{3.4}$$

where t is the time, $\lambda(t)$ the hazard rate, and $R(t)$ the reliability.

For a known reliability function of a unit, subsystem, or system, the hazard rate function can be obtained by using Eq. (3.4).

Example 3.3

The reliability function of an aircraft engine is

$$R(t) = \exp(-t^c/\theta) \tag{3.5}$$

where t is the time, c the shape parameter, and θ the scale parameter.

Obtain an expression for the engine's hazard rate. With the aid of Eqs. (3.4) and (3.5), we get

$$\lambda(t) = -\frac{1}{\exp(-t^c/\theta)} \cdot \left[-\frac{ct^{c-1}}{\theta} \cdot \exp\left(-\frac{t^c}{\theta}\right)\right]$$
$$= \frac{ct^{c-1}}{\theta} \tag{3.6}$$

Thus, the engine's hazard rate is given by Eq. (3.6).

General Expression for Failure Density Function

This is expressed as

$$f(t) = -\frac{\mathrm{d}R(t)}{\mathrm{d}t}$$

$$= \lambda(t) \cdot \exp\left[-\left(\int_0^t \lambda(x)\,\mathrm{d}x\right)\right] \tag{3.7}$$

where t is the time, $f(t)$ the failure density function, $\lambda(t)$ the hazard rate, and $R(t)$ the reliability.

For a defined hazard rate function, one can obtain the failure density function from Eq. (3.7).

Example 3.4

Assume that the hazard rate $\lambda(t)$ of a railway engine is defined by

$$\lambda(t) = 2t/\theta \tag{3.8}$$

where t is the time and θ the scale parameter.

Obtain an expression for the engine's failure density function. Using Eq. (3.8) in Eq. (3.7) yields

$$f(t) = \left(\frac{2t}{\theta}\right) \cdot \exp\left[-\left(\int_0^t \frac{2x}{\theta}\,\mathrm{d}x\right)\right]$$

$$= \left(\frac{2t}{\theta}\right) \cdot \exp\left[-\left(\frac{t^2}{\theta}\right)\right] \tag{3.9}$$

Thus, the failure density function of the railway engine is given by Eq. (3.9).

General Expression for Mean Time to Failure

The general equation for obtaining an item's mean time to failure (MTTF) is

$$\mathrm{MTTF} = \int_0^\infty R(t)\,\mathrm{d}t$$

$$= \int_0^\infty \exp\left[-\left(\int_0^t \lambda(x)\,\mathrm{d}x\right)\right] \cdot \mathrm{d}t \tag{3.10}$$

where t is the time, $R(t)$ the reliability, and $\lambda(t)$ the hazard rate.

For the known hazard rate function of a component, subsystem, or system, the mean time to failure can be obtained using Eq. (3.10).

Example 3.5

The hazard rate (failure rate) of a bus is 0.002 failures/h. Calculate the mean time to the failure (MTTF) of bus.

In this example, the failure rate of the bus is constant and thus its failure times are exponentially distributed. Substituting the given data for hazard rate (failure rate) into Eq. (3.10), we get

$$\mathrm{MTTF} = \int_0^{\infty} \exp\left[-\left(\int_0^t (0.002)\, \mathrm{d}x\right)\right] \cdot \mathrm{d}t$$

$$= \int_0^{\infty} \exp\left[-(0.002)t\right] \mathrm{d}t = \frac{1}{(0.002)} = 500 \text{ h}$$

3.3 GENERAL RELIABILITY MODELS

This section is concerned with the reliability evaluation of common reliability systems as given below.[2–4]

Series System

The block diagram of a k-unit series system is shown in Fig. 3.1. Each block in the diagram represents a unit or component. The symbol R_i in each block denotes unit i reliability, for $i = 1,2,3,...,k$. In this case, if any one of the series items fails, the total system fails.

For independent failures (i.e., the series units do not interact), the reliability of the series system shown in Fig. 3.1, R_{ss} is given by

$$R_{ss} = R_1 \cdot R_2 \cdot R_3 \cdot \ldots \cdot R_k \qquad (3.11)$$

where k is the number of independent units in the series and R_i the unit i reliability for $i = 1,2,3,...,k$.

For identical units, Eq. (3.11) becomes

$$R_{ss} = R^k \qquad (3.12)$$

where R is the reliability of a unit.

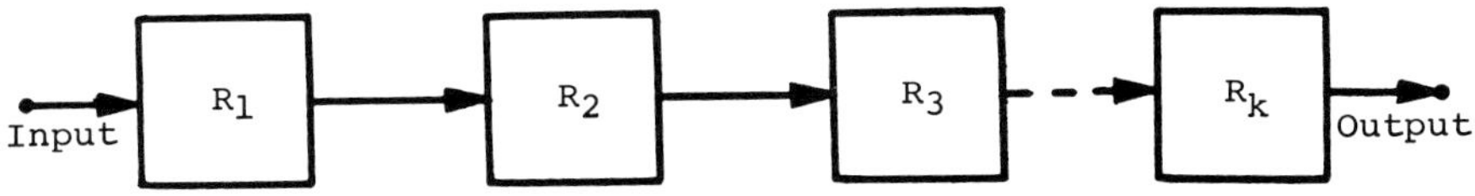

Fig. 3.1 Block diagram of a k-unit series system.

Example 3.6

A vehicle operates with four independent and identical tires and the reliability of each tire is 0.97. Calculate the vehicle's reliability with respect to tires.

In this case, all four tires are required for the operation of the vehicle. Thus, the tires form a series system. Using the given data into Eq. (3.12), we get

$$R_{ss} = (0.97)^4 = 0.8853$$

Thus, the reliability of the vehicle with respect to tires is 0.8853.

Parallel System

In this case, the system is composed of all active units. At least one unit must operate normally for the system's success. The system fails only when all of its units fail. The block diagram of a k active unit parallel system is shown in Fig. 3.2. Each block in the diagram denotes an active unit. The symbol R_i in each block denotes the reliability of unit i, for $i = 1,2,3,4,...,k$.

The reliability of the parallel system with independent units is

$$R_{ps} = 1 - (1 - R_1)(1 - R_2)(1 - R_3)(1 - R_4) \ldots (1 - R_k) \qquad (3.13)$$

where R_{ps} is the reliability of the parallel system, k the number of independent units in parallel, and R_i the reliability of unit i, for $i = 1,2,3,4,...,k$.

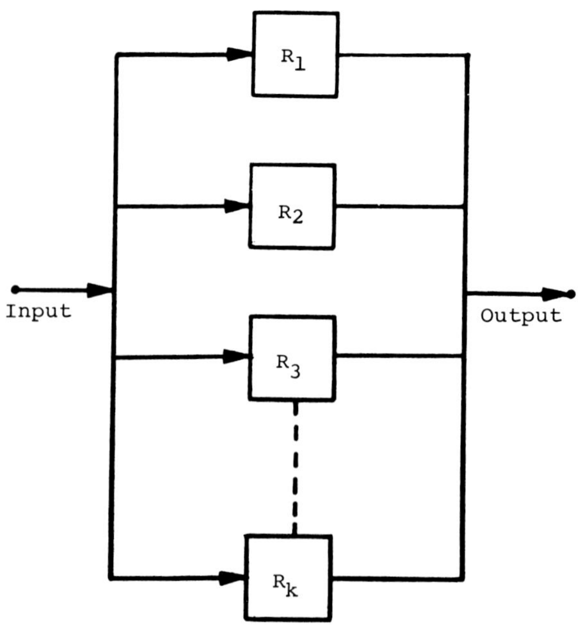

Fig. 3.2 Block diagram of the parallel system.

Since $R_i + F_i = 1$, Eq. (3.13) is rewritten in the form

$$R_{ps} = 1 - F_1 \cdot F_2 \cdot F_3 \cdot F_4 \cdot \ldots \cdot F_k \tag{3.14}$$

where F_i is the failure probability of unit i, for $i = 1,2,3,4,\ldots,k$.

For identical units, Eqs. (3.13) and (3.14) become

$$R_{ps} = 1 - (1 - R)^k \tag{3.15}$$

and

$$R_{ps} = 1 - F^k \tag{3.16}$$

where R is the unit's reliability and F the unit's failure probability.

Example 3.7

An aircraft has two independent engines. Both the engines are active and identical and the reliability of each engine is 0.83. At least one of the engines must work normally for the aircraft to fly successfully. Calculate the reliability of the aircraft flying with respect to its engines.

Substituting these data into Eq. (3.15), we get

$$R_{ps} = 1 - (1 - 0.83)^2 = 0.9711$$

The reliability of aircraft flying is 0.9711.

Standby System

The block diagram of a two-unit standby system is given in Fig. 3.3. Each block in the diagram represents a unit. In this case, it is assumed that one unit is active (operating) and the remaining one on standby. As soon as the operating unit fails, it is replaced with the standby unit. The system fails when both the units fail. More specifically, only one unit is required to operate for the system's success.

For perfect switch and identical units, the reliability of the two-unit standby system is

$$R_{sb} = R(1 - \ell n R) \tag{3.17}$$

where R_{sb} is the reliability of the two-unit standby system and R the single unit's reliability.

Example 3.8

A mechanical system is composed of two identical units (i.e., one operating, the other on standby). Only one unit is required to operate for system

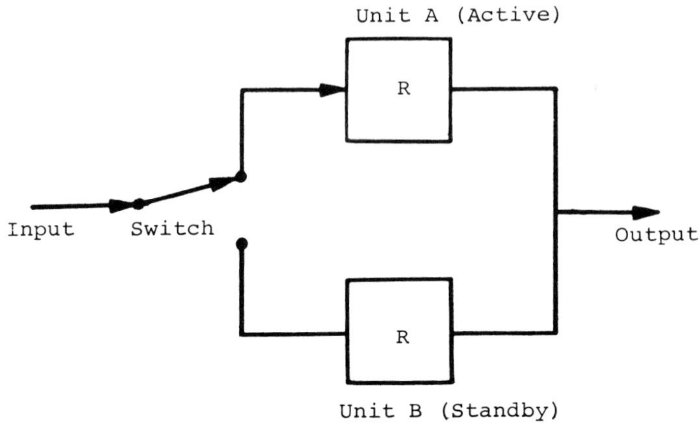

Fig. 3.3 Two-unit standby system (the symbol R denotes the unit's reliability).

success. As soon as the operating unit fails, the standby unit is switched in. The "switch in mechanism" is perfect. If the reliability of a unit is 0.75, calculate the reliability of the standby system. Using the given data into Eq. (3.17) yields

$$R_{sb} = (0.75)[1 - \ln(0.75)] = 0.9658$$

The reliability of the given standby system is 0.9658.

r-out-of-m Unit System

This basically is the same as the parallel system but with one exception, that is, at least r units must function normally for the system's success instead of at least one unit in the parallel case. Series and parallel systems are the special cases of this configuration when $r = m$ and $r = 1$, respectively. The system contains m number of active units. For independent and identical units, the reliability of the system in question is

$$R_{r/m} = \sum_{y=r}^{m} \binom{m}{y} R^y (1 - R)^{m-y} \tag{3.18}$$

where $R_{r/m}$ is the reliability of the r-out-of-m unit system, R the reliability of a unit, m the number of units in parallel, r the minimum number of units required for the system success, and

$$\binom{m}{y} = \frac{m!}{y!(m-y)!} \tag{3.19}$$

Example 3.9

Assume that an aircraft has four active, independent, and identical engines. The reliability of each engine is 0.9. At least two of the engines must function normally for the aircraft to fly successfully. Determine the aircraft reliability with respect to engines.

In this example, data are specified for the following items: $R = 0.9$, $r = 2$, and $m = 4$. Using this data in Eq. (3.18), we get

$$R_{2/4} = \sum_{y=2}^{4} \binom{4}{y} (0.9)^y (1 - 0.9)^{4-y}$$

$$= \binom{4}{2} (0.9)^2 (0.1)^2 + \binom{4}{3} (0.9)^3 (0.1) + \binom{4}{4} (0.9)^4$$

$$= 6(0.9)^2(0.1)^2 + 4(0.9)^3(0.1) + 1(0.9)^4$$

$$= 0.9963$$

The reliability of the aircraft flying is 0.9963.

3.4 TIME-DEPENDENT RELIABILITY MODELS

This section presents reliability analysis of systems whose units' failure times are represented by well-known statistical distributions.

Series System

This system is discussed in Sec. 3.3 and its block diagram is shown in Fig. 3.1. With the aid of Eq. (3.11), the time-dependent reliability of the series system is given by

$$R_{ss}(t) = R_1(t) \cdot R_2(t) \cdot R_3(t) \cdot \ldots \cdot R_k(t) \tag{3.20}$$

where $R_{ss}(t)$ is the reliability of the series system at time t, k the number of independent units in series, and $R_i(t)$ the reliability of the unit i at time t, for $i = 1,2,3,...,k$.

Substituting Eq. (3.1) into Eq. (3.20) yields

$$R_{ss}(t) = \prod_{i=1}^{k} \exp\left[-\left(\int_0^t \lambda_i(x)\, dx \right) \right] \tag{3.21}$$

where $\lambda_i(x)$ is the hazard rate of unit i, for $i = 1,2,3,...,k$.

Weibull Distributed Failure Times. For Weibull distributed failure times of each unit, the series system reliability from Eqs. (3.6) and (3.21) is

$$R_{ss}(t) = \prod_{i=1}^{k} \exp\left[-\left(\frac{t^{c_i}}{\theta_i} \right) \right] \tag{3.22}$$

where c_i is the Weibull shape parameter associated with unit i, for $i = 1,2,3,...,k$, and θ_i the Weibull scale parameter associated with unit i, for $i = 1,2,3,...,k$.

Exponentially Distributed Failure Times. For exponentially distributed failure times of a unit, the unit's hazard rate (failure rate) is constant. Thus, at $c = 1$, from Eq. (3.6) we get

$$\lambda = 1/\theta \tag{3.23}$$

In this case, θ is the mean time to failure (scale parameter) of the unit and λ the unit's constant failure rate.

Substituting Eq. (3.23) into Eq. (3.21) results in

$$R_{ss}(t) = \prod_{i=1}^{k} \exp\left(-\frac{t}{\theta_i}\right) \tag{3.24}$$

where θ_i is the mean time to failure (scale parameter) of unit i, for $i = 1,2,3,...,k$.

General Expression for Series System Mean Time to Failure. The mean time to failure (MTTF) of an item or a system is expressed as

$$\text{MTTF} = \int_0^{\infty} R(t)\,\mathrm{d}t = \lim_{s \to 0} R(s) \tag{3.25}$$

where $R(t)$ is the reliability of an item or a system and s the Laplace transform variable. $R(s)$ is the Laplace transform of the reliability function of an item or a system.

For a series system, this is given by

$$\text{MTTF}_{ss} = \int_0^{\infty} \left\{\prod_{i=1}^{k} \exp\left[-\left(\int_0^{t} \lambda_i(x)\,\mathrm{d}x\right)\right]\right\} \mathrm{d}t \tag{3.26}$$

where MTTF_{ss} is the mean time to failure of the series system.

The above equation was obtained by substituting Eq. (3.21) into Eq. (3.25).

For constant failure rate of unit i from Eq. (3.23), we get

$$\lambda_i = 1/\theta_i \tag{3.27}$$

where λ_i is the constant failure rate of unit i for $i = 1,2,3,...$ and θ_i the mean time to failure of unit i, for $i = 1,2,3,....$

Using Eq. (3.27) in Eq. (3.26) yields

$$\mathrm{MTTF}_{ss} = \int_0^\infty \left\{ \prod_{i=1}^{k} \exp\left[-\left(\int_0^t \frac{1}{\theta_i}\, dx \right) \right] \right\} dt$$

$$= \frac{1}{\sum_{i=1}^{k} (1/\theta_i)} \tag{3.28}$$

Rearranging Eq. (3.28) results in

$$\frac{1}{\mathrm{MTTF}_{ss}} = \sum_{i=1}^{k} \frac{1}{\theta_i} \tag{3.29}$$

Alternatively,

$$\frac{1}{\mathrm{MTTF}_{ss}} = \sum_{i=1}^{k} \lambda_i \tag{3.30}$$

General Expression for Series System Hazard Rate. This is obtained by substituting Eq. (3.21) into Eq. (3.4) as

$$\lambda_{ss}(t) = -\frac{1}{\prod_{i=1}^{k} \exp\left[-\left(\int_0^t \lambda_i(x)\, dx \right) \right]} \cdot \frac{d\left\{ \prod_{i=1}^{k} \exp\left[-\left(\int_0^t \lambda_i(x)\, dx \right) \right] \right\}}{dt}$$

$$= \sum_{i=1}^{k} \lambda_i(t) \tag{3.31}$$

where $\lambda_{ss}(t)$ is the hazard rate of the series system.

When unit i hazard rate is governed by the Weibull distribution, the series system hazard rate from Eqs. (3.31) and (3.6) is

$$\lambda_{ss}(t) = \sum_{i=1}^{k} \left(\frac{c_i t^{c_i - 1}}{\theta_i} \right) \tag{3.32}$$

where c_i is the Weibull shape parameter associated with unit i, for $i = 1,2,3,\ldots,k$; θ_i the Weibull scale parameter associated with unit i, for $i = 1,2,3,\ldots,k$; t the time; and k the number of independent units in series.

Setting $c_i = 1$ in Eq. (3.32), we get

$$\lambda_{ss} = \sum_{i=1}^{k} \frac{1}{\theta_i} \tag{3.33}$$

or

$$\lambda_{ss} = \sum_{i=1}^{k} \lambda_i \tag{3.34}$$

It is to be noted that Eqs. (3.33) and (3.34) are valid only when the unit i hazard rate is governed by the exponential distribution with mean value θ_i.

Example 3.10

A system has four independent and identical pulverizers. All of the pulverizers must work successfully for system success. The failure times of each pulverizer are exponentially distributed with a mean value of 1500 h. Calculate the reliability of the system for a 10 h mission.

In this case, the system units form a series system. Thus, substituting the given data into Eq. (3.24), we get

$$R_{ss}(10) = \left[\exp\left(-\frac{10}{1500}\right)\right]^4 = 0.9737$$

The reliability of the pulverizer system is 0.9737.

Example 3.11

Using the given data in Example 3.10, compute the mean time to failure and hazard rate (failure rate) of the pulverizer system.

Using the specified data in Eq. (3.28) results in

$$\text{MTTF}_{ss} = \frac{1}{(4/1500)} = 375 \text{ h}$$

Similarly, from Eq. (3.33) we get

$$\lambda_{ss} = \frac{1}{1500} + \frac{1}{1500} + \frac{1}{1500} + \frac{1}{1500}$$

$$= 0.0027 \text{ failures/h}$$

The mean time to failure and the hazard rate of the pulverizer system are 375 h and 0.0027 failures/h, respectively.

Parallel System

This system was described in detail in Sec. 3.3. The block diagram shown in Fig. 3.2 represents the parallel system. The time-dependent reliability of the parallel system is

$$R_{ps}(t) = 1 - [1 - R_1(t)][1 - R_2(t)][1 - R_3(t)][1 - R_4(t)]...[1 - R_k(t)] \quad (3.35)$$

where $R_{ps}(t)$ is the time-dependent reliability of the parallel system, t the time, $R_i(t)$ the time-dependent reliability of unit i, for $i = 1,2,3,4,...,k$, and k the number of independent units in parallel.

For identical units, the above equation reduces to

$$R_{ps}(t) = 1 - [1 - R(t)]^k \quad (3.36)$$

where $R(t)$ is the reliability of a unit.

Inserting Eq. (3.1) into Eqs. (3.35) and (3.36) leads to

$$R_{ps}(t) = 1 - \prod_{i=1}^{k} \left\{1 - \exp\left[-\int_0^t \lambda_i(x)\,dx\right]\right\} \quad (3.37)$$

and

$$R_{ps}(t) = 1 - \left\{1 - \exp\left[-\int_0^t \lambda(x)\,dx\right]\right\}^k \quad (3.38)$$

where $\lambda_i(x)$ is the hazard rate of unit i, for $i = 1,2,3,...,k$, and $\lambda(x)$ the hazard rate of a unit.

For Weibull governed hazard rate functions of units, with the aid of Eqs. (3.6), (3.37), and (3.38) we get

$$R_{ps}(t) = 1 - \prod_{i=1}^{k} \left[1 - \exp\left(-\frac{t^{c_i}}{\theta_i}\right)\right] \quad (3.39)$$

and

$$R_{ps}(t) = 1 - \left[1 - \exp\left(-\frac{t^c}{\theta}\right)\right]^k \quad (3.40)$$

where θ_i is the Weibull scale parameter associated with unit i and c_i the Weibull shape parameter associated with unit i.

Similarly, when each unit's hazard rate function is governed by the exponential distribution, we get the following expressions for system reliability using Eqs. (3.23), (3.37), and (3.38):

$$R_{ps}(t) = 1 - \prod_{i=1}^{k} \left[1 - \exp\left(-\frac{t}{\theta_i} \right) \right] \tag{3.41}$$

and

$$R_{ps}(t) = 1 - \left[1 - \exp\left(-\frac{t}{\theta} \right) \right]^k \tag{3.42}$$

where θ_i is the mean time to failure of unit i.

General Expression for Parallel System Mean Time to Failure. Inserting Eq. (3.37) into Eq. (3.25) yields the following general expression for the parallel system mean time to failure:

$$\begin{aligned} \mathrm{MTTF}_{ps} &= \int_0^\infty \left\{ 1 - \left[1 - \exp\left(-\frac{t}{\theta} \right) \right]^k \right\} dt \\ &= \theta \sum_{i=1}^{k} \frac{1}{i} \end{aligned} \tag{3.43}$$

where θ is the unit's mean time to failure.

It is to be noted that in Eq. (3.43) it was assumed that the unit's failure times are exponentially distributed with a mean value of θ. More specifically, the units' failure rates are constant.

Equation (3.43) is rewritten in the following form:

$$\mathrm{MTTF}_{ps} = \theta + \frac{\theta}{2} + \frac{\theta}{3} + \frac{\theta}{4} + \cdots + \frac{\theta}{k} \tag{3.44}$$

General Expression for Parallel System Hazard Rate. The following general expression for the parallel system hazard rate is obtained by substituting Eq. (3.37) into Eq. (3.4):

$$\lambda_{ps}(t) = -\frac{1}{\left(1 - \prod_{i=1}^{k} \left\{ 1 - \exp\left[-\int_0^t \lambda_i(x)\, dx \right] \right\} \right)} \cdot \frac{d\left(1 - \prod_{i=1}^{k} \left\{ 1 - \exp\left[-\int_0^t \lambda_i(x)\, dx \right] \right\} \right)}{dt} \tag{3.45}$$

where $\lambda_{ps}(t)$ is the hazard rate of the parallel system.

For constant failure rates of the units, inserting Eq. (3.27) into Eq. (3.45) yields

$$\lambda_{ps}(t) = -\frac{1}{\left\{1-\prod_{i=1}^{k}\left[1-\exp\left(-\frac{t}{\theta_i}\right)\right]\right\}}\cdot\frac{\mathrm{d}\left\{1-\prod_{i=1}^{k}\left[1-\exp\left(-\frac{t}{\theta_i}\right)\right]\right\}}{\mathrm{d}t} \tag{3.46}$$

For $k \geqslant 2$ and $t/\theta_i < 0.2$, we get the following approximate expression for $\lambda_{ps}(t)$:

$$\lambda_{ps}(t) \simeq kt^{k-1}\left(\prod_{i=1}^{k}\frac{1}{\theta_i}\right) \tag{3.47}$$

Similarly as $t \to \infty$, Eq. (3.46) becomes

$$\lambda_{ps}(\infty) = \min_i\,(1/\theta_i) \tag{3.48}$$

Since $\lambda_i = 1/\theta_i$, Eq. (3.48) is rewritten in the following form:

$$\lambda_{ps}(\infty) = \min_i\,\lambda_i \tag{3.49}$$

where λ_i is the failure rate of unit i.

Example 3.12

Using Eq. (3.40), obtain an expression for the parallel system mean time to failure. The system components obey the Weibull law in this case. Equation (3.40) is rewritten in the following form:

$$R_{ps} = \sum_{i=1}^{k}(-1)^{i+1}\binom{k}{i}\exp\left(-\frac{it^c}{\theta}\right) \tag{3.50}$$

In order to obtain the system mean time to failure, we substitute Eq. (3.50) in Eq. (3.25) as

$$\begin{aligned}\mathrm{MTTF}_{ps} &= \int_0^{\infty}\left[\sum_{i=1}^{k}(-1)^{i+1}\binom{k}{i}\exp\left(-\frac{it^c}{\theta}\right)\right]\mathrm{d}t \\ &= \sum_{i=1}^{k}(-1)^{i+1}\binom{k}{i}\left(\frac{1}{i}\right)^{-1/c}\left[\frac{\Gamma(1/c)}{c(\theta)^{-1/c}}\right] \end{aligned} \tag{3.51}$$

where MTTF_{ps} is the parallel system mean time to failure when its units obey Weibull law.

Example 3.13

Two active, independent, and identical pumps form a parallel system. Failure times of each pump are described by the Weibull distribution, with scale and shape parameters values of 1200 h and 2, respectively. Determine the system reliability for a 12 h mission.

Substituting the specified data into Eq. (3.40) yields

$$R_{ps}(12) = 1 - \left\{1 - \exp\left[-\frac{(12)^2}{1200}\right]\right\}^2$$

$$= 0.9872$$

The pump system reliability is 0.9872.

Example 3.14

Assume that the parallel pump system of Example 3.13 is composed of units whose failure times are described by an exponential distribution. The mean time to failure of each pump is 400 h. Compute the value of the system mean time to failure.

Substituting the given data into Eq. (3.44), we get

$$\mathrm{MTTF}_{ps} = \theta + \frac{\theta}{2} = 400 + \frac{400}{2} = 600 \text{ h}$$

The mean time to failure of the pump system is 600 h.

r-out-of-m Unit System

In a manner similar to the parallel system, general equations for r-out-of-m unit system reliability, mean time to failure, and hazard rate can be developed. The same results can also be obtained for other configurations such as series-parallel, bridge, and standby.

3.5 MEAN TIME TO FAILURE OF RELIABILITY NETWORKS

This section presents formulas to obtain the mean time to failure of independent unit series, parallel, r-out-of-m unit, standby, and bridge systems. In all of these networks, it is assumed that the failure rates of units (components) are constant.

Series Network

The mean time to failure of the series system is

$$\mathrm{MTTF}_{ss} = \frac{1}{\sum_{i=1}^{k} (1/\theta_i)} \tag{3.52}$$

where θ_i is the mean time to failure of unit i with exponentially distributed failure times and k the number of independent units in series.

Equation (3.52) can be rewritten in the following form:

$$\frac{1}{\mathrm{MTTF}_{ss}} = \frac{1}{\theta_1} + \frac{1}{\theta_2} + \frac{1}{\theta_3} + \cdots + \frac{1}{\theta_k} \tag{3.53}$$

For identical units, Eq. (3.52) simplifies to

$$\mathrm{MTTF}_{ss} = \theta / k \tag{3.54}$$

where θ is the mean time to failure of a unit with exponentially distributed failure times.

Parallel Network

The identical unit parallel network's mean time to failure MTTF_{ps} is given by

$$\begin{aligned}\mathrm{MTTF}_{ps} &= \theta \sum_{i=1}^{k} \frac{1}{i} \\ &= \theta + \frac{\theta}{2} + \frac{\theta}{3} + \frac{\theta}{4} + \cdots + \frac{\theta}{k}\end{aligned} \tag{3.55}$$

where k is the number of independent and identical units in parallel and θ the mean time to failure of a unit with exponentially distributed failure times.

r-out-of-m Unit Network

The system mean time to failure is expressed as follows:

$$\begin{aligned}\mathrm{MTTF}_{r/m} &= \theta \sum_{y=r}^{m} \frac{1}{y} \\ &= \frac{\theta}{r} + \frac{\theta}{r+1} + \frac{\theta}{r+2} + \cdots + \frac{\theta}{m}\end{aligned} \tag{3.56}$$

where $\mathrm{MTTF}_{r/m}$ is the mean time to failure of the identical and independent unit r-out-of-m system, r the minimum number of units required for the system success, m the total number of units in the system, and θ the mean time to failure of a unit with exponentially distributed failure times.

When values of r and m are reasonably large, the mean time to failure can be approximated as

$$\mathrm{MTTF}_{r/m} \simeq \theta \ell n \left(\frac{m + 0.5}{r - 0.5} \right) \tag{3.57}$$

Standby System

In this case, one unit is operating and k units are on standby. The system has a perfect switch in arrangement and identical units. The system fails when all of its units fail. The system's mean time to failure is given by

$$MTTF_{sb} = \theta(k + 1) \tag{3.58}$$

where k is the number of standbys and θ the mean time to failure of a unit with exponentially distributed failure times.

For nonidentical units, Eq. (3.58) becomes

$$MTTF_{sb} = \sum_{i=1}^{k+1} \theta_i \tag{3.59}$$

where θ_i is the mean time to failure of unit i with exponentially distributed failure times.

Bridge Network

In this case, it is assumed that the network is composed of five independent and identical units. In addition, each unit's failure times are exponentially distributed. The mean time to failure $MTTF_b$ of the bridge system is

$$MTTF_b = (49/60)\theta \tag{3.60}$$

where θ is the mean time to failure of a unit with exponentially distributed failure times.

Example 3.15

An engineering system is composed of five independent and identical units forming a bridge configuration. Each unit's failure times are exponentially distributed with a mean value of 250 h. Calculate the engineering system's mean time to failure.

Substituting the specified data into Eq. (3.60) we get

$$MTTF_b = (49/60)(250) = 204.167 \text{ h}$$

Thus, the mean time to failure of the engineering system is 204.167 h.

Example 3.16

A system is composed of two identical diesel generators forming the standby redundancy configuration. One generator is operating and the other is on standby. As soon as the operating generator fails, it is replaced with the standby. The "switch in mechanism" is perfect and the generator's failure times are exponentially distributed. The mean time to failure (MTTF) of a generator is 700 h. Determine the system mean time to failure.

Using the given data in Eq. (3.58) yields

$$\mathrm{MTTF}_{\mathrm{sb}} = \theta(k + 1) = 700(2) = 1400 \text{ h}$$

The mean time to failure ($\mathrm{MTTF}_{\mathrm{sb}}$) of the standby system is 1400 h.

3.6 RELIABILITY ANALYSIS TECHNIQUES

This section presents techniques concerned with reliability analysis of engineering systems. Three of these methods are presented below.

Network Reduction Method

This is a straightforward approach used for determining the reliability of engineering systems composed of series and parallel subsystems. The systems containing bridge configurations can also be handled using decomposition or delta-star techniques.[5] The series and parallel configurations of the system are reduced in sequence to equivalent single units until the entire system becomes an equivalent single-unit system. The method is demonstrated in the following example.

Example 3.17

The block diagram representing a hypothetical mechanical system is shown in Fig. 3.4. Each block in the diagram represents an independent mechanical unit. The symbol R_i used in the diagram denotes reliability of unit i, for $i = 1,2,...,6$. The units are identified by numerals as shown in the diagram. Develop an expression for the mechanical system's reliability by applying the network reduction technique.

Using Eq. (3.11), the reliabilities of subsystems A and B shown in Fig. 3.4 are

$$R_{\mathrm{A}} = R_4 \cdot R_5 \tag{3.61}$$

and

$$R_{\mathrm{B}} = R_1 \cdot R_2 \cdot R_3 \tag{3.62}$$

where R_{A} is the reliability of subsystem A, R_{B} the reliability of subsystem B, and R_i the reliability of unit i, for $i = 1,2,3,4,5$.

Thus, subsystems A and B of Fig. 3.4 can be replaced with single equivalent units of reliabilities R_{A} and R_{B}, respectively, as shown in Fig. 3.5. Utilizing Eq. (3.13), the reliability of the subsystem shown in Fig. 3.5 is

$$R_{\mathrm{C}} = 1 - (1 - R_{\mathrm{A}})(1 - R_{\mathrm{B}}) \tag{3.63}$$

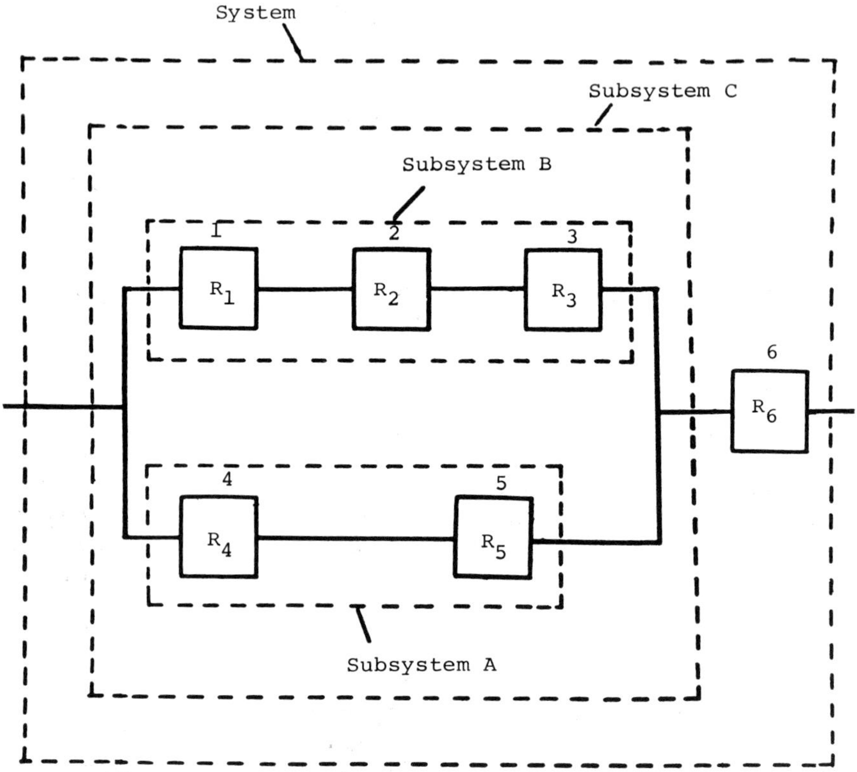

Fig. 3.4 Block diagram of a mechanical system.

where R_C is the reliability of subsystem C. Now, one can replace the subsystem C of Fig. 3.5 with an equivalent unit of reliability R_C as shown in Fig. 3.6.

With the aid of Eq. (3.11), the reliability R_s of the system shown in Fig. 3.6 is

$$R_s = R_C \cdot R_6 \tag{3.64}$$

where R_6 is the reliability of unit 6.

As shown in Fig. 3.7, the system of Fig. 3.6 is replaced with an equivalent unit of reliability R_s. Thus, the single unit of Fig. 3.7 represents the reliability of network shown in Fig. 3.4. More specifically, the network of Fig. 3.4 is reduced to an equivalent single unit. Substituting Eqs. (3.61) and (3.62) into Eq. (3.63) leads to

$$R_C = 1 - (1 - R_4R_5)(1 - R_1R_2R_3) \tag{3.65}$$

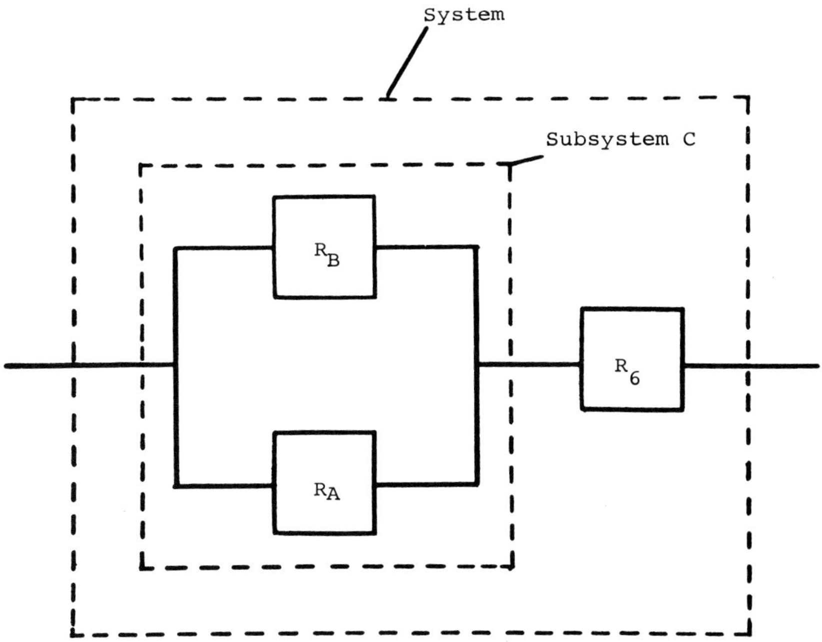

Fig. 3.5 Three-unit reduced network.

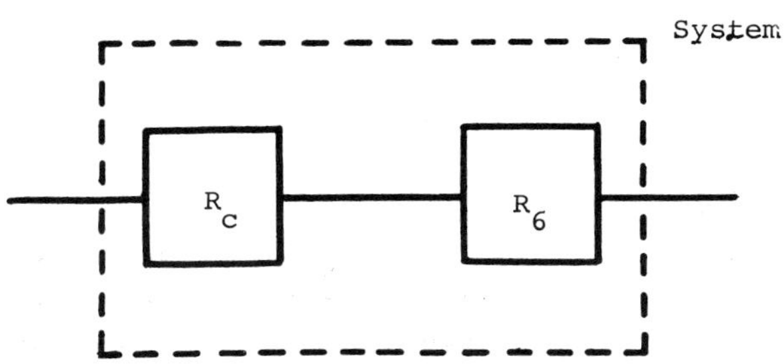

Fig. 3.6 Two-unit reduced network.

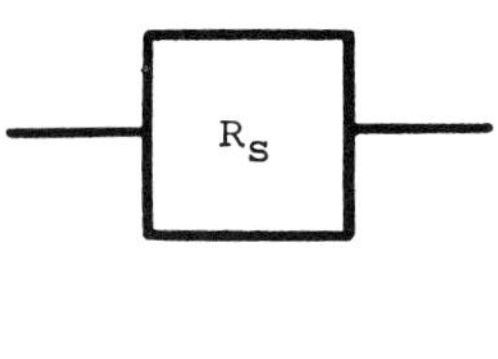

Fig. 3.7 Equivalent single unit of reliability R_s for network shown in Fig. 3.4.

Finally, from Eqs. (3.64) and (3.65), we get the following expression for the mechanical system's reliability:

$$R_s = [1 - (1 - R_4R_5)(1 - R_1R_2R_3)]R_6 \tag{3.66}$$

where R_s is the reliability of the mechanical system whose block diagram is shown in Fig. 3.4.

Example 3.18

Assume that in Example 3.17 the reliabilities of mechanical units shown in Fig. 3.4 are as follows:

$$R_1 = 0.85 \qquad R_4 = 0.7$$
$$R_2 = 0.9 \qquad R_5 = 0.75$$
$$R_3 = 0.8 \qquad R_6 = 0.87$$

Calculate the mechanical system's reliability using Eq. (3.66).

Using the specified data in Eq. (3.66) we get

$$\begin{aligned} R_s &= \{1 - [1 - (0.7)(0.75)][1 - (0.85)(0.9)(0.8)]\}(0.87) \\ &= \{1 - [1 - 0.5250][1 - 0.6120]\}(0.87) \\ &= [1 - (0.4750)(0.3880)](0.87) \\ &= (1 - 0.1843)(0.87) \\ &= (0.8157)(0.87) \\ &= 0.7097 \end{aligned}$$

The reliability of the mechanical system is 0.7097.

Fault Tree Method

This method was developed in the early 1960's.[5] Fault tree analysis starts with the identification of the undesirable event in the system. This event is also known as the top event. Fault events that can cause the occurrence of the undesirable event are connected and generated by logic gates such as OR and AND. The construction of fault tree proceeds by generation of fault events in a successive manner until the point where there is no need to develop fault events any further. This involves successively asking the question, "how could this event occur?." The fault tree method makes use of various symbols. Some of those are given below.

Symbols Used in Fault Tree Construction. This section presents four commonly used symbols in fault tree construction. Symbols for both logic gates and events are described below.

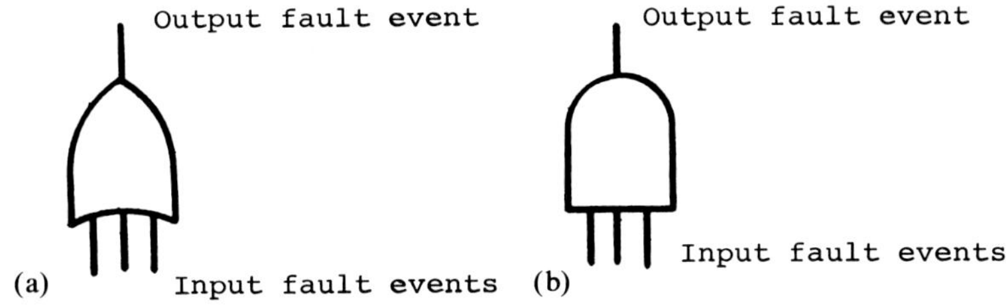

Fig. 3.8 Symbols for OR and AND gates. a) OR gate, b) AND gate.

AND gate. Shown in Figure 3.8b, this symbol denotes that an output fault event will occur only if all of the input fault events occur.

OR gate. Shown in Fig. 3.8a, this symbol denotes that an output fault event will occur if one or more of the input fault events occur.

Basic fault event. This represents the failure of an elementary part and is denoted by a circle. The failure parameters (of the event) such as failure and repair rates, unavailability, and probability are obtained from empirical data.

Resultant event. This event results from the combination of fault events through the input of a logic gate such as AND or OR. It is represented by a rectangle.

Other fault tree symbols are presented in Ref. 6. The next example demonstrates fault tree construction by using all of the above symbols.

Example 3.19

Develop a fault tree for the network representing a mechanical system shown in Fig. 3.4. The top event is mechanical system failure. Each block in Fig. 3.4 represents an independent mechanical unit and is identified with a numeral. The failure probability of unit i is given by $F_i = (1 - R_i)$, for $i = 1,...,6$. Thus, F_1, F_2, F_3, F_4, F_5, and F_6 are the failure probabilities of units 1, 2, 3, 4, 5, and 6, respectively.

Using symbols for the AND gate, OR gate, basic fault event, and the resultant event, a fault tree for the network of Fig. 3.4 was developed and is shown in Fig. 3.9.

Probability Evaluation of a Fault Tree. This section presents equations to determine the probability of occurrence of output events at the AND and OR gates. In the case of the AND gate, the probability of occurrence of an output fault event is

$$P(x_1 \cdot x_2 \cdot x_3 \cdot \ldots) = P(x_1) \cdot P(x_2) \cdot P(x_3) \ldots \qquad (3.67)$$

where x_i is the AND gate's input fault event, for $i = 1,2,3,...$, and $P(x_i)$ the probability of occurrence of an independent input fault event x_i, for $i = 1,2,3,....$

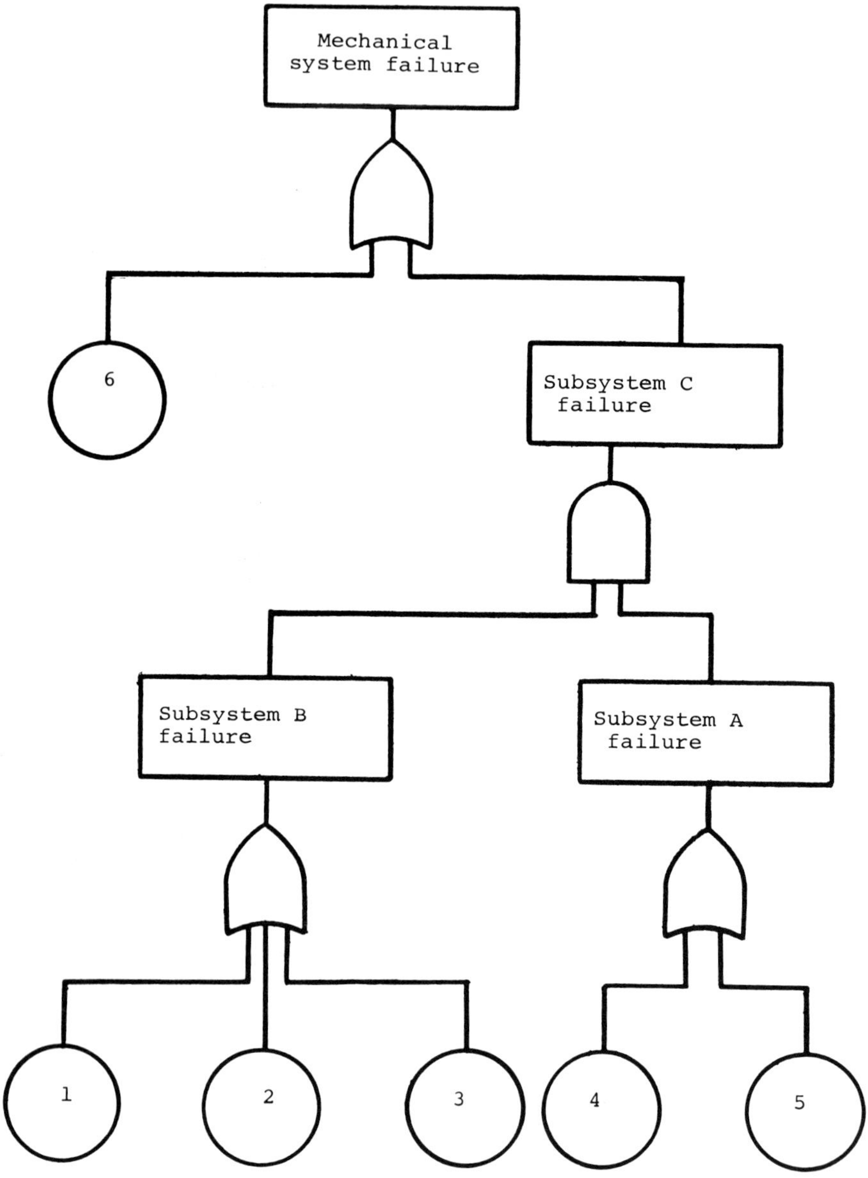

Fig. 3.9 Fault tree of a hypothetical mechanical system.

Similarly, the probability of occurrence of the output fault event of the OR gate is

$$P(Y_1 + Y_2 + Y_3 + \ldots) = 1 - [1 - P(Y_1)][1 - P(Y_2)][1 - P(Y_3)] \ldots \quad (3.68)$$

where Y_i is the OR gate's input fault event, for $i = 1,2,3,\ldots$; $P(Y_i)$ the probability of occurrence of an independent input fault event Y_i, for $i = 1,2,3,\ldots$; and + the symbol used to denote the union of sets (i.e., events $Y_1, Y_2, Y_3,\ldots$).

For very small values of $P(Y_i)$, Eq. (3.68) can be approximated as

$$P(Y_1 + Y_2 + Y_3 + \ldots) \simeq P(Y_1) + P(Y_2) + P(Y_3) + \ldots \quad (3.69)$$

Example 3.20

Assume that in Example 3.19 the values of F_1, F_2, F_3. F_4, F_5, and F_6 are 0.15, 0.1, 0.2, 0.3, 0.25, and 0.13, respectively. Calculate the probability of occurrence of the mechanical system failure (i.e., the top event) using the fault tree of Fig. 3.9. The fault tree of Fig. 3.9 is redrawn in Fig. 3.10 with specified values of basic fault events' occurrence probabilities.

The probabilities of occurrence of events (see Fig. 3.9), i.e., failure of subsystem A, B, or C and/or mechanical system failure, were obtained using Eqs. (3.67) and (3.68) as follows:

Probability of occurrence of subsystem B failure $= 1 - (1 - 0.15)(1 - 0.1)(1 - 0.2)$
$= (0.3880)$

Probability of occurrence of subsystem A failure $= 1 - (1 - 0.3)(1 - 0.25)$
$= 0.4750$

Probability of occurrence of subsystem C failure $= (0.3880)(0.4750)$
$= 0.1843$

Probability of occurrence of mechanical system failure $= 1 - (1 - 0.1843)(1 - 0.13)$
$= 0.2903$

These values are shown in Fig. 3.10. The probability of the top event (mechanical system failure) occurrence is 0.2903. Subtracting 0.2903 from unity, we get 0.7097. Thus, the reliability of the mechanical system is 0.7097. This result is the same as given in Example 3.18.

Failure Mode and Effects Analysis

This technique was developed in the early 1950's for use in the development of flight control systems.[7] Failure mode and effects analysis (FMEA)

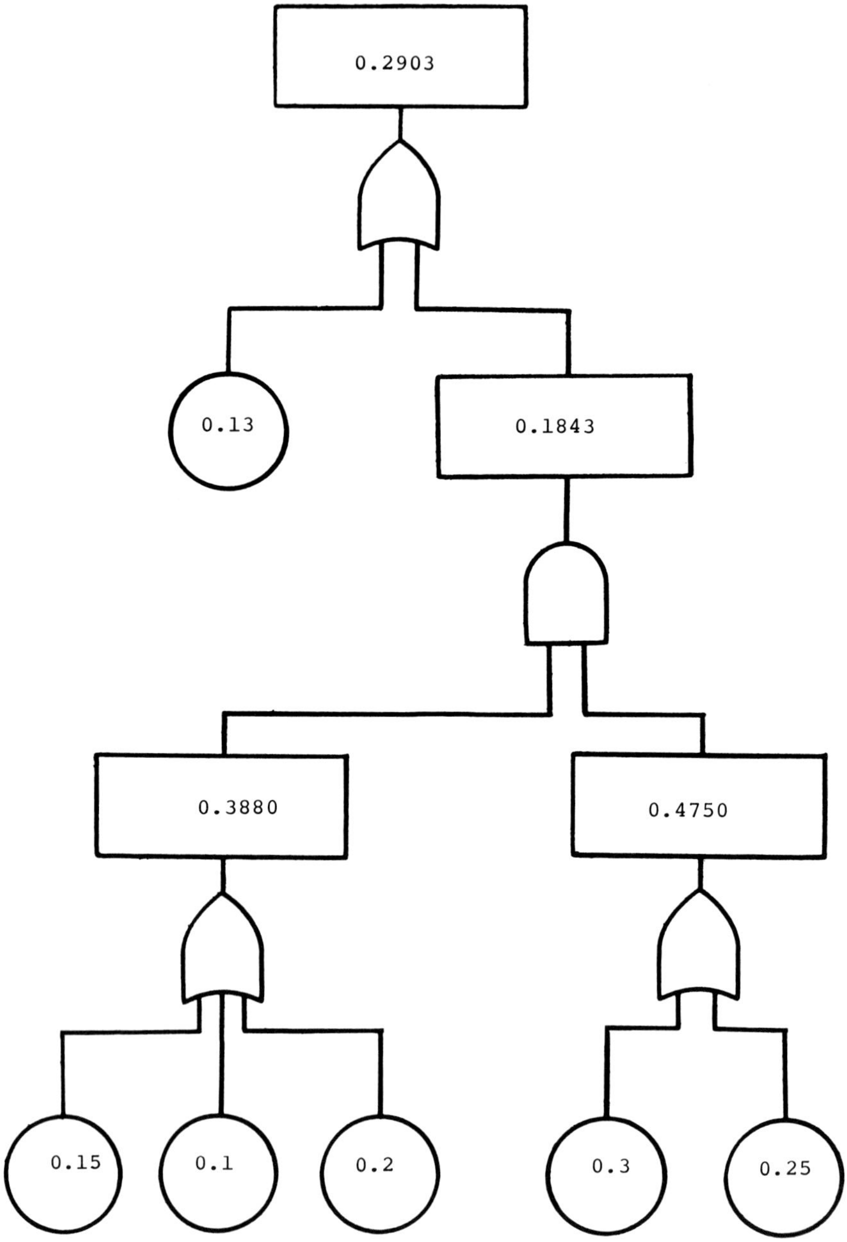

Fig. 3.10 Fault tree with occurrence probabilities of fault events.

is one of the most effective design reliability analysis methods and is basically concerned with identification of potential failure modes and assessment of their effects throughout the system in question. More specifically, it helps to gain an insight into any potential reliability problems. The basic difference between the fault tree method and this technique is that the former is event oriented and the latter malfunction oriented.

Some of the major steps involved in performing FMEA are defining the system boundaries and requirements; listing all of the components and subsystems in the system; identifying and describing the components and their failure modes; assigning a failure rate to each failure mode; listing the effects of each failure mode on the subsystems, plant, etc.; entering remarks for each failure mode; reviewing each crucial failure mode; and initiating any necessary action.

The method is described in detail in Ref. 8.

3.7 A METHOD FOR ALLOCATING FAILURE RATES

In order to meet the specified failure rate (or MTBF) of an engineering system, the general approach is to allocate the system failure rate to its subsystems and components. This way, when the components' failure rate targets are fully satisfied, the overall specified failure rate of the system will be achieved automatically. This section presents a method for allocating failure rates to system units or components.

The method is based on the following assumptions: 1) the system and component failure rates are constant, 2) the components form a series system, and 3) the components fail independently.

From Equation (3.34), the series system failure rate is

$$\lambda_{ss} = \sum_{i=1}^{k} \lambda_i \tag{3.70}$$

where λ_{ss} is the constant failure rate of the series system, λ_i the constant failure rate of the component i, and k the number of independent components in the series.

If the specified rate of the system is λ_{sps}, then we allocate failure rates $\hat{\lambda}_i$ for component i such that

$$\sum_{i=1}^{k} \hat{\lambda}_i \leqslant \lambda_{sps} \tag{3.71}$$

The failure rate allocation can be accomplished in the following five steps:

1) Specify the system failure rate λ_{sps}.

2) Obtain estimates for failure rates, $\lambda_1, \lambda_2, \lambda_3, \ldots, \lambda_k$ of components associated with the system. These estimates can be obtained from field data or other sources.

3) Using the failure rate estimates of step 1, compute the value of λ_{fss} with the aid of the following equation:

$$\lambda_{fss} = \sum_{i=1}^{k} \lambda_i \tag{3.72}$$

where λ_{fss} is the estimate for field failure rate of the series system, λ_i the estimate for field failure rate of component i, and k the number of independent components in the series.

4) Compute ith component's relative weight w_i from the following relationship:

$$w_i = \frac{\lambda_i}{\lambda_{fss}} = \frac{\lambda_i}{\sum_{i=1}^{k} \lambda_i}, \qquad \text{for } i = 1,2,3,4,\ldots,k \tag{3.73}$$

One should note here that the sum of the relative weights is equal to unity, as

$$\sum_{i=1}^{k} w_i = 1 \tag{3.74}$$

5) Allocate failure rate to component i as follows:

$$\hat{\lambda}_i = w_i \lambda_{sps}, \qquad \text{for } i = 1,2,3,\ldots,k \tag{3.75}$$

where $\hat{\lambda}_i$ is the allocated failure rate of unit i.

The above equation is subject to Eq. (3.71).

Example 3.21

The specified failure rate of a mechanical system is 0.008 failures/h. The system is composed of five independent units in series. The failure times of each unit are assumed to be exponentially distributed. The field failure rates of units 1, 2, 3, 4, and 5 are, in failures per hour, $\lambda_1 = 0.001$, $\lambda_2 = 0.0015$, $\lambda_3 = 0.002$, $\lambda_4 = 0.0025$, and $\lambda_5 = 0.003$, respectively. Allocate the specified failure rate of the mechanical system to its units.

Substituting the specified data into Eq. (3.72), we get

$$\lambda_{fss} = (0.001) + (0.0015) + (0.002) + (0.0025) + (0.003)$$

$$= 0.01 \text{ failures/h}$$

Using this result and the specified data for units' failure rates into Eq. (3.73) yields

$$w_1 = \frac{0.001}{0.01} = 0.1 \qquad w_4 = \frac{0.0025}{0.01} = 0.25$$

$$w_2 = \frac{0.0015}{0.01} = 0.15 \qquad w_5 = \frac{0.003}{0.01} = 0.3$$

$$w_3 = \frac{0.002}{0.01} = 0.2$$

Using the above values of w_i (for $i = 1,2,3,4,5$) and the specified value of the mechanical system's failure rate in Eq. (3.75), we get the following allocated values for the failure rates:

$$\hat{\lambda}_1 = (0.1)(0.008) = 0.0008 \text{ failures/h}$$

$$\hat{\lambda}_2 = (0.15)(0.008) = 0.0012 \text{ failures/h}$$

$$\hat{\lambda}_3 = (0.2)(0.008) = 0.0016 \text{ failures/h}$$

$$\hat{\lambda}_4 = (0.25)(0.008) = 0.002 \text{ failures/h}$$

$$\hat{\lambda}_5 = (0.3)(0.008) = 0.0024 \text{ failures/h}$$

where $\hat{\lambda}_1$, $\hat{\lambda}_2$, $\hat{\lambda}_3$, $\hat{\lambda}_4$, and $\hat{\lambda}_5$ are the allocated failure rates for system units 1, 2, 3, 4, and 5, respectively.

Several other reliability allocation methods are presented in Ref. 8.

3.8 FAILURE RATE EVALUATION MODELS FOR MECHANICAL ITEMS

This section presents a number of mathematical models for estimating failure rates of some mechanical items.[9]

Failure Rate Evaluation Model for Pumps

The following equation can be used to predict failure rate of a pump[9]:

$$\lambda_{\text{pump}} = \lambda_{\text{pfd}} + \lambda_{\text{pb}} + \lambda_{\text{pc}} + \lambda_{\text{ps}} + \lambda_{\text{psh}} \text{ failures}/10^6 \text{ cycles} \qquad (3.76)$$

where

λ_{pfd} = constant failure rate of the pump fluid driver
λ_{pb} = constant failure rate of all pump bearings

λ_{pc} = constant failure rate of the pump casing
λ_{ps} = constant failure rate of all pump seals
λ_{psh} = constant failure rate of the pump shaft
λ_{pump} = constant total failure rate of the pump

One should note here that all of the above failure rates are expressed in failures/10^6 cycles.

Failure Rate Evaluation Model for Pump Seals

The failure rate of the pump seal can be expressed as

$$\lambda_{ps} = \lambda_{bps} \prod_{i=1}^{7} f_i \tag{3.77}$$

where λ_{ps} is the constant failure rate of the pump seal, λ_{bps} the constant base failure rate of the pump seal, and f_i the ith modifying factor; $i = 1$ represents the effects of seal smoothness, $i = 2$ of fluid viscosity, $i = 3$ of temperature, $i = 4$ of contaminates, $i = 5$ of surface finish, $i = 6$ of the pressure/velocity factor, and $i = 7$ the casing thrust load.

Thus, Eq. (3.77) can be used to predict failure rate of the pump seal.

Failure Rate Evaluation Model for Pump Shafts

The pump shaft failure rate is given by

$$\lambda_{psh} = \lambda_{bpsh} \prod_{i=1}^{6} f_i \tag{3.78}$$

where λ_{psh} is the constant failure rate of the pump shaft, λ_{bpsh} the constant base failure rate of the pump shaft, and f_i the ith modifying factor. $i = 1$ is the casing thrust load, $i = 2$ the shaft surface finish, $i = 3$ the contamination, $i = 4$ the material temperature, $i = 5$ the pump displacement, and $i = 6$ the material endurance limit.

Using Eq. (3.78), one can predict the failure rate of the pump shaft. This model is described in detail in Ref. 9.

Failure Rate Evaluation Model for Filters

The filter failure rate can be predicted by using the following equation:

$$\lambda_{ft} = \lambda_{bft} \prod_{i=1}^{6} f_i \tag{3.79}$$

where λ_{ft} is the constant failure rate of the filter, λ_{bft} the constant base failure rate of the filter, and f_i the ith modifying factor. $i = 1$ represents the effects of a cold start, $i = 2$ of vibration, $i = 3$ of water contamination, $i = 4$ of cyclic flow effects, $i = 5$ of temperature, and $i = 6$ of differential pressure.

The detailed description of this model can be found in Ref. 9.

References

[1]Rau, J. G., *Optimization and Probability in Systems Engineering*, Van Nostrand Reinhold, New York, pp. 145–186.

[2]*Reliability Prediction of Electronic Equipment*, prepared by U.S. Department of Defense, Washington D.C., MIL-HDBK-217B, 1974, pp. A1–A26. (Available from Rome Air Development Center, Griffiss Air Force Base, NY, 13441.)

[3]Ang, A. H. S. and Tang, W. H., *Probability Concepts in Engineering Planning and Design*, Vol. II, Wiley, New York, 1984, pp. 448–504.

[4]Pages, A. and Gondran, M., *System Reliability*, Springer-Verlag, New York, 1986, pp. 91–96.

[5]Dhillon, B. S., *Reliability Engineering in Systems Design and Operation*, Van Nostrand Reinhold, New York, 1983, pp. 32–42.

[6]Dhillon, B. S. and Singh, C., *Engineering Reliability: New Techniques and Applications*, Wiley, New York, 1981, pp. 48–93.

[7]Countinho, J. S., "Failure-Effect Analysis," *Transactions of the New York Academy of Sciences*, Vol. 26, 1964, pp. 564–584.

[8]Dhillon, B. S., *Systems Reliability, Maintainability and Management*, Petrocelli Books, New York, 1983, pp. 223–237.

[9]Raze, J. D., Nelson, J. J., Simar, D. J., and Bradley, M., Reliability Models for Mechanical Equipment, "*Proceedings of the Annual Reliability and Maintainability Symposium*, 1987, pp. 130–134.

Exercises

3.1 Assume that the failure times of a pump are Rayleigh distributed. Obtain an expression for the pump hazard rate by utilizing the Rayleigh probability density function.

3.2 The Weibull hazard rate function $\lambda(t)$ is given by

$$\lambda(t) = \beta t^{\beta - 1}/\theta \tag{3.80}$$

where β is the shape parameter, θ the scale parameter, and t the time. Obtain an expression for the reliability function.

3.3 An aircraft has two independent and identical engines. Both the engines must work normally for the aircraft to fly successfully. The reliability of each engine is 0.99. Calculate the reliability of the aircraft flying with respect to engines.

3.4 Assume that an aircraft has three active, independent, and identical engines. At least two engines must work normally for the aircraft to fly successfully. The reliability of an engine is 0.86. Compute the reliability of the aircraft flying with respect to engines.

3.5 Assume that in Exercise 3.4, the engine's constant failure rate, $\lambda = 0.00005$ failures/h, is given instead of its reliability (0.86). Determine the following: 1) the aircraft's reliability with respect to engines for a 10 h mission and 2) the aircraft's mean time to failure.

3.6 Prove that the mean time to failure of a r-out-of-m unit system is given by Eq. (3.56).

3.7 Assume that the failure times of independent mechanical units shown in Fig. 3.4 are Weibull distributed. Develop an expression for reliability of the mechanical system.

3.8 Assume that in Exercise 3.7 the failure times of mechanical units are exponentially distributed instead of Weibull distributed. Develop an expression for the mean time to failure of the mechanical system.

3.9 Assume that the specified failure rate of an engineering system is 0.007 failures/h. The system has four independent units in series. Each unit's failure times are represented by an exponential distribution. The failure data collected from the field indicate that the failure rates of units 1, 2, 3, and 4 in failure per hour are $\lambda_1 = 0.002$, $\lambda_2 = 0.003$, $\lambda_3 = 0.004$, and $\lambda_4 = 0.001$. Allocate the specified failure rate of the system to its units.

4
MECHANICAL RELIABILITY: Failure Modes, Causes, Defects, and Safety Factors

This chapter is concerned with several areas related to mechanical reliability. A number of justifications for the increased attention on mechanical reliability are presented. The commonly observed modes of mechanical failure are listed, including failures caused by corrosion, fretting, buckling, wear, fatigue, impact, brittle fracture, thermal shock, thermal relaxation, stress corrosion, and so on. The failure modes of structural and contactual elements are discussed. The factors affecting the structural element's basic modes of failure are listed. Failure modes of thermodynamic, structural, hydraulic-actuation, kinematic, and fluid systems are presented. The causes of failure are grouped into seven distinct categories of defective design, wrong application, manufacturing defect, wear-out, incorrect installation, failure of other parts, and gradual deterioration in performance. Gear failures and their causes are discussed. Casting and weld defects in pressure vessels and piping are described in detail. Definitions of the safety factor are presented along with those of the safety margin. A number of guidelines for the designer to achieve a high degree of reliability are listed.

4.1 INTRODUCTION

In comparison to electronic reliability, the field of mechanical reliability is relatively less developed. The early years of the 1960's witnessed the beginning of in-depth efforts in this field. There have been various compelling reasons for the development of the mechanical reliability field that extend beyond the boundaries of the utilization of good design practices. The following are some of the justifications for the increased attention on mechanical reliability[1]:

1) Increasing costs.

2) Demanding product development schedules (meaning that the designer has no time to learn from his or her errors; the cut-and-try procedure cannot be tolerated).

3) More stringent requirements (which lower the margin for error).

4) Influence of advances made in electronic reliability.

5) More severe environments.

6) Lack of design experience (technological advances are accelerating faster than the design experience is being accumulated by designers).

7) Optimization of a workable design for maximum value (i.e., tradeoffs between reliability, weight, cost, performance, and size have to be performed).

This chapter discusses failure modes, causes, defects, and safety factors.

4.2 MODES OF MECHANICAL FAILURE

Before discussing the modes of mechanical failure, let us define a mechanical failure. Thus, according to Ref. 2, one may define mechanical failure as any change in the shape, size, or material properties of a structure, piece of equipment, or equipment part that renders it unfit to carry out its specified function adequately. Similarly, a failure mode might be expressed as the physical process or processes that occur or combine their effects to generate failure. There are several modes of mechanical failure. The commonly observed ones are as follows[2]:

1) Corrosion failures. There are several forms of corrosion such as pitting, stress, cavitation, direct chemical attack, erosion, and galvanic.

2) Fretting failures. There are three types of fretting failures: fretting fatigue, fretting wear, and fretting corrosion.

3) Creep and buckling failures. These include three types of failures: creep, buckling, and creep buckling.

4) Corrosion fatigue and wear failures. These include corrosion fatigue and corrosion wear.

5) Fatigue failures. There are many forms of fatigue such as thermal, high-cycle, fretting, low-cycle, and surface.

6) Brinelling, yielding, and spalling failures.

7) Wear failures. There are various forms of wear: impact, adhesive, fretting, deformation, abrasive, corrosive, and surface fatigue.

8) Ductile rupture and stress rupture failures.

9) Impact failures. The subcategories of these failures are impact fatigue, impact fracture, impact fretting, impact deformation, and impact wear.

10) Brittle fracture failures.

11) Combined creep and fatigue failures.

12) Thermal shock failures.

13) Thermal relaxation failures.

14) Stress corrosion failures.

15) Galling and seizure failures.

16) Force-induced elastic deformation failures.

17) Temperature-induced elastic deformation failures.

18) Combined force- and temperature-induced elastic deformation failures.

19) Radiation damage failures.

A detailed description of these modes of mechanical failures is given in Ref. 2.

Failure Modes of Structural and Contactual Elements

The solid parts belonging to a mechanical device may be considered of having structural and contactual elements. According to Ref. 3, one may describe a structural element as a portion of a part whose basic purpose is to withstand loads, which can either be cyclic, thermal, intermittent, impact, steady, centrifugal, or vibrational. Similarly, a contactual element may be described[3] as a portion of a part whose basic purpose is to provide a contact surface with another contactual element that normally belongs to another part. In this case, the overall purpose is to provide for static bearing or relative motion of the rolling, impacting, sliding, pivoting, or sealing varieties (i.e., with or without the application of an intermediate lubricant).

The structural element's basic modes of failure are given in Fig. 4.1. There are several factors that affect these failure modes. Some of these are environments, size of the part, material, manner of loading, shape of the part, properties of the material, and condition of the part's surfaces.

The failure modes of a structural element that depend on time (or cycle) are fatigue, creep buckling, creep, creep rupture, etc. The contactual element also has several modes of failure, some of which are seizure, fretting, stress corrosion, wear, galling, corrosion, and surface fatigue. One should note here that these failure modes are time dependent.

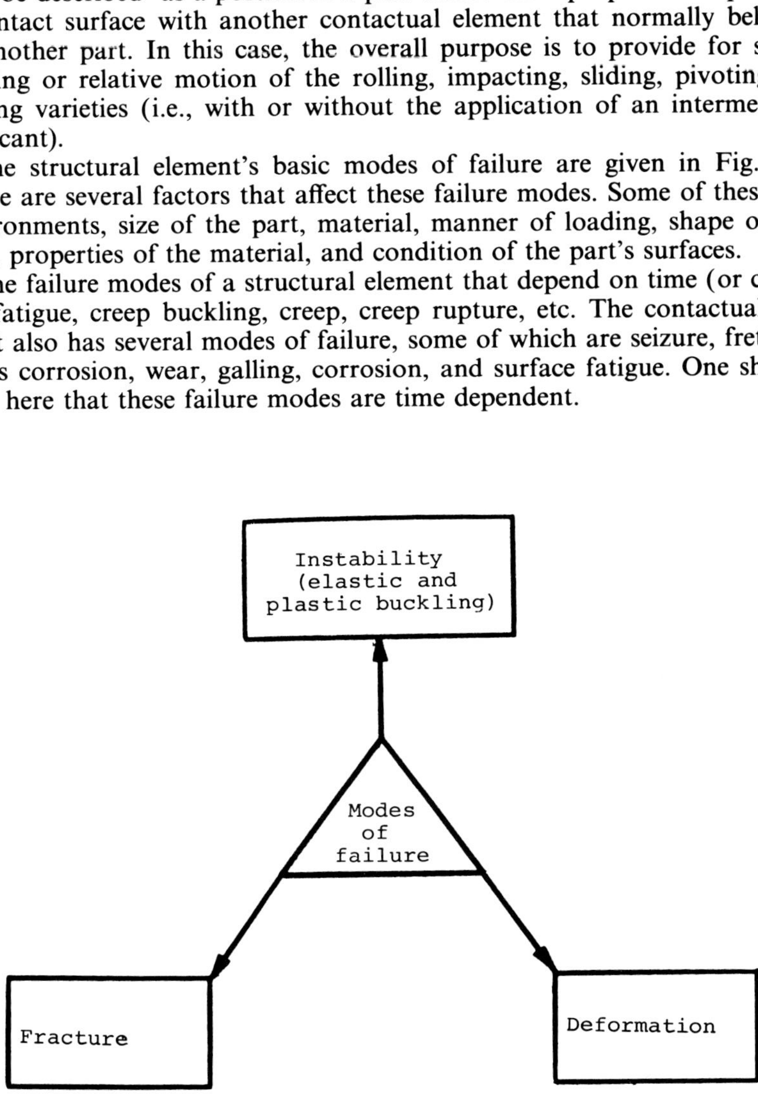

Fig. 4.1. Basic modes of failure of a structural element.

Table 4.1 Failure Modes of Some Systems

No.	System	Failure modes
1	Thermodynamic	Overheating Reduction in efficiency
2	Structural	Excessive deflection Fracture
3	Hydraulic-actuation	Fitting leakage Rupture of actuator cylinder Fluid dirt contamination Servo amplifier drift Static seal leak
4	Kinematic	Bearing seizure Reduction in accuracy of relative movement
5	Fluid	Distorted flow Leakage

Failure Modes of Some Systems

This section presents typical failure modes associated with the following systems[1,4]: thermodynamic, structural, hydraulic-actuation, kinematic, and fluid. Examples of failure modes associated with the above systems are presented in Table 4.1.

4.3 FAILURE CAUSES

There are many causes of failure. In Ref. 5, such causes have been grouped into seven distinct categories: 1) defective design, 2) wrong application, 3) manufacturing defect, 4) wear-out, 5) incorrect installation, 6) failure of other parts, and 7) gradual deterioration in performance.

A more thorough inspection, adjustment, or preventive maintenance can be practiced to correct failures in categories 2–7. However, it is not possible to readily detect failures due to a design defect because they are often of a fatigue nature and happen without admonition.

The sources of responsibility for product failure in the following areas[5] are:

1) Steel industry. Manufacturing and metallurgy (34%), design (58%), and service (8%).

2) Engines. Manufacturing (20%), engineering (40%), misuse in field (30%), and other causes (10%).

3) Truck industry. Manufacturing (12%), design (55%), and materials (33%).

4) Electrical industry. Manufacturing (37%), design (37%), and service (26%).

A study[5] of 931 gear failures occurring over a 35 year period reported that breakage, surface fatigue, wear, and plastic flow accounted for 61.2, 20.3, 13.2, and 5.3% of the total failures, respectively. In addition, the study reported the following causes: heat-treatment related (16.2%), manufacturing related (1.4%), service related (74.7%), material related (0.8%), and design related (6.9%).

The heat treatment-related causes included distortion (0.2%), inadequate case depth (4.8%), excessive core hardness (0.5%), incorrect hardening (5.9%), inadequate core hardness (2.0%), excessive case depth (1.3%), and improper tempering (1.0%). The manufacturing-related causes were grinding burns (0.7%) and tool marks or notches (0.7%). The improper assembly (21.2%), operator errors (0.3%), impact loading (13.9%), incorrect lubrication (11.0%), bearing failure (0.7%), abusive handling (1.2%), continual overloading (25.0%), and foreign material (1.4%) belong to the category of service-related causes. There were three material-related causes: steel defects (0.5%), forging defects (0.1%), and mixed steel or incorrect composition (0.2%). Finally, the design-related causes included incorrect material selection (1.6%), specification of unsuitable heat treatment (2.5%), and wrong design (2.8%).

4.4 CASTING AND WELD DEFECTS IN PRESSURE VESSELS AND PIPING

A defect may be described as the deficiency or absence of something vital for completeness or perfection.[6] This section concentrates on two types of defects in pressure vessels and piping, i.e., casting and weld defects.

Casting Defects

There are various types of casting defects. The commonly known are shown in Fig. 4.2. Five of the defects shown in Fig. 4.2 are briefly discussed below.

Unfused chaplets. These are metal spacers or supports employed in molds, generally for the purpose of upholding the cores during casting in their correct locations. Chaplets may cause service failures.

Inclusions. During the solidification oxides, dross or scale entrapped within the framework of the casting are represented by inclusions. They become critical only when their applicable specification limits are bypassed. Another factor for the criticality can be their occurrence, in a fair size, at or close to the surface.

Hot tears and cracking. During the early stages of cooling, the hot tears occur and frequently contain oxide or scale. On the other hand, the cracks generally occur at or close to room temperature. In appearance, cracks and hot tears are quite similar.

Gas and blow holes. These result because of gases dissolved in the molten metal. Such holes on the inside of castings rarely cause any service failure.

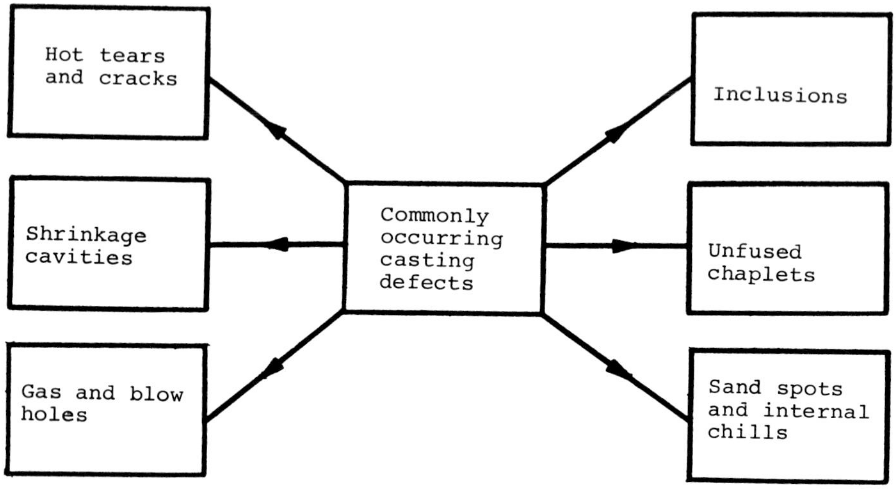

Fig. 4.2. Commonly occurring casting defects.

Shrinkage cavities. These occur quite commonly in steel castings and have caused service failures. The causes for large shrinkage cavities are risers of inadequate size, improper gating, etc. Similarly, an insufficient number and the position of risers can be the causes for the occurrence of smaller cavities.

Weld Defects

Numerous publications have appeared on the study of weld defects. Just as in the case of base metals, weld deposits are not defect free.[6] Some weld defects are more severe than others and in the past have led to failures. Some of the weld defect classifications are as follows[6]: burn through, cracks, high/low, porosity, sink or concavity, slag inclusions, weld spatter, wagon tracks, slugging, arc strikes, tungsten inclusions, undercut, crater pits, incomplete fusion, lack of penetration, and overlap. All of these items are described in detail in Ref. 6. Here, we will describe briefly some of the important ones with respect to reliability.

Slag inclusions. In this case, slag represents solid material of a nonmetallic nature that is entrapped in the weld deposit or between the base and weld metals. Usually, slag inclusions do not cause service failures, unless they are of a significant size or occur along the surface.

Tungsten inclusions. In inert-gas, tungsten-arc welding, the tungsten electrode produces particles that may in turn deposit in the weld metal. These are known as tungsten inclusions. They can be harmful when their number and size become large.

Burn through. A coalescence of weld metal beyond the root of the weld is referred to as burn through. The burn-through conditions can be harmful when they are continuous and very severe.

Cracks. These may be described as linear ruptures of metal under stress. In welded joints, the cracking occurs in various forms. Usually, these are base-metal cracking, hot cracking in weld deposits, microfissuring, and cold cracking in weld deposits. Both the weld- and the base-metal crackings are considered to be critical and have resulted in service failures.

Lack of penetration. This means that the penetration of the weld through the thickness of the joint is incomplete. In the past, service failures have resulted from lack of penetration.

Sink or concavity. The concavity may be described as sink in a root weld bead. A number of service failures have resulted from excessive sink.

Slugging. This may be described as the addition of a separate piece or pieces of material in a weld joint prior to or during the welding process. Slugging may lead to very harmful failures.

Overlap. This is usually associated more with fillet welds than with butt welds and may be described as an excess of weld metal extending beyond the boundaries of fusion over the base-metal surface. Failures may be caused by the severe overlap in fillet welds.

Incomplete fusion. This may be described as a deficiency in the total melting and coalescence (fusion) of some part of the metal in the joint of a weld. The incomplete fusion may lead to service failures.

4.5 SAFETY FACTORS

The factor used to determine the safeness of a member is known as the factor of safety. In cases where the safety factors are established from the previous experience, they may provide satisfactory design. However, design based only on such factors can be misleading.

There have been several definitions of the safety factor, some of which are given below.[7–16]

Definition I

The safety factor is defined by[8,9]

$$SF = \frac{US}{WS} \tag{4.1}$$

where SF is the safety factor, WS the working stress in lb/in.^{2} (psi), and US the ultimate strength in psi.

Definition II

In this definition the safety factor is expressed as[10]

$$SF = \frac{MSL}{NSL} \tag{4.2}$$

where SF is the safety factor, MSL the maximum safe load, and NSL the normal service load.

This is considered to be a good definition, especially in the case when the loads are distributed.

Definition III

The safety factor is expressed as[11]

$$SF = \frac{MS}{ML} \tag{4.3}$$

where SF is the safety factor, MS the mean strength, and ML the mean load.

Definition IV

This safety factor is defined as[8]

$$SF = \frac{YS}{WS} \tag{4.4}$$

where SF is the safety factor, YS the yield strength, and WS the actual or allowable working stress.

Definition V

In this case, the safety factor is defined by[12]

$$SF = \frac{DS}{MNS} \tag{4.5}$$

where SF is the safety factor, DS the damaging stress (i.e., fatigue limit), and MNS the maximum known working stress.

Definition VI

In this case to compute the factor of safety for gears is defined as[15]

$$FS_g = F_{OL} \cdot F_{LD} \,.\, FS_U \tag{4.6}$$

where FS_g is the factor of safety for gears, FS_U the usual factor of safety, F_{OL} the overload correction factor, and F_{LD} the load-distribution factor.

In order to guard against fatigue failure, the value of FS_U used is two or more. The tabulated values of F_{OL} and F_{LD} are presented in Ref. 15.

Definition VII

In this case, it is considered that the factor of safety is the product of several independently determined factors as[16]

$$FS = \prod_{i=1}^{3} f_i \tag{4.7}$$

where FS is the factor of safety, f_1 the shock factor, f_2 the elastic ratio, and f_3 the real margin of safety.

The shock factor f_1 is given by

$$f_1 = S_{sh}/S_g \tag{4.8}$$

where S_{sh} is the stress generated by a shock or by a suddenly applied load and S_g the stress generated by the same load when it is applied in a gradual manner.

According to Ref. 16 the mean values of the shock factor for the following three cases are 1.25–1.50 (light shock), 2–3 (very heavy shock), and 5 (a few extreme cases).

For static loads, the elastic ratio f_2 is defined by

$$f_2 = US/EL \tag{4.9}$$

where US is the ultimate strength and EL the elastic limit. This is normally taken as the yield stress.

The real margin of safety factor is used by designers to allow for unreliable material, temperature effects, costly repairs, etc. The mean values of this factor for some materials are 1.5–2 (ductile materials of uniform structure, e.g., iron and steel), 2.5–3 (brittle materials, e.g., cast iron), and 3–4 (timber).

Selection of an Appropriate Value of Safety Factor

Factors such as the following play an important role in the selection of an appropriate value of safety factor[14]: cost, loading uncertainty, uncertainty of material strength, failure consequence, and degree of uncertainty involved in relating applied loads to item strength.

4.6 SAFETY MARGINS

In many publications the basic definition of safety margin is given as

$$SM = (\text{safety factor}) - 1 \tag{4.10}$$

where SM is the safety margin.

Equation (4.10) states that the safety margin is given simply by the value of the safety factor minus unity or more specifically, the quantity by which the value of the safety factor surpasses unity. In addition, it should be noted that the above definition does not take into consideration the fact that the strength and stress are distributed. In order to overcome this problem, for normally distributed stress and strength, we define the safety margin as[7]

$$SM = \frac{m_{sth} - MST_{max}}{\sigma_{sth}} \tag{4.11}$$

where σ_{sth} is the standard deviation of the strength, m_{sth} the mean value of the strength, and MST_{max} the maximum value of the stress. The maximum value of the stress is given by

$$MST_{max} = m_{st} + k\sigma_{st} \tag{4.12}$$

where m_{st} is the mean value of the stress and σ_{st} the standard deviation of the stress.

The usual value of k used is between 3 and 6, i.e., $k = 3$–6. Substituting Eq. (4.12) into Eq. (4.11) yields

$$SM = \frac{m_{sth} - m_{st} - k\sigma_{st}}{\sigma_{sth}} \tag{4.13}$$

Example 4.1

Suppose that we have the following given data: $m_{st} = 8000$ psi, $\sigma_{st} = 400$ psi, $m_{sth} = 16{,}000$ psi, and $\sigma_{sth} = 1000$ psi. Calculate the values of the safety margin for $k = 3$, 3.5, 4, 4.5, 5, 5.5, and 6.

Using the above given data in Eq. (4.13), the following values of the safety margin result:

For $k = 3$, $$SM = \frac{16{,}000 - 8000 - (3)(400)}{1000} = 6.8$$

For $k = 3.5$, $$SM = \frac{16{,}000 - 8000 - (3.5)(400)}{1000} = 6.6$$

For $k = 4$, $$SM = \frac{16{,}000 - 8000 - (4)(400)}{1000} = 6.4$$

For $k = 4.5$, $$SM = \frac{16{,}000 - 8000 - (4.5)(400)}{1000} = 6.2$$

For $k = 5$, $$SM = \frac{16{,}000 - 8000 - (5)(400)}{1000} = 6.0$$

For $k = 5.5$, $$SM = \frac{16{,}000 - 8000 - (5.5)(400)}{1000} = 5.8$$

For $k = 6$, $$SM = \frac{16{,}000 - 8000 - (6)(400)}{1000} = 5.6$$

It can be seen from these results that the value of the safety margin decreases with the corresponding increasing values of k.

In Ref. 17, the margin of safety is defined by

$$\mathrm{MS} = \left(\frac{\mathrm{MAS}}{\mathrm{MDS}} - 1\right)(100) \tag{4.14}$$

where MS is the margin of safety, MAS the minimum allowable stress, and MDS the maximum design stress.

According to Ref. 18, the design margin of safety is given by

$$\mathrm{DMS} = \left[\frac{\mathrm{US}}{(\mathrm{LS})(\mathrm{SF})} - 1\right] \tag{4.15}$$

where DMS is the design margin of safety, US the ultimate strength, LS the limit stress, and SF the safety factor.

4.7 GUIDELINES FOR THE DESIGNER TO ACHIEVE A HIGH DEGREE OF RELIABILITY

This section presents a number of useful guidelines for a designer to achieve a high level of reliability. Guidelines such as given below directly or indirectly contribute in achieving a high degree of reliability.

1) Make design as simple as possible.
2) Avoid introducing cost-saving procedures at the expense of reliability.
3) Make allowances for human error.
4) Make use of well-tried parts and materials.
5) Carefully test new designs.
6) Perform analyses of data received from the field.
7) Pay attention to data received from the field when making modifications and improvements.
8) In the case of critical parameters, consider using safety margins of 3–6 standard deviations.
9) Consider diagnosis of critical items.
10) Include facilities for inspection in the design.
11) Introduce redundancy whenever necessary.
12) Pay attention to the maintenance aspect with respect to reliability.
13) Take into account the effects of transport, handling, and storage.
14) Make use of standard parts as much as possible.
15) Pay attention to the manufacturing aspect with respect to reliability.

References

[1]McCrory, R. J., "Elements of Realism in Mechanical Reliability," *Mechanical Reliability Concepts, ASME Design Engineering Conference Proceedings*, ASME, New York, 1965, pp. 1–16.

[2]Collins, J. A., *Failure of Materials in Mechanical Design*, Wiley, New York, 1981, pp. 6–14.

[3]Mittenbergs, A. A., "Fundamental Aspects of Mechanical Reliability," *Mechanical Reliability Concepts, ASME Design Engineering Conference Proceedings*, ASME, New York, 1965, pp. 17–34.

[4]Martin, P., "Reliability in Mechanical Design and Production," *Proceedings of Generic Techniques in Systems Reliability Assessment*, Noordhoff-Leyden, Amsterdam, 1976, pp. 267–271.

[5]Lipson, C., *Analysis and Prevention of Mechanical Failures*, Course Notes 8007, University of Michigan, Ann Arbor, June 1980.

[6]Thielsch, H., *Defects and Failures in Pressure Vessels and Piping*, R. E. Krieger Publishing Co., Huntington, NY, 1977, pp. 210–255.

[7]Kececioglu, D. and Haugen, E. B., "A Unified Look at Design Safety Factors, Safety Margins and Measures of Reliability," *Proceedings of Annual Reliability and Maintainability Conference*, ASME, New York, 1968, pp. 520–530.

[8]Faires, V. M., *Design of Machine Elements*, MacMillan, New York, 1955, p. 205.

[9]Howell, G. M., "Factors of Safety," *Machine Design*, July 12, 1956, pp. 76–81.

[10]Phelan, M. R., *Fundamentals of Machine Design*, McGraw-Hill, New York, 1962, p. 208.

[11]McCalley, R. B. "Nomogram for Selection of Safety Factors," *Design News*, Sept. 1957, pp. 138–141.

[12]Schoof, R., "How Much Safety Factor?" *Allis-Chalmers Electrical Review*, 1960, pp. 21–24.

[13]MacGregor, J. G., "Safety and Limit States Design For Reinforced Concrete," *Canadian Journal of Civil Engineering*, Vol. 3, 1976, pp. 484–513.

[14]Juvinall, R. C., *Fundamentals of Machine Component Design*, Wiley, New York, 1983, pp. 71–73.

[15]Shigley, J. E. and Mitchell, L. D., *Mechanical Engineering Design*, McGraw-Hill, New York, 1983, pp. 610–611.

[16]Vallance, A. and Doughtie, V. L., *Design of Machine Members*, McGraw-Hill, New York, 1943, pp. 71–73.

[17]Dicks, C. and Wilson, S., "Structural System Design Technique for Reliability and Weight Tradeoffs," *Proceedings of Annual Reliability and Maintainability Conference*, ASME, New York, July 1965, pp. 421–430.

[18]Baker, H. C., "Reliability Analysis and Test of Aluminum Honeycomb Structures," *Proceedings of Annual Reliability and Maintainability Conference*, ASME, New York, July 1965, pp. 421–430.

[19]Furman, T. T., *Approximate Methods in Engineering Design*, Academic Press, London, 1981, pp. 162–164.

Exercises

4.1 What are the reasons for the increased attention on mechanical reliability?

4.2 List at least 10 modes of mechanical failure.

4.3 Describe the following: 1) structural element and 2) contactual element.

4.4 What are the basic modes of failure of a structural element?

4.5 List the failure modes of a hydraulic-actuation system.

4.6 What are the commonly occurring casting defects?

4.7 Describe at least five classifications of weld defects.

5
RELIABILITY EVALUATION OF MECHANICAL COMPONENTS AND SYSTEMS

This chapter is concerned with reliability evaluation of mechanical components and systems. A number of basic objectives of structural design criteria are listed. Factors for setting a structural reliability level are given. Stress and strength distributions are described. Steps for determining failure governing stress and strength distributions are presented. Thirteen steps of "design-by-reliability" methodology are listed. General reliability and unreliability expressions concerned with stress-strength modeling are developed. Eight interference theory models are presented for

1) Exponentially distributed stress and normally distributed strength.
2) Exponentially distributed stress and strength.
3) Normally distributed stress and strength.
4) Normally distributed stress and exponentially distributed strength.
5) Maxwellian distributed stress and strength.
6) Exponentially distributed stress and gamma distributed strength.
7) Exponentially distributed stress and Maxwellian distributed strength.
8) Gamma distributed stress and exponentially distributed strength.

Equations for performing stress-strength analysis of redundant mechanical systems are presented. A parallel mechanical system is analyzed (i.e., for exponentially distributed stress and strength). A graphical method is outlined for calculating reliability of a mechanical item using Mellin transforms.

5.1 INTRODUCTION

Having a high degree of reliability in each mechanical item that is designed and manufactured is one's expectation. It can generally be made possible simply by overdesigning mechanical items when factors such as cost and size are not the problems to be considered and, in addition, when the design problem is within the boundaries of the present state-of-the-art. However, today's competitive economic environment dictates a high reliability of mechanical items at the lowest life-cycle cost possible. In addition, the designer may also have to deal with requirements for minimum weight and limited space. These are the kinds of problems a designer has to overcome when designing a mechanical item. Such requirements impose demands to tradeoff reliability with factors such as performance, cost, and weight. Because these factors can be expressed in quantitative terms, the reliability has to be quantified in order to make the tradeoffs meaningful.

A reliability prediction may be described as the process by which a numerical estimate is made in terms of the ability of a design to carry out its specified function without failure. Frequently, in evaluating the reliability of an electronic item, the assumption of a constant failure rate is made. This assumption is based on the belief that electronic product failures follow the "bathtub" hazard rate curve. In the case of mechanical items, this may or may not be true. In many cases, it is found that mechanical components generally follow an increasing failure rate pattern. This chapter is concerned with reliability evaluation of mechanical parts and systems.

5.2 STRUCTURAL RELIABILITY DESIGN FACTORS

This section briefly discusses the basic objectives of structural design criteria and the considerations in setting a structural reliability level for aerospace vehicles.

Objectives of Structural Design Criteria

There are a number of basic objectives listed for aircraft and space vehicles in Ref. 1 as follows:

1) Define the vehicle's satisfactory level of strength by taking into consideration its specified mission.

2) Specify the operating limitations to the user.

3) Outline an administrative tool for making decisions regarding the issue of structural compliance or noncompliance with the specifications for the vehicle.

4) Outline a decision-making tool for allocating the responsibility of causing any structural failure.

5) Provide a mechanism for deciding any changes in the structural design criteria.

6) Define only criteria that are possible to implement with the existing state-of-the art.

7) Define requirements that will lead to a structural system with acceptable reliability.

These basic objectives may also be applicable to other mechanical items.

Factors for Setting a Structural Reliability Level

In setting a desired level of structural reliability for space vehicles, one has to consider a number of factors. Some of those are prestige, psychological effects of concerned failures, mission type (e.g., manned or unmanned), and incorporation of testing, analysis, maintenance, and operation errors in the final reliability figure. Such factors may also be considered for other mechanical products.

5.3 STRENGTH AND STRESS DISTRIBUTIONS

Strength Distribution

Before we discuss the strength distribution, let us look into strength itself. Strength may be defined as the ability of an item to resist failure due to external environment and loading.[2,3] The strengths of nominally identical components subject to the same conditions normally vary from component to component. This variation in strengths takes place because of variabilities in material properties, production processes, geometric dimensions, etc. All of the important variabilities must be considered when estimating the expected strength distribution of an item. The techniques to obtain the resulting strength distribution are given in Refs. 4–7. According to Ref. 8, the failure governing strength distribution can be determined in five steps. These steps are concerned with developing the failure governing strength criterion, evaluating the nominal strength, making use of the necessary strength factors to modify the nominal strength, obtaining the nominal strength distribution as well as of each modifying factor and parameter associated with the strength equation, and synthesizing the distribution functions into the strength distribution. Three methods of synthesizing the failure governing strength distribution are[9] Monte Carlo simulation, generating system moments, and binary synthesis.

It is always desirable to aim for narrow strength distributions. A narrow strength distribution leads to a higher reliability than its counterpart, the wide strength distribution, for the same average value of strength. In addition, it is also true for a load level less than the average strength. Another advantage of having a narrow strength distribution is that, for the same reliability, it allows one to make use of components with lower average strength.

Stress Distribution

The term stress (load) may be simply described as any factor that tends to produce a failure. Some of the examples of the factor in question are temperature, environment, and electric current. The stress changes under different conditions. Factors such as environment, maintenance, and use pattern lead to differences in the degree of stress imposed. Generally, it can be said that the larger the spread in the stress distribution, the greater the spread in the failure distribution.[10] Reference 8 presents a number of steps for determining the failures governing stress distributions. These steps are concerned with identifying the significant failure modes, performing stress probing when the fracture is one of the significant failure modes, computing nominal stress components, making use of the appropriate stress-modifying factors in obtaining each stress component's maximum value, computing the principal stresses, combining the stresses into the failure governing stress at each location, determining each nominal stress's distribution and of each modifying factor and parameter associated with the failure-governing stress equation, and synthesizing distributions into the failure-governing stress distribution. The frequently used methods of synthesizing the failure governing stress distribution are the moment-generating functions, Monte Carlo simulation, binary synthesis, and central limit theorem.

5.4 DESIGN METHODOLOGY

This section presents a "design-by-reliability" methodology composed of the following steps:

1) State and define the design problem requiring solution.
2) Identify and list all the appropriate design variables and parameters.
3) Carry out failure mode and effect analysis (FMEA) and list the critical items.
4) Verify the selection of the important design parameters.
5) Establish the relationship between the crucial parameters and the failure governing criteria.
6) Determine the failure-governing stress function and then the failure-governing stress distribution.
7) Determine the failure-governing strength function and then the failure-governing strength distribution.
8) Compute the reliability associated with the above failure-governing stress and strength distributions for each crucial failure mode in question.
9) Iterate the design in question to meet the assigned reliability goals.
10) Perform the design optimization in terms of factors such as reliability, cost, performance, volume, weight, safety, and maintainability.
11) Repeat the above design optimization for all of the crucial parts and subsystems of the system in question.
12) Compute reliability of the system.
13) Iterate the design in question until the system's reliability goal is fully satisfied.

5.5 STRESS-STRENGTH ANALYSIS

In some cases, the failure of certain items is not necessarily dependent upon their usage time—their failures may be directly traceable to some other stress variable. In situations such as this, the item's reliability can be ascertained by comparing its strength to the stress. An item's stress and strength are random variables. If the probability density functions of stress and strength are known, the reliability of a component may be determined through analytical means.

General Reliability and Unreliability Expressions

From the point of view of failure-governing stress and strength, the reliability may be expressed as the probability that strength exceeds stress as

$$R = P(x > y) = P(y < x) = P(x - y > 0) \tag{5.1}$$

where x is the strength random variable, y the stress random variable, R the reliability, and P the probability.

Similarly, the unreliability F is given by

$$F = P(y > x) = P(x < y) = P(y - x > 0) \tag{5.2}$$

Figure 5.1 shows the left and right tails of the strength and stress distributions, respectively. The stress-strength interference area is also shown in this figure. The probability that a stress of value y_0 exists in a small interval of width dy is given by

$$P\left(y_0 - \frac{\mathrm{d}y}{2} \leqslant y \leqslant y_0 + \frac{\mathrm{d}y}{2}\right) = f(y_0) \cdot \mathrm{d}y \tag{5.3}$$

One should note that the right-hand side of Eq. (5.3) is equal to the area α of the element dy. The probability of strength exceeding a stress of value y_0 is given by

$$P(x > y_0) = \int_{y_0}^{\infty} f(x)\,\mathrm{d}x \tag{5.4}$$

where $f(x)$ is the strength probability density function.

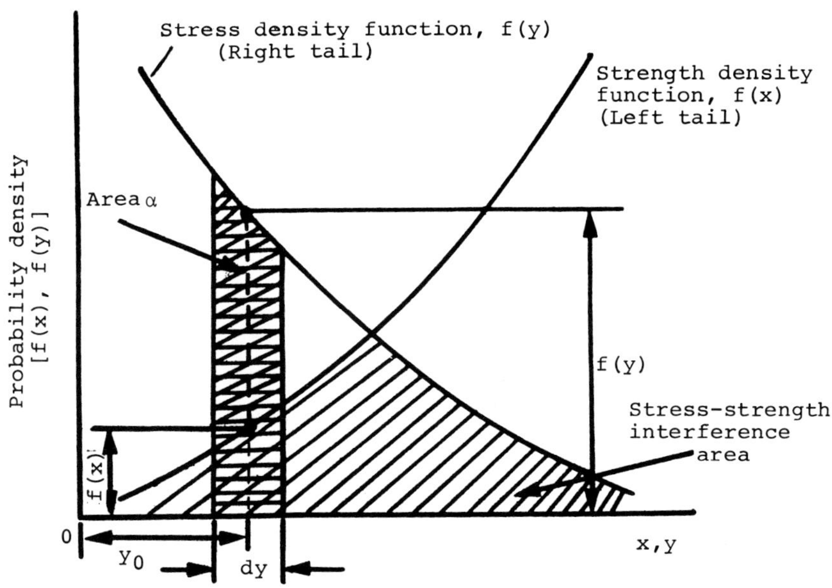

Fig. 5.1 Plots of the left and right tails of strength and stress distributions, respectively.

If the stress is in the infinitesimal neighborhood of y_0, the reliability (at y_0) is given by

$$\left[f(y_0)\cdot \mathrm{d}y\right]\cdot\left[\int_{y_0}^{\infty} f(x)\,\mathrm{d}x\right] \tag{5.5}$$

Note here that the above reliability is given by the product of the right-hand sides of Eqs. (5.3) and (5.4).

If y_0 is allowed to range over all possible positive values, then the probability that strength x is greater than stress y (i.e., all possible values of stress) is given by

$$R=\int_{-\infty}^{\infty}\left[\int_{y}^{\infty} f(x)\,\mathrm{d}x\right]f(y)\cdot \mathrm{d}y \tag{5.6}$$

where R is the item's reliability and $f(y)$ the stress probability density function.

Three alternative forms of the above relationship are

$$R=\int_{-\infty}^{\infty}\left[\int_{-\infty}^{x} f(y)\cdot \mathrm{d}y\right]f(x)\,\mathrm{d}x \tag{5.7}$$

$$R=\int_{-\infty}^{\infty} f(y)\left[1-\int_{-\infty}^{y} f(x)\,\mathrm{d}x\right]\mathrm{d}y \tag{5.8}$$

since

$$\int_{-\infty}^{y} f(x)\,\mathrm{d}x+\int_{y}^{\infty} f(x)\,\mathrm{d}x=1 \tag{5.9}$$

and

$$R=\int_{-\infty}^{\infty} f(x)\left[1-\int_{x}^{\infty} f(y)\,\mathrm{d}y\right]\mathrm{d}x \tag{5.10}$$

since

$$\int_{-\infty}^{x} f(y)\,\mathrm{d}y+\int_{x}^{\infty} f(y)\,\mathrm{d}y=1 \tag{5.11}$$

Subtracting Eq. (5.6) from unity, we get the following expression for an item's unreliability:

$$F=1-\int_{-\infty}^{\infty}\left[\int_{y}^{\infty} f(x)\,\mathrm{d}x\right]f(y)\,\mathrm{d}y \tag{5.12}$$

where F is the item's unreliability.

Using Eq. (5.9) in Eq. (5.12) yields

$$F = 1 - \int_{-\infty}^{\infty} \left[1 - \int_{-\infty}^{y} f(x)\, \mathrm{d}x \right] f(y)\, \mathrm{d}y$$

$$= 1 - \int_{-\infty}^{\infty} f(y)\, \mathrm{d}y + \int_{-\infty}^{\infty} \left[\int_{-\infty}^{y} f(x)\, \mathrm{d}x \right] f(y)\, \mathrm{d}y$$

$$= 1 - 1 + \int_{-\infty}^{\infty} \left[\int_{-\infty}^{y} f(x)\, \mathrm{d}x \right] f(y)\, \mathrm{d}y$$

$$= \int_{-\infty}^{\infty} \left[\int_{-\infty}^{y} f(x)\, \mathrm{d}x \right] f(y)\, \mathrm{d}y \tag{5.13}$$

Similarly, subtracting Eq. (5.7) from unity, we get

$$F = 1 - \int_{-\infty}^{\infty} \left[\int_{-\infty}^{x} f(y)\, \mathrm{d}y \right] f(x)\, \mathrm{d}x \tag{5.14}$$

where F is the item's unreliability.

Using Eq. (5.11) in Eq. (5.14) yields

$$F = 1 - \int_{-\infty}^{\infty} \left\{ 1 - \left[1 - \int_{-\infty}^{x} f(y)\, \mathrm{d}y \right] \right\} f(x)\, \mathrm{d}x$$

$$= 1 - \int_{-\infty}^{\infty} f(x)\, \mathrm{d}x + \int_{-\infty}^{\infty} \left[1 - \int_{-\infty}^{x} f(y)\, \mathrm{d}y \right] f(x)\, \mathrm{d}x$$

$$= 1 - 1 + \int_{-\infty}^{\infty} \left[1 - \int_{-\infty}^{x} f(y)\, \mathrm{d}y \right] f(x)\, \mathrm{d}x$$

$$= \int_{-\infty}^{\infty} \left[1 - \int_{-\infty}^{x} f(y)\, \mathrm{d}y \right] f(x)\, \mathrm{d}x \tag{5.15}$$

Differentiating Eqs. (5.13) and (5.15) with respect to y and x, respectively, yields the following failure functions:

$$\frac{\mathrm{d}F}{\mathrm{d}y} = f(y) \left[\int_{-\infty}^{y} f(x)\, \mathrm{d}x \right] \tag{5.16}$$

$$\frac{\mathrm{d}F}{\mathrm{d}x} = f(x) \left[1 - \int_{-\infty}^{x} f(y)\, \mathrm{d}y \right] \tag{5.17}$$

Interference Theory Models

This section is concerned with determining an item's reliability when the stress and strength probability density functions are known or, more specifically, when the stress and strength probability density functions associated with the item in question are defined. A number of specific interference theory models have been presented in Refs. 11–14.

Model I. In this case, an item's stress and strength are described by the exponential and normal probability density functions, respectively, as

$$f(y) = \theta e^{-\theta y}, \qquad y > 0 \tag{5.18}$$

$$f(x) = \frac{1}{\sigma\sqrt{2\pi}} \exp\left[-\frac{1}{2}\left(\frac{x-\mu}{\sigma}\right)^2\right], \qquad -\infty < x < \infty \tag{5.19}$$

where θ is the reciprocal of the mean value of stress $\bar{y}$, μ the average value of the strength, and σ the standard deviation of the strength.

Substituting Eqs. (5.18) and (5.19) into Eq. (5.7) results in

$$\begin{aligned}
R &= \int_{-\infty}^{\infty}\left(\int_0^x \theta e^{-\theta y}\,\mathrm{d}y\right)\left\{\frac{1}{\sigma\sqrt{2\pi}}\exp\left[-\frac{1}{2}\left(\frac{x-\mu}{\sigma}\right)^2\right]\right\}\mathrm{d}x \\
&= \int_{-\infty}^{\infty}(1-e^{-\theta x})\left\{\frac{1}{\sigma\sqrt{2\pi}}\exp\left[-\frac{1}{2}\left(\frac{x-\mu}{\sigma}\right)^2\right]\right\}\mathrm{d}x \\
&= \int_{-\infty}^{\infty}\frac{1}{\sigma\sqrt{2\pi}}\exp\left[-\frac{1}{2}\left(\frac{x-\mu}{\sigma}\right)^2\right]\mathrm{d}x \\
&\quad - \int_{-\infty}^{\infty}\frac{1}{\sigma\sqrt{2\pi}}\exp\left\{-\left[\frac{1}{2}\left(\frac{x-\mu}{\sigma}\right)^2+\theta x\right]\right\}\mathrm{d}x \\
&= 1 - \int_{-\infty}^{\infty}\frac{1}{\sigma\sqrt{2\pi}}\exp\left\{-\left[\frac{1}{2}\left(\frac{x-\mu}{\sigma}\right)^2+\theta x\right]\right\}\mathrm{d}x
\end{aligned} \tag{5.20}$$

Since

$$\frac{1}{2}\left(\frac{x-\mu}{\sigma}\right)^2 + \theta x = \frac{2\mu\theta\sigma^2 + (\sigma^2\theta - \mu + x)^2 - \sigma^4\theta^2}{2\sigma^2} \tag{5.21}$$

Eq. (5.20) is rewritten as

$$R = 1 - \exp\left[-\tfrac{1}{2}(2\mu\theta - \sigma^2\theta^2)\right]\cdot A \tag{5.22}$$

where

$$A = \int_{-\infty}^{\infty}\frac{1}{\sigma\sqrt{2\pi}}\exp\left[-\frac{1}{2}\left(\frac{\sigma^2\theta-\mu+x}{\sigma}\right)^2\right]\mathrm{d}x \tag{5.23}$$

Since, $A = 1$, Eq. (5.22) becomes[13]

$$R = 1 - \exp[-\tfrac{1}{2}(2\mu\theta - \sigma^2\theta^2)] \tag{5.24}$$

Example 5.1

The stress and strength of a mechanical component are exponentially and normally distributed, respectively. The mean value of the stress is 18,000 psi. The values of the mean and standard deviation associated with the normal distribution are 27,000 and 2000 psi, respectively. Obtain the mechanical component's reliability using Eq. (5.24).

In this case, we have $\theta = 1/18{,}000$, $\mu = 27{,}000$ psi, and $\sigma = 2000$ psi. After substituting these data in Eq. (5.24), we get

$$R = 1 - \exp\left\{-\frac{1}{2}\left[\frac{2(27{,}000)}{(18{,}000)} - \frac{(2000)^2}{(18{,}000)^2}\right]\right\}$$

$$= 1 - \exp[-\tfrac{1}{2}(3 - 0.0123)]$$

$$= 0.7755$$

Thus, the mechanical component's reliability is 0.7755.

Model II. In this case, both stress and strength are represented by exponential probability density functions as

$$f(y) = \theta_s e^{-\theta_s y}, \qquad 0 \leqslant y < \infty \tag{5.25}$$

and

$$f(x) = \theta_{sh} e^{-\theta_{sh} x}, \qquad 0 \leqslant x < \infty \tag{5.26}$$

where θ_s and θ_{sh} are the parameters associated with stress and strength probability density functions, respectively.

Substituting the above functions into Eq. (5.6) yields

$$R = \int_0^\infty \left[\int_y^\infty \theta_{sh} e^{-\theta_{sh} x}\, dx\right] \theta_s e^{-\theta_s y}\, dy \tag{5.27}$$

$$= \int_0^\infty \theta_s e^{-(\theta_{sh} + \theta_s)y}\, dy$$

$$= \frac{\theta_s}{\theta_{sh} + \theta_s} \tag{5.28}$$

For $\theta_s = 1/\bar{y}$ and $\theta_{sh} = 1/\bar{x}$, the above equation becomes

$$R = \frac{\bar{x}}{\bar{x} + \bar{y}} \tag{5.29}$$

where $\bar{x}$ is the mean strength and $\bar{y}$ the mean stress.

Example 5.2

A component's stress and strength are distributed exponentially with mean values of $\bar{y} = 15{,}000$ psi and $\bar{x} = 25{,}000$ psi, respectively. Calculate reliability of the component. Substituting these data into Eq. (5.29) yields

$$R = \frac{25{,}000}{25{,}000 + 15{,}000} = 0.625$$

Thus, the reliability of the component is 0.625.

Model III. In this case, it is assumed that both the stress and strength of an item are represented by normal distributions, respectively, as follows:

$$f(y) = \frac{1}{\sigma_y\sqrt{2\pi}} \exp\left[-\frac{1}{2}\left(\frac{y-\mu_y}{\sigma_y}\right)^2\right], \qquad -\infty < y < \infty \tag{5.30}$$

$$f(x) = \frac{1}{\sigma_x\sqrt{2\pi}} \exp\left[-\frac{1}{2}\left(\frac{x-\mu_x}{\sigma_x}\right)^2\right], \qquad -\infty < x < \infty \tag{5.31}$$

where μ_x is the mean strength, y the stress random variable, μ_y the mean stress, σ_y the standard deviation of the stress, x the strength random variable, and σ_x the standard deviation of the strength.

If we let $z = x - y$, the difference z is a random variable with the following normal probability density function:

$$f(z) = \frac{1}{\sigma_z\sqrt{2\pi}} \exp\left[-\frac{1}{2}\left(\frac{z-\mu_z}{\sigma_z}\right)^2\right], \qquad -\infty < z < \infty \tag{5.32}$$

where

$$\mu_z = \mu_x - \mu_y \tag{5.33}$$

$$\sigma_z = (\sigma_x^2 + \sigma_y^2)^{\frac{1}{2}} \tag{5.34}$$

The reliability R is given by

$$R = P(z > 0) = \int_0^\infty \frac{1}{\sigma_z\sqrt{2\pi}} \exp\left[-\frac{1}{2}\left(\frac{z-\mu_z}{\sigma_z}\right)^2\right] dz \tag{5.35}$$

In the above relationship, by letting $T = (z - \mu_z)/\sigma_z$, we get

$$\frac{dT}{dz} = \frac{1}{\sigma_z} \quad \text{or} \quad dz = \sigma_z \, dT \tag{5.36}$$

$$T = \frac{0 - \mu_z}{\sigma_z}, \quad \text{for } z = 0 \tag{5.37}$$

and

$$T = \infty, \quad \text{for } z = \infty \tag{5.38}$$

Substituting for μ_z and σ_z from Eqs. (5.33) and (5.34) results in

$$T = -\left[\frac{\mu_x - \mu_y}{(\sigma_x^2 + \sigma_y^2)^{\frac{1}{2}}}\right] \tag{5.39}$$

Thus, from Eq. (5.35), we get

$$R = \frac{1}{\sqrt{2\pi}} \int_{-[(\bar{x} - \bar{y})/(\sigma_x^2 + \sigma_y^2)^{\frac{1}{2}}]}^{\infty} \exp\left[-\frac{T^2}{2}\right] dT \tag{5.40}$$

The reliability R can be obtained by using the standardized normal distribution tables.

Example 5.3

An item's stress and strength are defined by normal distributions. The values of mean μ_y and standard deviation σ_y associated with the stress distribution are 25,000 and 1000 psi, respectively. Similarly, for the strength distribution, the values of mean μ_x and standard deviation σ_x are 30,000 and 2000 psi, respectively. Calculate the reliability of the item.

Using these data in Eqs. (5.33) and (5.34) yields

$$\mu_z = 30{,}000 - 25{,}000 = 15{,}000 \text{ psi}$$

and

$$\sigma_z = [(2000)^2 + (1000)^2]^{\frac{1}{2}}$$

$$= 2236.07 \text{ psi}$$

Utilizing the above results and Eq. (5.39), we get

$$T = -5000/2236.07 = -2.2361$$

Thus, for $T = -2.2361$, the component reliability is 0.98745.

Model IV. In this case, the component's stress and strength are defined by normal and exponential probability density functions, respectively, as

$$f(y) = \frac{1}{\sigma_y\sqrt{2\pi}} \exp\left[-\frac{1}{2}\left(\frac{y-\mu_y}{\sigma_y}\right)^2\right], \qquad -\infty < y < \infty \tag{5.41}$$

$$f(x) = \theta_{sh} e^{-\theta_{sh} x}, \qquad x \geqslant 0 \tag{5.42}$$

where μ_y is the mean stress, σ_y the standard deviation of the stress, and θ_{sh} the parameter associated with strength probability density function.

Using Eqs. (5.41) and (5.42) in Eq. (5.6) yields

$$R = \int_{-\infty}^{\infty}\left[\int_{y}^{\infty} \theta_{sh} e^{-\theta_{sh} x}\, dx\right] \frac{1}{\sigma_y\sqrt{2\pi}} \exp\left[-\frac{1}{2}\left(\frac{y-\mu_y}{\sigma_y}\right)^2\right] dy$$

$$= \int_{-\infty}^{\infty} \frac{1}{\sigma_y\sqrt{2\pi}} \exp\left\{-\left[\frac{1}{2}\left(\frac{y-\mu_y}{\sigma_y}\right)^2 + \theta_{sh} y\right]\right\} dy \tag{5.43}$$

Since

$$\frac{(y-\mu_y)^2}{2\sigma_y^2} + \theta_{sh} y = \frac{2\mu_y\theta_{sh}\sigma_{sh}^2 - \theta_{sh}^2\sigma_y^4 + (y - \mu_y + \theta_{sh}\sigma_y^2)^2}{2\sigma_y^2} \tag{5.44}$$

Eq. (5.43) yields

$$R = \exp\left[-\tfrac{1}{2}(2\mu_y\theta_{sh} - \theta_{sh}^2\sigma_y^2)\right] \tag{5.45}$$

Example 5.4

A mechanical item's strength is described by an exponential probability distribution with a mean value of 20,000 psi. The stresses acting on the item are represented by the normal distribution with mean $\mu_y = 15{,}000$ psi and standard deviation $\sigma_y = 3000$ psi. Determine the mechanical item's reliability by using Eq. (5.45).

In this example, the data for the following items are given $\theta_{sh} = 1/20{,}000$, $\mu_y = 15{,}000$ psi, and $\sigma_y = 3000$ psi.

Using these values in Eq. (5.45) results in

$$R = \exp\left\{-\frac{1}{2}\left[\frac{2(15{,}000)}{20{,}000} - \frac{(3000)^2}{(20{,}000)^2}\right]\right\}$$

$$= \exp\left[-\tfrac{1}{2}(1.5 - 0.0225)\right]$$

$$= 0.4777$$

Thus, the reliability of the mechanical item is 0.4777.

Model V. In this case, stress and strength of an item are Maxwellian distributed. The probability density functions are defined as

$$f(y) = \frac{4}{\sqrt{\pi}} \frac{1}{\alpha_y^3} y^2 \exp\left(-\frac{y^2}{\alpha_y^2}\right), \qquad \alpha_y > 0, y \geqslant 0 \tag{5.46}$$

$$f(x) = \frac{4}{\sqrt{\pi}} \frac{1}{\alpha_x^3} x^2 \exp\left(-\frac{x^2}{\alpha_x^2}\right), \qquad \alpha_x > 0, x \geqslant 0 \tag{5.47}$$

where α_y is the stress parameter and α_x the strength parameter.

Substituting Eqs. (5.46) and (5.47) into Eq. (5.6) yields

$$R = \frac{4}{\sqrt{\pi}} \frac{1}{\alpha_y^3} \int_0^\infty y^2 \exp\left(-\frac{y^2}{\alpha_y^2}\right) \left[\int_y^\infty \frac{4}{\sqrt{\pi}} \cdot \frac{1}{\alpha_x^3} x^2 \exp\left(-\frac{x^2}{\alpha_x^2}\right) \mathrm{d}x\right] \mathrm{d}y \tag{5.48}$$

At $\alpha_x = 1$, from Eq. (5.48), we get

$$R = \frac{1}{\alpha_y^3}\left[\frac{1}{\sqrt{\pi}}\left(\frac{\alpha_y^2+1}{\alpha_y^2}\right)^{-2} \Gamma(2) + \frac{1}{2}\left(\frac{1}{\alpha_y^2}\right)^{-3/2} \Gamma\left(\frac{3}{2}\right) - \int_0^\infty y^2 \exp\left(-\frac{y^2}{\alpha_y^2}\right) \cdot \operatorname{erf}(y) \cdot \mathrm{d}y\right] \tag{5.49}$$

Model VI. In this case, it is assumed that the component's stress and strength are described by exponential and gamma distributions, respectively. Their probability density functions are

$$f(y) = \theta e^{-\theta y}, \qquad 0 \leqslant y < \infty \tag{5.50}$$

$$f(x) = \frac{\mu^k x^{k-1} \exp(-\mu x)}{\Gamma(k)}, \qquad 0 \leqslant x < \infty \tag{5.51}$$

where μ is the (strength) scale parameter, θ the (stress) scale parameter, and k the shape parameter.

The gamma function is defined by

$$\Gamma(k) = \int_0^\infty t^{k-1} \exp(-t)\, \mathrm{d}t \tag{5.52}$$

The component's reliability R is given by

$$R = 1 - \left(\frac{\mu}{\mu+\theta}\right)^k = 1 - \left(\frac{1}{1+\lambda}\right)^k \tag{5.53}$$

where $\lambda = \theta/\mu$.

Model VII. In this case, the mechanical component's stress and strength follow exponential and Maxwellian distributions, respectively. The probability density functions are defined as

$$f(y) = \theta e^{-\theta y}, \qquad y \geqslant 0 \tag{5.54}$$

$$f(x) = \frac{4}{\sqrt{\pi}} \frac{1}{\alpha^3} x^2 \exp\left(-\frac{x^2}{\alpha^2} \right), \qquad x \geqslant 0 \tag{5.55}$$

where θ is the stress parameter and α the strength parameter.

Using Eqs. (5.54) and (5.55) in Eq. (5.6) results in

$$R = \theta \int_0^{\infty} e^{-\theta y} \, \mathrm{d}y \left[\int_y^{\infty} \frac{4}{\sqrt{\pi}} \frac{1}{\alpha^3} x^2 \exp\left(-\frac{x^2}{\alpha^2} \right) \mathrm{d}x \right] \tag{5.56}$$

From this equation, we get

$$R = \frac{\sqrt{\pi}}{4} \exp\left(\frac{\theta^2}{4} \right) \cdot \operatorname{erf}\left(\frac{\theta}{2} \right) + \frac{\sqrt{\pi}}{4} + \frac{\theta}{4} - \frac{\sqrt{\pi}}{4} \exp\left(\frac{\theta^2}{4} \right)$$
$$- \sqrt{\pi}\theta^2 \cdot \exp\left(\frac{\theta^2}{4} \right) \cdot \left[1 - \operatorname{erf}\left(\frac{\theta}{2} \right) \right] \tag{5.57}$$

Model VIII. In this model, the mechanical item's stress and strength are represented by gamma and exponential distributions, respectively, as

$$f(y) = \frac{\lambda^{\beta} y^{\beta-1} \exp(-\lambda y)}{\Gamma(\beta)}, \qquad 0 \leqslant y < \infty \tag{5.58}$$

$$f(x) = \theta e^{-\theta x}, \qquad 0 \leqslant x < \infty \tag{5.59}$$

where λ is the scale parameter associated with the stress distribution, β the shape parameter associated with the stress distribution, and θ the scale parameter associated with the strength distribution.

The gamma function is given by

$$\Gamma(\beta) = \int_0^{\infty} t^{\beta-1} \exp(-t) \, \mathrm{d}t \tag{5.60}$$

With the aid of Eqs. (5.58) and (5.59), the following expression is obtained for the item's reliability:

$$R = \left(\frac{\lambda}{\theta + \lambda} \right)^{\beta} = \left(\frac{1}{1 + p} \right)^{\beta} \tag{5.61}$$

where $p = \theta/\lambda$.

Example 5.5

For the following given data, determine the component's reliability using Eq. (5.61): $\beta = 1$, $1/\theta = 10{,}000$ psi, and $1/\lambda = 5000$ psi. Substituting these data into Eq. (5.61), we get

$$R = \left[\frac{(5000)^{-1}}{(10{,}000)^{-1} + (5000)^{-1}}\right]^1 = 0.6667$$

Thus, the component's reliability is 0.6667.

Stress-Strength Analysis of Redundant Mechanical Systems

This section is concerned with the reliability determination of independent and identical component redundant mechanical systems using the interference theory.[15] The probability of an item's strength exceeding a stress of value y_0 is given by

$$P(x > y_0) = \int_{y_0}^{\infty} f(x)\,\mathrm{d}x \tag{5.62}$$

where y_0 is the fixed stress level and $f(x)$ the strength probability density function.

The overall reliability of a redundant system under stress is given by

$$R_s = \int_0^{\infty} f(y)\cdot R_s(y)\,\mathrm{d}y \tag{5.63}$$

where $f(y)$ is the probability density function of stress and $R_s(y)$ is the system reliability at stress y.

Parallel System. The independent component parallel system's[16] reliability is given by

$$R_s(y) = 1 - [1 - P(x > y_0)]^k \tag{5.64}$$

where k is the number of independent and identical items in parallel and $R_s(y)$ the parallel system's reliability at stress y.

Substituting Eq. (5.62) into Eq. (5.64) yields

$$R_s(y) = 1 - \left[1 - \int_y^{\infty} f(x)\,\mathrm{d}x\right]^k \tag{5.65}$$

Inserting Eq. (5.65) into Eq. (5.63) results in

$$R_s = \int_0^{\infty}\left\{1 - \left[1 - \int_y^{\infty} f(x)\,\mathrm{d}x\right]^k\right\} f(y)\,\mathrm{d}y \tag{5.66}$$

where R_s is the overall reliability of the parallel system under stress.

Special Case Model (Parallel System). In this case, it is assumed that the stress and strength are exponentially distributed. Their probability density functions are

$$f(y) = \theta_y e^{-\theta_y y}, \qquad 0 \leqslant y < \alpha \tag{5.67}$$

$$f(x) = \theta_x e^{-\theta_x x}, \qquad 0 \leqslant x < \infty \tag{5.68}$$

where θ_y is the parameter associated with the stress distribution and θ_x the parameter associated with the strength distribution.

Using Eqs. (5.67) and (5.68) in Eq. (5.66) results in

$$R_s = \int_0^\infty \left[1 - \left(1 - \int_y^\infty \theta_x e^{-\theta_x x}\,dx\right)^k\right] \theta_y e^{-\theta_y y}\,dy$$

$$= \int_0^\infty \left\{1 - \left[1 - \exp\left(-\theta_x y\right)\right]^k\right\} \theta_y e^{-\theta_y y}\,dy$$

$$= 1 - \theta_y \sum_{i=0}^{k} \frac{\binom{k}{i}(-1)^i}{(\theta_y + i\theta_x)} \tag{5.69}$$

Example 5.6

Using Eq. (5.69) and $k = 2$, obtain an expression for overall reliability of the two unit parallel system under stress. Thus, from Eq. (5.69), we get

$$R_s = 1 - \theta_y \sum_{i=0}^{2} \frac{\binom{2}{i}(-1)^i}{(\theta_y + i\theta_x)}$$

$$= \frac{2\theta_y}{\theta_y + \theta_x} - \frac{\theta_y}{\theta_y + 2\theta_x} \tag{5.70}$$

Example 5.7

Assume that we have the following data: $k = 1$, $1/\theta_y = 10{,}000$ psi, and $1/\theta_x = 12{,}000$ psi. Calculate the value of R_s using Eq. (5.69). For $k = 1$, Eq. (5.69) yields

$$R_s = 1 - \theta_y \sum_{i=0}^{1} \frac{\binom{1}{i}(-1)^i}{(\theta_y + i\theta_x)}$$

$$= 1 - \theta_y \left[\frac{\binom{1}{0}}{\theta_y} - \frac{\binom{1}{1}}{\theta_y + \theta_x}\right] = \frac{\theta_y}{\theta_y + \theta_x} \tag{5.71}$$

Substituting the specified values for θ_y and θ_x into Eq. (5.71) yields

$$R_s = \frac{(10{,}000)^{-1}}{(10{,}000)^{-1} + (12{,}000)^{-1}}$$

$$= 0.5455$$

A Graphical Method for Calculating Reliability of a Mechanical Item

This method employs Mellin transforms[12] in evaluating the mechanical component's reliability. The method is extremely useful for calculating reliability when the stress and strength distributions cannot be assumed, but one has a sufficient amount of empirical data. The Mellin transforms for Eq. (5.6) are

$$A = \int_y^{\infty} f(x)\,\mathrm{d}x = 1 - F_x(y) \tag{5.72}$$

and

$$B = \int_0^{y} f(y)\,\mathrm{d}y = F_y(y) \tag{5.73}$$

where $F_x(y)$ and $F_y(y)$ are the cumulative distribution functions.

Differentiating Eq. (5.73) with respect to y, we get

$$\frac{\mathrm{d}B}{\mathrm{d}y} = f(y) \tag{5.74}$$

Rearranging the above equation results in

$$\mathrm{d}B = f(y)\cdot \mathrm{d}y \tag{5.75}$$

It is obvious from Eq. (5.73) that B takes values of 0–1 (i.e., at $y = 0$, $B = 0$ and at $y = \alpha$, $B = 1$).

Substituting Eqs. (5.72) and (5.75) into Eq. (5.6) results in

$$R = \int_0^{1} A\,\mathrm{d}B \tag{5.76}$$

The above equation indicates that the reliability of the item is given by the area under the A vs B plot. Simpson's or the trapezoidal rule can be used to compute the area in question. Both of these rules are presented below.

Simpson's Rule. For even m, this is expressed as

$$\int_c^d f(z)\,\mathrm{d}z \simeq \frac{d-c}{3m}(Y_0 + 4Y_1 + 2Y_2 + 4Y_3 + \cdots + 2Y_{m-2} + 4Y_{m-1} + Y_m) \tag{5.77}$$

where $f(z)$ is a function defined in an interval $c \leqslant z \leqslant d$, m the even number of equal subdivided parts of interval (c,d), and $(d-c)/m = h$ the width of the subdivided part (i.e., the width of the subinterval).

At $c = z_0, z_1, z_2, z_3, \ldots, z_{m-1}, z_m = d$; the corresponding values of Y are $Y_0 = f(z_0), Y_1 = f(z_1), Y_2 = f(z_2), Y_3 = f(z_3), \ldots, Y_m = f(z_m)$.

Trapezoidal Rule. This is expressed as

$$\int_c^d f(z)\,\mathrm{d}z \simeq \frac{d-c}{2m}(Y_0 + 2Y_1 + 2Y_2 + \cdots + 2Y_{m-1} + Y_m) \tag{5.78}$$

Example 5.8

As in Example 5.2, a component's stress and strength are exponentially distributed with mean values of $\bar{y} = 10{,}000$ psi and $\bar{x} = 20{,}000$ psi, respectively. Determine reliability of the component using the graphical method. Verify the end result by using the analytical (interference theory) approach.

Thus, the probability density functions of stress and strength are defined by

$$f(y) = \frac{1}{10{,}000}\exp\left(-\frac{y}{10{,}000}\right) \tag{5.79}$$

and

$$f(x) = \frac{1}{20{,}000}\exp\left(-\frac{x}{20{,}000}\right) \tag{5.80}$$

From Eqs. (5.72) and (5.80), we get

$$A = \int_y^\infty \frac{1}{20{,}000}\exp\left(-\frac{x}{20{,}000}\right)\mathrm{d}x \tag{5.81}$$

$$A = \exp\left(-\frac{y}{20{,}000}\right) \tag{5.82}$$

Substituting Eq. (5.79) into Eq. (5.73) results in

$$B = \int_0^y \frac{1}{10{,}000} \exp\left(-\frac{y}{10{,}000}\right) dy$$
$$= 1 - \exp\left(-\frac{y}{10{,}000}\right) \tag{5.83}$$

For several different values of stress y, the values of A and B with the aid of Eqs. (5.82) and (5.83), respectively, are tabulated in Table 5.1. The plot of data given in Table 5.1 is shown in Fig. 5.2. In order to estimate the area (i.e., component reliability) under the curve shown in Fig. 5.2, use Eq. (5.77) as follows:

$$R \simeq \frac{(1-0)}{3(4)} (Y_0 + 4Y_1 + 2Y_2 + 4Y_3 + Y_4)$$
$$\simeq \frac{1}{12} (1 + 4(0.87) + 2(0.71) + 4(0.5) + 0)$$
$$\simeq 0.6583$$

Table 5.1 Calculated Values of A and B for Several Different Values of Stress y

No.	Stress y, psi	$A = \exp\left(-\frac{y}{20{,}000}\right)$	$B = 1 - \exp\left(-\frac{y}{10{,}000}\right)$
1	0	1	0
2	1,000	0.95	0.1
3	4,000	0.82	0.33
4	6,000	0.74	0.45
5	8,000	0.67	0.55
6	12,000	0.55	0.70
7	16,000	0.45	0.80
8	20,000	0.37	0.87
9	25,000	0.29	0.92
10	30,000	0.22	0.95
11	40,000	0.14	0.98
12	∞	0	1

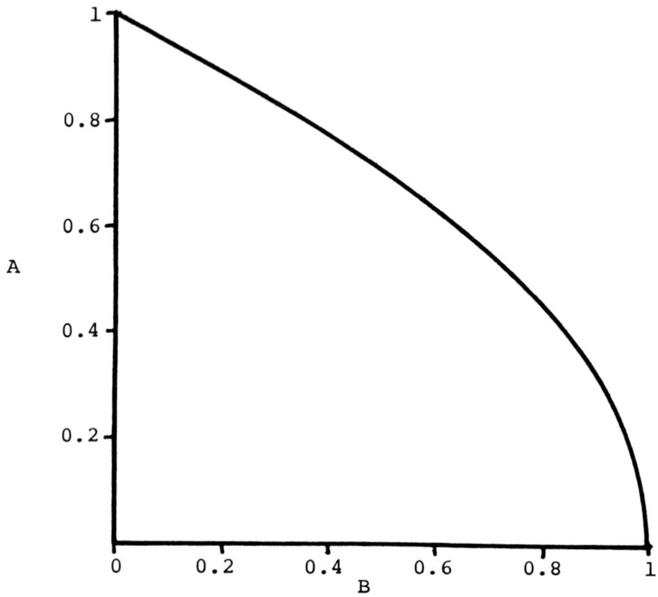

Fig. 5.2 Plot of *A* and *B* values given in Table 5.1.

Substituting the specified data for the mean values of stress and strength into Eq. (5.29) yields

$$R = \frac{\bar{x}}{\bar{x} + \bar{y}} = \frac{(20{,}000)}{(20{,}000) + (10{,}000)} = 0.6667$$

The above two results for component reliability (i.e., 0.6583 and 0.6667) obtained through two different means are quite close.

References

[1]Bouton, I., Trent, D. J., and Chenoweth, H. B., "Design Factors for Structural Reliability," *Proceedings of the Fifth Annual Reliability and Maintainability Conference*, ASME, New York, July 1966, pp. 229–235.

[2]Mittenbergs, A. A., "Fundamental Aspects of Mechanical Reliability," *Mechanical Reliability Concepts*, ASME, New York, 1965, pp. 17–34.

[3]Mittenbergs, A. A., "The Materials Problem in Structural Reliability," *Proceedings of the Fifth Annual Reliability and Maintainability Conference*, ASME, New York, July 1966, pp. 148–158.

[4]Mark, D. G., "Choosing the Best Method for a Variability Analysis," *Electronic Design*, Vol. 11, Nov. 1963, pp. 39–49.

[5]Kececioglu, D. and Cormier, D., "Designing a Specified Reliability Directly into a Component," *Proceedings of the Aerospace Reliability and Maintainability Conference*, ASME, New York, 1964, pp. 546–565.

[6]Broding, W. C., Diederich, F. W., and Parker, P. S., "Structural Optimization and Design Based on a Reliability Design Criterion," *Journal of Spacecraft and Rockets*, Vol. 1, 1964, pp. 56–61.

[7]Mesloh, R. E. and Mark, D. G., "Variability Analysis of Precision Mechanical Devices," *Proceedings of the Aerospace Reliability and Maintainability Conference*, ASME, New York, 1964, pp. 576–588.

[8]Kececioglu, D., "Reliability Analysis of Mechanical Components, and Systems," *Nuclear Engineering and Design*, Vol. 19, 1972, pp. 259–290.

[9]Hahn, G. J. and Shapiro, S. S., *Statistical Models in Engineering*, Wiley, New York, 1967, pp. 225–259.

[10]Bompas-Smith, J. H., *Mechanical Survival: The Use of Reliability Data*, McGraw-Hill, London, 1973, pp. 27–30.

[11]Kapur, K. C. and Lamberson, L. R., *Reliability in Engineering Design*, Wiley, New York, 1977, pp. 122–163.

[12]*Quality Assurance (Reliability Handbook)*, Headquarters, U.S. Army Material Command, Washington, DC, Rept. AMCP 702–3, 1968, pp. A.55–A.67 (available from the National Technical Information Service, Springfield, VA).

[13]Kececioglu, D. and Li, D., "Exact Solutions for the Prediction of the Reliability of Mechanical Components and Structural Members," *Proceedings of the Failure Prevention and Reliability Conference*, ASME, New York, 1985, pp. 115–122.

[14]Dhillon, B. S., "Mechanical Reliability: Interference Theory Models," *Proceedings of the Annual Reliability and Maintainability Symposium*, IEEE, New York, 1980, pp. 462–467.

[15]"Practical Reliability, Vol. IV—Prediction," NASA CR-1129, 1968, pp. 98–105.

[16]Dhillon, B. S., "Interference Theory Application to Mechanical Systems," *Proceedings of the Annual Reliability and Maintainability Symposium*, IEEE, New York, 1981, pp. 24–27.

Exercises

5.1 Discuss steps to be taken for determining the failure governing stress distribution.

5.2 Describe the "design-by-reliability" methodology developed for use in mechanical design.

5.3 Show the stress-strength interference area with the aid of a diagram.

5.4 Prove that the reliability of a mechanical item is given by

$$R = \int_{-\infty}^{\infty} \left[\int_{-\infty}^{x} f(y)\, \mathrm{d}y \right] f(x)\, \mathrm{d}x \qquad (5.84)$$

where R is the item reliability, x the strength random variable, y the stress random variable, $f(y)$ the stress probability density function, and $f(x)$ the strength probability density function.

5.5 Assume that stress and strength of a mechanical component are distributed lognormally. Develop an expression for the component's reliability.

5.6 Assume that stress and strength of an item are exponentially and normally distributed, respectively. The mean stress is equal to 14,000 psi and the values of the mean and standard deviation associated with the normal distribution are 21,000 and 3000 psi, respectively. Calculate the item's reliability.

5.7 For the specified values of $\theta_x = (19{,}000 \text{ psi})^{-1}$ and $\theta_y = (14{,}000 \text{ psi})^{-1}$, calculate the two unit parallel system reliability using Eq. (5.70).

6 MECHANICAL RELIABILITY MODELING

This chapter presents various aspects of mechanical-reliability-related modeling. Basically, the chapter is divided into four areas, i.e., general models, optimization in structural reliability, modeling of an earthquake protection system, and modeling of fire-related failures in buildings. Under the general models topic, eight mathematical models are presented. These models are concerned with determining the average depth of wear in the system composed of two flat plates with one resting on the other, estimating the rate of corrosion penetration, calculating the total specimen weight loss due to fretting, determining the depth of erosion, etc. The reliability optimization of a mechanical component is presented for two cases when stress and strength are independent and described by the normal distribution. In one case, the total cost subject to specified component reliability is minimized and in the other case the component reliability is maximized subject to resource constraints. A Markov model representing an earthquake protection system with k reserve elements is presented. Several models concerned with fire-related failures in buildings are presented. These models deal with estimating the structural failure probability during the planned service life of a building, the monetary fire loss expectation, the life loss expectation; and the monetary fire cost expectation.

6.1 INTRODUCTION

Even though the advances made in mechanical reliability may not be as much as in the field of electronic reliability, there has been considerable progress made in various directions in mechanical reliability modeling. Examples of such directions are time-dependent stress-strength modeling, general models, reliability modeling of structural systems with earthquake protection, and optimization in structural reliability. The purpose of this chapter is to make readers aware of developments in various selective areas related to mechanical reliability. These areas cover a wide spectrum, i.e., from mechanical reliability optimization models to safety factors for fire loads. The advances made in each of these areas varies from one area to another. However, the relative amount of coverage given to each of these areas in this chapter in no way reflects the respective amount of advances made on those topics. More specifically, it does not necessarily mean that the greater coverage of a certain topic indicate that the advances made in that subject are higher than the one has less coverage. The following areas are covered in this chapter: general models, optimization in structural reliability, modeling of an earthquake protection system, and modeling of fire-related failures in buildings.

The chapter focuses more on the structure of the concepts than the minute details.

6.2 GENERAL MODELS

This section presents a number of mathematical models directly or indirectly related to mechanical reliability. These models are presented below.

Model I

This model is concerned with determining the average depth of wear in a system composed of two flat plates, one resting on the other. When one of the plates is slid over the other, a specific volume of material is worn away. This volume is proportional to the sliding distance, the nature of the material, and the actual contact.[1] The mean depth of wear β_m in centimeters is given by

$$\beta_m = V_m/\alpha_n \tag{6.1}$$

where V_m is the volume of material worn given in cm^3 and α_n the nominal contact area.

The volume of material removed V_m is expressed as

$$V_m = (\alpha_t) \cdot (D_{tv}) \cdot (\lambda_w) \tag{6.2}$$

where α_t is the true contact area in cm^2, D_{tv} the distance traveled in cm, and λ_w the wear coefficient.

The true contact area α_t is given by

$$\alpha_t = L_a/h_i \tag{6.3}$$

where h_i is the indentation hardness, given in kg/cm^2, of softer material and L_a the load applied in kg.

Substituting Eqs. (6.2) and (6.3) into Eq. (6.1) results in

$$\beta_m = \frac{(\alpha_t) \cdot (D_{tv}) \cdot \lambda_w}{\alpha_n} = \frac{L_a \cdot D_{tv} \cdot \lambda_w}{h_i \alpha_n} \tag{6.4}$$

Thus, the above equation can be used to calculate the average depth of wear.

Model II

This model is related to model I and is concerned with estimating the wear depth of harder material when the plates are made of differing materials. In this case, the wear depth β_{hm} of the harder material[1] is given by

$$\beta_{hm} = \frac{S_{ys}^2 \cdot \beta_{sm}}{S_{yh}^2} \tag{6.5}$$

where S_{ys} is the softer material's yield strength, S_{yh} the harder material's yield strength, and β_{sm} the softer material's wear depth.

Model III

This model is sometimes known as the empirical model for zero wear. The term "zero wear" simply means that the wear process does not significantly alter the surface finish because the wear magnitude is so small. In this case, the depth of wear is in the neighborhood of half the peak-to-peak-surface finish dimension. The following equation can be used to calculate the number of passes that can be tolerated without bypassing the level of zero wear[2]:

$$m = \frac{2000(\lambda \cdot Y_{sp})^9}{(S_{s\max})^9} \tag{6.6}$$

where m is the number of passes that can be accommodated without bypassing the level of zero wear, λ a constant associated with the specific combination of lubricant and materials, Y_{sp} the shear yield point of the given material, and $S_{s\max}$ the maximum shearing stress produced by sliding.

Model IV

This model is concerned with determining the corrosion rate. The most common type of corrosion is probably the direct chemical attack.[2,3] Normally, it is possible to estimate the rate of corrosion due to direct attack from relatively straightforward laboratory tests. The following equation can be used to compute rate of corrosion penetration μ_c in mils/year[3]:

$$\mu_c = \frac{(543)w_L}{d \cdot T \cdot \alpha_s} \tag{6.7}$$

where α_s is the specimen's exposed area in in.^{2}, d the specimen's density in g/cm^3, w_L the weight loss in mg, and T the exposure time expressed in hours.

Model V

This model is basically concerned with fretting wear. This kind of wear (fretting corrosion) may be described simply as accelerated surface damage occurring at the interface of contacting materials (when such materials are under small oscillatory displacements). Factors[4,5] such as temperature, frequency, slip, applied load, humidity, lubricants, atmosphere, material hardness, surface finish, and coefficient of friction influence the fretting wear. The total specimen weight loss WT_L due to fretting[6] is given by

$$WT_L = Af_c + \theta S_{pp} \cdot L_n f_c \tag{6.8}$$

$$A \equiv [\alpha(L_n)^{\frac{1}{2}} - \beta L_n]/f_f \tag{6.9}$$

where f_c is the total number of fretting cycles, f_f the fretting frequency, S_{pp} the slip between fretting surfaces (peak-to-peak), L_n the contact load, and α, β, and θ empirically determined constants.

Model VI

This model is concerned with estimating the depth of erosion on the basis that the energy absorbed by the material eroded from a surface is a characteristic of the material itself. The following equation[1,7] is proposed to estimate the depth of erosion d_E:

$$d_E = v/a_E = (270)P_I T_E/E_s \tag{6.10}$$

where a_E is the exposed area in in.^{2}, v the erosion volume in in.^{3}, E_s the erosion strength in psi, P_I the power level intensity in W/ft^2, and T_E the exposure time given in hours.

Model VII

This model is concerned with fatigue design criteria. For safe operation, the endurance limit of the material selected must be adequately high. The following criterion is proposed for use in initial design[8]:

$$A_1 + A_2 \leqslant 0.5 \tag{6.11}$$

$$A_1 \equiv (SH_{s\max} - SH_{\min})/2E_L$$

$$A_2 \equiv (SH_{\max} + SH_{\min})/2S_Y$$

or

$$f_{sf} \cdot F_s A_1 + f_{st} \cdot T_{sr} A_2 \leqslant 0.5 \tag{6.12}$$

where f_{sf} is the fatigue safety factor (the purpose of this factor is to assure that no failure occur due to fatigue), F_s and T_{sr} the fatigue strength and tensile strength reduction factors, respectively, S_Y the yield strength associated with simple tension, f_{st} the tension safety factor (the purpose of this factor is to assure that no tension failure occurs), $SH_{\max}$ the expected maximum shear stress, E_L the endurance limit, and $SH_{\min}$ the expected minimum shear stress.

It is to be noted that one can make use of either Eq. (6.11) or (6.12). However, Eq. (6.12) is considered to be more practical.[1]

Model VIII

This model is concerned with determining the reliability of an equipment, $R_{ew}(t)$, during wearout period. For normal distribution wearout, the reliability is given by:

$$R_{ew}(t) = \frac{1}{(\sqrt{2\pi})s} \int_t^{\infty} \exp\left[-\frac{(t-m_w)^2}{2s^2}\right] \cdot \mathrm{d}t \tag{6.13}$$

where $s = [\Sigma\,(t - m_w)^2/z]^{\frac{1}{2}}$, the standard deviation; t the accumulated equipment operating time since new (i.e., age); m_w the average wearout life; z the total number of equipment failure events over which the summation $\Sigma\,(t - m_w)^2$ is to be made.

This model is described in detail in Ref. 9.

6.3 OPTIMIZATION IN STRUCTURAL RELIABILITY

Optimization is an invaluable tool and is frequently being used in many engineering studies. Numerous publications have appeared on reliability optimization.[10] Many of these publications are concerned with optimization of redundant systems subject to constraints such as cost and weight. Traditional operations research methods such as linear programming and Lagrange multiplier are used. In mechanical reliability work, the concept of optimization is also used for purposes such as to minimize total cost or maximize reliability.[11,12] The reliability optimization of a mechanical component is demonstrated for the two cases presented below.

Case I

In this case, it is assumed that the component stress and strength are independent and described by the normal distribution. Thus, the probability density functions for stress and strength, respectively, are

$$p_{st}(x) = \frac{1}{s_{st}\sqrt{2\pi}} \exp\left[-\frac{1}{2}\left(\frac{x - m_{st}}{s_{st}}\right)^2\right], \qquad -\infty < x < \infty \tag{6.14}$$

and

$$p_{sth}(y) = \frac{1}{s_{sth}\sqrt{2\pi}} \exp\left[-\frac{1}{2}\left(\frac{y - m_{sth}}{s_{sth}}\right)^2\right], \qquad -\infty < y < \infty \tag{6.15}$$

where x is the stress random variable, y the strength random variable, $p_{st}(x)$ the probability density function for the stress x, $p_{sth}(y)$ the probability density function for the strength y, s_{st} the standard deviation of the stress, s_{sth} the standard deviation of the strength, m_{st} the mean stress, and m_{sth} the mean strength.

For normally distributed stress and strength, the component reliability R_c, is given by[12]

$$R_c = \frac{1}{\sqrt{2\pi}} \int_{-\theta}^{\infty} e^{-t^2/2}\,\mathrm{d}t \tag{6.16}$$

where

$$\theta = (m_{sth} - m_{st})(s_{sth}^2 + s_{st}^2)^{-\frac{1}{2}}$$

In order to maximize the component reliability R_c, the value of θ should be as low as possible. For desired level of component reliability, the total cost C_t can be minimized as

$$C_t = c_1(m_{st}) + k_1(m_{sth}) + k_2(s_{sth}) + c_2(s_{st}) \tag{6.17}$$

subject to

$$\frac{(m_{sth} - m_{st})}{(s_{sth}^2 + s_{st}^2)^{\frac{1}{2}}} \geqslant t \tag{6.18}$$

where t is obtained from normal tables for a desired level of item reliability, $c_1(m_{st})$ the monotonically decreasing cost function associated with the mean stress m_{st}, $k_1(m_{sth})$ the monotonically increasing cost function associated with the mean strength m_{sth}, $c_2(s_{st})$ the monotonically decreasing cost function associated with the standard deviation of the stress, and $k_2(s_{sth})$ the monotonically decreasing cost function associated with the standard deviation of the strength.

The following Lagrangian function results from Eqs. (6.17) and (6.18):

$$L(m_{st}, m_{sth}, s_{st}, s_{sth}, \lambda) = c_1(m_{st}) + k_1(m_{sth}) + k_2(s_{sth})$$

$$+ c_2(s_{st}) + \lambda[m_{sth} - m_{st} - t(s_{sth}^2 + s_{st}^2)^{\frac{1}{2}}] \tag{6.19}$$

After differentiating the above equation with respect to m_{st}, m_{sth}, s_{st}, s_{sth}, and λ, we get

$$\frac{\partial L(.)}{\partial m_{st}} = \frac{\partial c_1(m_{st})}{\partial m_{st}} - \lambda \tag{6.20}$$

$$\frac{\partial L(.)}{\partial m_{sth}} = \frac{\partial k_1(m_{sth})}{\partial m_{sth}} + \lambda \tag{6.21}$$

$$\frac{\partial L(.)}{\partial s_{st}} = \frac{\partial c_2(s_{st})}{\partial s_{st}} - ts_{st}\lambda(s_{sth}^2 + s_{st}^2)^{-\frac{1}{2}} \tag{6.22}$$

$$\frac{\partial L(.)}{\partial s_{sth}} = \frac{\partial k_2(s_{sth})}{\partial s_{st}} - ts_{sth}\lambda(s_{sth}^2 + s_{st}^2)^{-\frac{1}{2}} \tag{6.23}$$

$$\frac{\partial L(.)}{\partial \lambda} = m_{sth} - m_{st} - t(s_{sth}^2 + s_{st}^2)^{\frac{1}{2}} \tag{6.24}$$

In order to find the locally optimal solutions, equate each differentiation to zero as

$$\frac{\partial c_1(m_{st})}{\partial m_{st}} - \lambda = 0 \tag{6.25}$$

$$\frac{\partial k_1(m_{sth})}{\partial m_{sth}} + \lambda = 0 \tag{6.26}$$

$$\frac{\partial c_2(s_{st})}{\partial s_{st}} - ts_{st}\lambda(s_{sth}^2 + s_{st}^2)^{-\frac{1}{2}} = 0 \tag{6.27}$$

$$\frac{\partial k_2(s_{sth})}{\partial s_{sth}} - ts_{sth}\lambda(s_{sth}^2 + s_{st}^2)^{-\frac{1}{2}} = 0 \tag{6.28}$$

$$m_{sth} - m_{st} - t(s_{sth}^2 + s_{st}^2)^{\frac{1}{2}} = 0 \tag{6.29}$$

To obtain all of the local optima, the above simultaneous equations have to be solved. In addition to finding the globally optimal solution, the objective function has to be evaluated for all the local solutions.[12]

Case II

In this case, the component stress and strength are again independent and described by the normal distribution. However, instead of minimizing the total cost subject to specified component reliability, here we are concerned with maximizing component reliability subject to resource constraints. In some sense, this problem is the dual of the problem in case I. The problem is formulated as[12]

$$\text{Maximize} \qquad t = \frac{(m_{sth} - m_{st})}{(s_{sth}^2 + s_{st}^2)^{\frac{1}{2}}} \tag{6.30}$$

subject to

$$c_1(m_{st}) + k_1(m_{sth}) + k_2(s_{sth}) + c_2(s_{st}) \leqslant z \tag{6.31}$$

where z is the quantity of resources available. The remaining symbols are the same as defined in case I.

The Lagrangian function of the problem is given by

$$L(m_{st},m_{sth},s_{st},s_{sth};\lambda) = \frac{m_{sth} - m_{st}}{(s_{sth}^2 + s_{st}^2)^{\frac{1}{2}}} + \lambda B \tag{6.32}$$

where

$$B = c_1(m_{st}) + k_1(m_{sth}) + k_2(s_{sth}) + c_2(s_{st}) - z$$

Differentiating Eq. (6.32) with respect to the variables, we get

$$\frac{\partial L(.)}{\partial m_{st}} = -\frac{1}{A^{\frac{1}{2}}} + \frac{\lambda\, \partial c_1(m_{st})}{\partial m_{st}} \tag{6.33}$$

$$A = (s_{sth}^2 + s_{st}^2)$$

$$\frac{\partial L(.)}{\partial m_{sth}} = \frac{1}{A^{\frac{1}{2}}} + \frac{\lambda\, \partial k_1(m_{sth})}{\partial m_{sth}} \tag{6.34}$$

$$\frac{\partial L(.)}{\partial s_{st}} = \frac{-(m_{sth} - m_{st})s_{st}}{A^{3/2}} + \frac{\lambda\, \partial c_2(s_{st})}{\partial s_{st}} \tag{6.35}$$

$$\frac{\partial L(.)}{\partial s_{sth}} = \frac{-(m_{sth} - m_{st})s_{sth}}{A^{3/2}} + \frac{\lambda\, \partial k_2(s_{sth})}{\partial s_{sth}} \tag{6.36}$$

$$\frac{\partial L(.)}{\partial \lambda} = c_1(m_{st}) + k_1(m_{sth}) + k_2(s_{sth}) + c_2(s_{st}) - z \tag{6.37}$$

To find the locally optimal solutions, we equate each differentiation to zero, as

$$\frac{\lambda\, \partial c_1(m_{st})}{\partial m_{st}} - A^{-\frac{1}{2}} = 0 \tag{6.38}$$

$$\frac{\lambda\, \partial k_1(m_{sth})}{\partial m_{sth}} + A^{-\frac{1}{2}} = 0 \tag{6.39}$$

$$\frac{\lambda\, \partial c_2(s_{st})}{\partial s_{st}} - \frac{(m_{sth} - m_{st})s_{st}}{A^{3/2}} = 0 \tag{6.40}$$

$$\frac{\lambda\, \partial k_2(s_{sth})}{\partial s_{sth}} - \frac{(m_{sth} - m_{st})s_{sth}}{A^{3/2}} = 0 \tag{6.41}$$

$$c_1(m_{st}) + k_1(m_{sth}) + k_2(s_{sth}) + c_2(s_{st}) - z = 0 \tag{6.42}$$

To obtain all the local optima, solve Eqs. (6.38–6.42) with five unknowns.

The globally optimal solution can be selected after evaluating the objective function for all these locally optimal solutions.

6.4 MODELING OF AN EARTHQUAKE PROTECTION SYSTEM

Many buildings have displayed very poor performance during earthquakes. Nowadays in order to make building structures more earthquake resistant, earthquake protection systems are developed. These systems employ reserve elements so that, during the earthquake, such elements can disengage (partly or totally) from the action of the structure's load-carrying system.[13] Once the reserve elements have disengaged, the structure is capable of independently resisting the design value of the horizontal load and the permanent vertical load because of its different dynamic characteristics as compared to the original. Reserve-element earthquake protective systems are already being used in many building structures.

This Markov model[14] is concerned with an earthquake protection system with k reserve elements. It is assumed that these elements disengage one by one in consecutive order. The system state space diagram is shown in Fig. 6.1. The following symbols are associated with this model:

k = number of reserve elements
T_i = system state; $i = 0$ (initial condition of the system, no reserve element disengaged), $i = 1$ (one of the system reserve elements disengaged), $i = 2$ (two of the system reserve elements disengaged), $i = k - 1$ [$(k - 1)$ of the system reserve elements disengaged], $i = k$ (all of the system reserve elements disengaged)
s = Laplace transform variable
$P_{T_i}(t)$ = probability that the protection system is in state T_i at time t, for $i = 0,1,2,(k - 1),k$
μ_i = constant system transition rate from state T_i to T_{i+1}, $i = 0,1,\ldots,(k - 1)$.

The following set of differential equations[14] is associated with Fig. 6.1:

$$\frac{dP_{T_i}(t)}{dt} - \mu_{i-1}P_{T_{i-1}}(t) + \mu_i P_{T_i}(t) = 0$$

$$\text{for } \mu_{-1} = \mu_k = 0 \text{ and } 0 \leqslant i \leqslant k \tag{6.43}$$

The initial condition is that when $t = 0$, $P_{T_0}(0) = 1$ and $P_{T_i}(0) = 0$ (for $1 \leqslant i \leqslant k$).

From Eq. (6.43), the Laplace transform[13] of the solution is

$$P_i(s) = \frac{\mu_0\mu_1\mu_2\ldots\mu_{i-1}}{(s + \mu_0)(s + \mu_1)\ldots(s + \mu_{i-1})(s + \mu_i)} \tag{6.44}$$

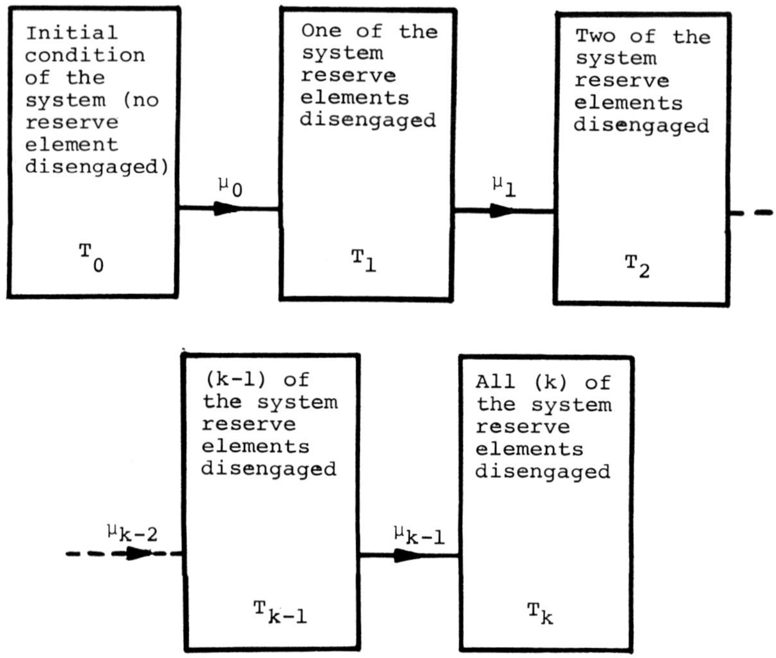

Fig. 6.1 Transition diagram of an earthquake protection system with *k* reserve elements.

With the aid of the above equation, the probability of the system being in state T_k is given by[14]

$$P_{T_k}(t) = 1 - \sum_{i=0}^{k-1} P_{T_i}(t) \tag{6.45}$$

where

$$\sum_{i=0}^{k-1} P_{T_i}(t) = \prod_{i=0}^{k-1} \mu_i \sum_{j=0}^{k-1} \left[\frac{\exp(-\mu_j t)}{\mu_j \prod\limits_{n=0,\, n \neq j} (\mu_n - \mu_j)} \right]$$

Further discussion on this model may be found in Ref. 13.

6.5 MODELING OF FIRE-RELATED FAILURES IN BUILDINGS

Fire in a building may lead to serious consequences, such as loss of life, monetary loss, and collapse of the building. The building may collapse because of the exposure of load-bearing members such as walls, columns, and beams to fire. The exposure reduces the strength of these load-bearing members. However, it is possible to prevent excessive loss of strength in these members with adequate fire protection.

When fire in a building reaches the fully developed stage (large fire), it is reasonable to assume that the failure of the building depends on two factors, i.e., the building's fire resistance and fire load. The building structure will fail in a situation when the fire load exceeds the fire resistance of the structure. The following are assumptions in relating the failure probability to the fire load and to the fire resistance of the building structure[15]:

1) The fire load and fire resistance are two independent variables.

2) It is possible to express the fire resistance F in terms of the fire load it (resistance) can withstand.

3) The fire load the building structure can withstand is described by a lognormal distribution. The mean value and the standard deviation associated with this distribution are represented by F_m and σ_F, respectively.

4) The fire load L is described by a lognormal distribution. The mean value and the standard deviation associated with this distribution are represented by L_m and σ_L, respectively.

During the planned service life of a building, the structural failure probability[16,17] is given by

$$P_s = 1 - e^{-\lambda_s t_s} \tag{6.46}$$

$$\lambda_s \equiv f_{cs} \cdot \theta_m \cdot B_a \tag{6.47}$$

where P_s is the structural failure probability during the planned service life of a building, λ_s the average number of failures/unit time, t_s the building's expected service life, f_{cs} the conditional probability associated with structural failure (given that a large fire occurs), B_a the building's floor area, and θ_m the average number of large fires (given in per unit area of the concerned building/unit time).

The conditional probability of structural failure f_{cs} can be estimated from

$$f_{cs} = W/2 \tag{6.48}$$

where

$$W = \text{erfc}\beta = G \int_{-\infty}^{\beta} e^{-t^2/2} \cdot dt$$

and $G \equiv \sqrt{2/\pi}$. The value of β can be obtained from

$$\beta = [(\ell n F)_m - (\ell n L)_m]/Y \tag{6.49}$$

where

$$Y = [2(\sigma^2_{\ell nF} + \sigma^2_{\ell nL})]^{\frac{1}{2}}$$

and where $\sigma_{\ell nF}$ is the standard deviation of ℓnF, $(\ell nF)_m$ the average (mean) of ℓnF, $\sigma_{\ell nL}$ the standard deviation of ℓnL, and $(\ell nL)_m$ the average (mean) of ℓnL.

Alternatively, the β can also be estimated[18,19] from

$$\beta = \frac{\ell n[m(G_L/G_F)^{\frac{1}{2}}]}{[2\ell n(G_L \cdot G_F)]^{\frac{1}{2}}} \tag{6.50}$$

where $G_L \equiv 1 + k_L^2$ and $G_F \equiv 1 + k_F^2$. Also, k_F is the coefficient of variation (for fire resistance), k_L the coefficient of variation (for the fire load), and m the fire load safety factor. This is given by the mean fire resistance (expressed in terms of fire load) over the mean value of the fire load expected.

Model I

This model is concerned with estimating monetary fire loss expectation for a building. Some of the assumptions associated with the model are as follows: 1) the losses occurred are in proportion to the area enveloped in the fire and 2) a part or the entire building has to be replaced in the event of a structural failure. The monetary fire loss expectation M_{fe} is given[15] by

$$M_{fe} = C_1 + C_2[1 - e^{-(Q+I)t_s}] + C_3[1 - e^{-(\lambda_s + I)t_s}] \tag{6.51}$$

where

$$C_1 = \frac{C_m}{C_t} \cdot q \cdot B_a k_t, \qquad q = \frac{\lambda_s}{\lambda_s + 1}, \qquad k_t = k_b + k_{fs}$$

$$C_2 = \frac{1}{C_t} \frac{Q}{(Q+1)} (1 - f_{cs})[(r + r_b)B_a k_t]$$

$$C_3 = \frac{C_m}{C_t} \cdot q \left[r - \frac{B_a \cdot k_t}{t_s(\lambda_s + 1)} \right]$$

$$Q = B_a \cdot \theta_m, \qquad I = I_i - I_{if}, \qquad r = \frac{L_c}{B_c}, \qquad r_b = \frac{C_R}{B_C}$$

λ_s, B_a, f_{cs}, t_s, and θ_m are as defined previously in this section. The remaining symbols used in the above equations are

I_i = interest rate
I_{if} = average inflation rate

C_m = number of compartments lost due to fire
C_t = total number of compartments in the building in question
K_b = building cost/unit floor area (this cost does not include the cost of fire safety measures)
K_{fs} = cost of fire safety measures (associated with the building) given in per unit floor area
L_c = sum of loss of contents and indirect losses
B_c = building cost
C_R = repair cost

Model II

This model is proposed[15] to estimate life loss expectation during the building's planned service life. Thus, the life loss expectation is given by

$$L_{ex} = N_p \cdot P_s \tag{6.52}$$

where

$N_p = \alpha \cdot N_{Tp}$
$N_{Tp} = L_{od} S_a N_s$
$P_s = 1 - e^{-\lambda_s t_s}$
N_p = number of people in the building when the failure occurs
N_{Tp} = number of people in the entire building when the fire starts
α = proportion of population still remaining in the building when the failure occurs
P_s = structural failure probability during the planned service life of a building
L_{od} = occupant load (number of people per unit floor area)
S_a = area of the building's single story
N_s = total number of stories in the building
λ_s = average number of failures/unit time
t_s = building's expected service life

Model III

This model is concerned with estimating the total fire cost expectation. This cost expectation is composed of fire loss expectation plus the cost of fire safety measures. Thus, the monetary fire cost expectation F_{mc} is expressed as[15]

$$F_{mc} = M_{fe} + B_a K_b (m k_{pm} + k_{am}) \tag{6.53}$$

where

k_{am} = active measures (e.g., sprinkler protection) cost, expressed as fraction of the building cost
m = fire load safety factor

k_{pm} = passive measures cost (e.g., structural fire protection cost) per unit increase of m (fire load safety factor), expressed as fraction of the building cost
M_{fe} = monetary fire loss expectation
B_a = building's floor area
k_b = building cost/unit floor area (does not include the cost of fire safety measures)

References

[1]Kivenson, G., *Durability and Reliability in Engineering Design*, Hayden Book Co., New York, 1971.

[2]Collins, J. A., *Failure of Materials in Mechanical Design*, Wiley, New York, 1981.

[3]Fontana, M. G. and Greene, N. D., *Corrosion Engineering*, McGraw-Hill, New York, 1967.

[4]Braunovic, M., "Tribology: Fretting Wear," *Engineering Digest*, Oct. 1981, pp. 48–54.

[5]Bloch, H. P. and Geitner, F. K., *Practical Machinery Management: Machinery Failure Analysis and Troubleshooting*, Gulf Publishing Co., Houston, TX, 1983.

[6]Uhlig, H. H., "Mechanism of Fretting Corrosion," *Transactions of ASME, Journal of Applied Mechanics*, Vol. 76, 1954, pp. 401–407.

[7]Thiruvengadam, A., "Now There is a Way to Work out Erosion Strength of Materials," *Product Engineering*, Vol. 37, Aug. 1966, p. 55.

[8]Johnson, R. C., "Predicting Part Failures," *Machine Design*, Vol. 37, Jan. 7, 1965, p. 137.

[9]Bazovsky, I., *Reliability Theory and Practice*, Prentice-Hall, Englewood Cliffs, NJ, 1961.

[10]Dhillon, B. S., *Reliability Engineering in Systems Design and Operation*, Van Nostrand Reinhold, New York, 1983.

[11]Taraman, S. I. and Kapur, K. C., "Optimization Considerations in Design Reliability by Stress-Strength Interference Theory," *IEEE Transactions on Reliability*, Vol. 24, June 1975, pp. 136–138.

[12]Kapur, K. C. and Lamberson, L. R., *Reliability in Engineering Design*, Wiley, New York, 1977.

[13]Eisenberg, I. M., "Safety of Structural Systems with Reserve Elements for Earthquake Protection," *Earthquake Engineering and Structural Dynamics*, Vol. 5, 1977, pp. 305–318.

[14]Kozlov, B. A. and Ushakov, I. A., *Reliability Handbook*, Holt, Rinehart and Winston, New York, 1970.

[15]Lie, T. T., "Safety Factors for Fire Loads," *Canadian Journal of Civil Engineering*, Vol. 6, 1979, pp. 617–628.

[16]Lie, T. T., "Optimum Fire Resistance of Structure," *ASCE Journal of the Structural Division*, Vol. 98, 1972, pp. 215–232.

[17]Burros, R. H., "Probability of Failure of Building from Fire," *ASCE Journal of the Structural Division*, Vol. 101, 1975, pp. 1947–1960.

[18]Allen, D. E., "Limit States Design—A Probabilistic Study," *Canadian Journal of Civil Engineering*, Vol. 2, 1975, pp. 36–49.

[19]Benjamin, J. R. and Cornell, C. A., *Probability, Statistics and Decision for Civil Engineers*, McGraw-Hill, New York, 1970.

Exercises

6.1 Discuss the term "zero wear."

6.2 Describe the following: fretting wear, corrosion, erosion, fatigue, fatigue safety factor, and tension safety factor.

6.3 Write an essay on usage of earthquake protection systems in buildings.

6.4 Discuss the meaning of the following two terms: fire load and fire resistance.

6.5 What are the assumptions associated with model I [i.e., Eq. (6.51)]. Comment on their practicality.

6.6 What is the fire load safety factor?

7 HUMAN ERRORS IN MECHANICAL SYSTEMS

This chapter discusses several aspects of human errors in mechanical systems. The findings of various studies concerned with human errors are briefly listed. The history of human reliability is reviewed. Six categories of human errors are discussed along with consequences and causes of human errors. The types of human errors made in operating equipment are reviewed. The topic of gross errors is discussed in detail. In addition, the causes of 120 structural failures are presented along with three mathematical models associated with human errors. Two approaches to reduce human error occurrence are presented. The historical background of human reliability data banks is reviewed briefly. Six human reliability data banks are described. These are American Institute for Research (AIR) Data Store, Operational Recording and Data System (OPREDS), Nuclear Plant Reliability Data System (NPRDS), Aerojet General method, Aviation Safety Reporting System (ASRS), and Bunker-Ramo tables. Selected statistics for U.S. Air Force aircraft accidents are presented. Human error rates for selected tasks are given. Two Markov models representing mechanical systems with human error are presented. System mean time to failure formulas are developed.

7.1 INTRODUCTION

Human errors are an important factor to be considered in mechanical systems or equipment reliability. Their impact on an item's reliability is much greater than many of us realize. Findings such as the following witness this belief:

1) Human error contributed 20–53% of system failures according to a survey of nine U.S. Air Force missile systems conducted by Stanford Research Institute.[1]

2) A total of 401 human errors occurred from June 1, 1973 to June 30, 1975 in U.S. commercial light-water nuclear reactors.[2]

3) Over 90% of the documented air traffic control system errors were due to human operators.[3]

4) A total of 82% of the production defects analyzed at the plant of a company were due to human errors.[4]

5) A study conducted in 1960 reported that 23–45% of the failures were due to humans.[5]

6) Human error contributed 50–70% of failures in electronic equipment.[6]

7) A study reported in Refs. 7 and 8 indicate that at least 20–30% of equipment failures were caused by humans.

8) A study of 213 maintenance events reported that 25.8% of the malfunctions were partially or wholly due to human error.[9]

9) About one in five defects in products are missed by inspectors.[10]

10) A study of critical incidents involving aircrews with the U.S. Strategic Air Command revealed that 16% of the time the critical actions of aircrews made the situation worse or did not provide any relief.[11]

For some applications, human errors and their rates can be predicted within an acceptable margin of error. Human error may be described as the failure to perform a given task (or the performance of a forbidden action) that can result in disruption of scheduled operations or result in damage to property and item (equipment). This chapter presents several different aspects associated with human errors.

7.2 A BRIEF HUMAN RELIABILITY HISTORICAL REVIEW

The mid-1950's is regarded as the period during which interest in human reliability grew. During this time, the first human-engineering standards for the U.S. Air Force were developed.[12,13] The need for human reliability was pointed out by Williams[14] in 1958. In the same year, Hettinger[8] and Meister[15] documented results of their findings related to human reliability in two Convair-Astronautics reports. Two years later in 1960, Shapero et al.[1] and LeVan[5] prepared reports of their findings for Wright-Patterson Air Force Base and Bell Aerosystems Company, respectively. Both of these documents pointed out that a large proportion of failures were due to human beings.

An important data base (Data Store) containing time and human performance reliability estimates for human engineering design features was published in 1962.[16] During the early 1960's, Rook,[4,17] Swain,[18] Meister,[19] and others contributed to human reliability. Since then, many other researchers, authors, and editors have contributed to the subject of human reliability. Recently a book entitled *Human Reliability: With Human Factors*[20] was published. Selected references, covering the period from 1958–1978, on the subject can be found in Ref. 21.

7.3 HUMAN ERRORS

This section describes several aspects of human errors.

Breakdowns of Human Errors

Human errors or failures can be categorized[22] into several types as operating, design, maintenance, fabrication, inspection, or contributory.

Operating errors occur almost solely in field use (of equipment) and result from humans operating the equipment incorrectly. Some of the operating error subcategories are motivational error, use of wrong procedures, failure to follow procedures, etc. Design errors are another important type of human errors and result from inadequate design. Maintenance errors result generally from wrong repair or installation of the item under consideration; thus, some of the subcategories of these errors are calibration, repair, and installation errors. Fabrication errors result from poor workmanship and have subcategories such as the use of an incorrect component, poor workmanship, and failure to follow a blueprint. Inspection errors result from poor inspection; accepting out-of-tolerance item(s) is one example. Finally, the failures that are difficult to classify exactly as either human or equipment fall into the category of contributory errors.

Consequences and Causes of Human Errors

The human error failure consequences may be classified into three distinct categories, as shown in Fig. 7.1.

There are several causal factors for human error failures. Six of them are poor skill level or training of personnel, inadequate equipment design, poor job environment, inadequate motivation, improper tools, and inadequate procedures to do the job.

Types of Human Errors Made in Operating Equipment

Errors made in operating equipment can be classified into five distinct types.[23] These are attention, memory, interpretation, operation, and identification errors.

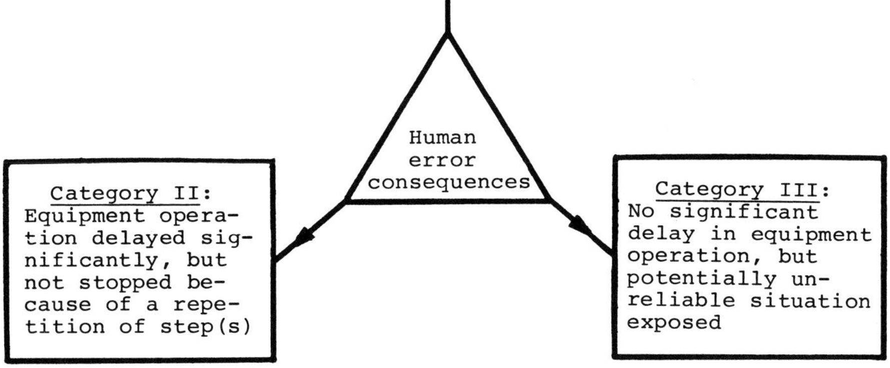

Fig. 7.1 Human error failure consequence categories.

Both attention and memory errors are also known as the errors of omission. In these errors, the human operator plays a passive role and errors result from things the human operator fails to carry out. In contrast to errors of omission, the interpretation, operation, and identification errors are known as the errors of commission. In all of them, the operator plays a more active role. These errors occur from things the human operator does carry out.

Each of the above five types of human errors can be defined briefly as

1) Attention errors occur when the operator fails to pay attention to things requiring attention.

2) Memory errors occur when the human operator's memory fails to perform a required task or when the operator forgets to carry out the task in the specific manner it is required.

3) Interpretation errors happen when the human operator misunderstands the meaning of the given information and then takes action incorrectly.

4) Operation errors are associated with operation. For our purpose, the error may be described simply as one in which the control movement is incorrect to achieve the desired effect.

5) Identification errors occur when an item is identified incorrectly and then considered to be the right one.

7.4 GROSS ERRORS IN STRUCTURES

In the published civil engineering literature, the term "gross errors" is widely used to denote human errors. There is no widely accepted definition of a gross error. However, it may be described as a basic or significant mistake in some aspect of the processes associated with a structure that has the potential for inducing failure.[24] The processes are planning, design and construction, use, or maintenance. The sources of gross errors may be classified as[24]:

1) Inspection. In this case, the nature of error involved is the failure to detect gross defects (e.g., crack in weld).

2) Construction. This is source for errors such as wrong construction, incorrectly interpretating drawings, and wrong fabrication.

3) Conceptual design. In this case, the kind of errors involved are the wrong assumptions regarding the nature of use and nonrecognition of potential failure modes.

4) Use. This is source for errors such as accidental loading and overlooking the requirement for special maintenance.

5) Design and analysis. This is home for errors such as geotechnical data misinterpretation, load omission, computational error in the analytical work, and incorrect interpretation of units.

A study of 120 structural failures in buildings[25] classified their causes into eight distinct categories, as shown in Fig. 7.2. The study revealed that 60% of the failures were due to inadequate experience.

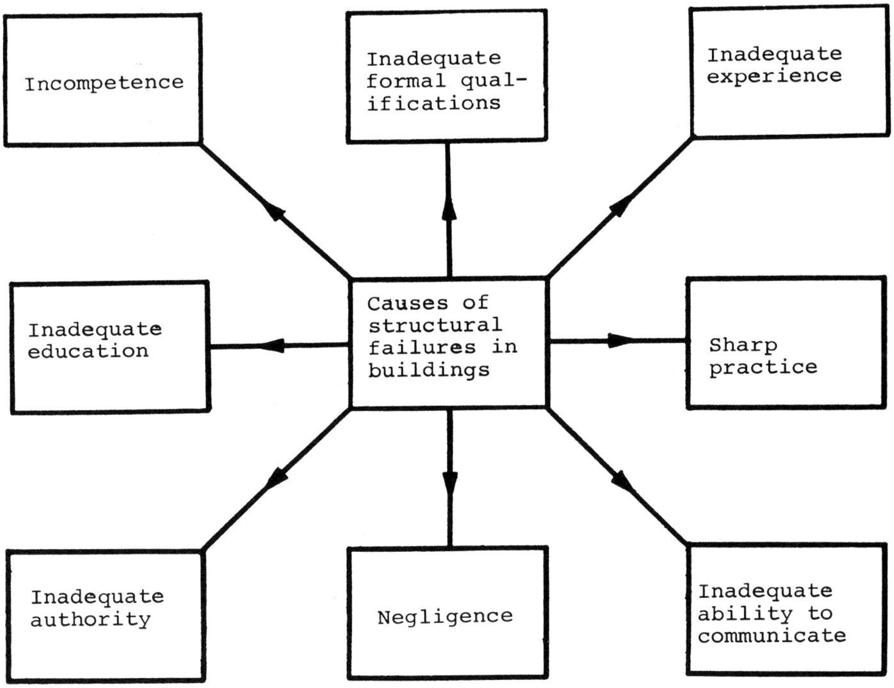

Fig. 7.2 Causes of 120 structural failures.

Checking and inspection is the common method of detecting gross errors during design and construction. According to a study reported in Ref. 26, the cost of inspection is less than 1.2% of the total cost for the entire United Kingdom. The breakdowns were as given in Table 7.1.

Model I

This is the model of the binary case (i.e., the gross error occurs or not) and is concerned with estimating the probability of failure Q.[27] In this case, we let parameter X represent structure reliability such that $X < 0$ for failure and $X > 0$ for a safe realization, e.g.,

$$X = Y - L \tag{7.1}$$

where L is the effect of the total load and Y the resistance.

For the binary case (i.e., the gross error occurs or not), one can consider two conditional probability density functions of the parameter X: as $g(x)$ is for the realization with error and $m(x)$ for the realization without error. Thus, we may write

$$F_c(x) = (1 - p)\int_{-\infty}^{x} m(x_1)\,\mathrm{d}x_1 + p\int_{-\infty}^{x} g(x_1)\,\mathrm{d}x_1 \tag{7.2}$$

Table 7.1 Costs of Inspection Breakdowns in the United Kingdom

No.	Type of inspector	Percentage of the total cost
1	Contractor	0.2
2	Designer	0.4–0.6
3	Checking authority	0.2–0.4

where $F_c(x)$ is the cumulative distribution function of x, p the gross error occurrence probability, and $(1-p)$ the gross error nonoccurrence probability.

After letting $M = \int_{-\infty}^{x} m(x_1)\, dx_1$ and $G = \int_{-\infty}^{x} g(x_1)\, dx_1$ in Eq. (7.2), the probability of failure Q is given by

$$Q = F_c(X=0) = (1-p)M + pG = p(G-M) + M \tag{7.3}$$

Model II

This model is concerned with estimating p, the gross error occurrence probability,

$$p = P_c \prod_{j=1}^{k} (1 - F_{dj}) \tag{7.4}$$

where k is the number of independent control steps, F_{dj} the error detection probability for step j, and P_c the error occurrence probability with no checks.

In order to develop a relationship between the gross error occurrence probability and the inspection and checking cost, we assume[27] that the checking cost is proportional to the number of checks and that the error detection probability for step j is constant.

Thus, considering K as a continuous variable, the gross error occurrence probability is expressed as

$$p = P_c e^{-(K_{h1}/K_{h2})} \tag{7.5}$$

where K_{h1} is the cost of checking, K_{h2} the attenuation cost of checking (i.e., the checking cost corresponding to $p = P_c/2.718$), and P_c the error occurrence probability when the value of K_{h1} is zero.

Model III

This model is concerned with finding the optimum cost of checking and the optimum error occurrence probability associated with the structure. The life-cycle cost LCC_{st} of the structure is given by[27]

$$LCC_{st} = K_f Q + \sum_{i=1}^{3} K_i + K_{h1} \tag{7.6}$$

where K_f is the failure cost, Q the probability of failure, K_1 the initial cost (i.e., design, fabrication, and construction), K_{h1} the cost of checking (control), K_2 the maintenance cost, K_3 the demolition cost, and $K_f Q$ the expected failure cost.

Substituting Eq. (7.5) into Eq. (7.3) results in

$$Q = M + (G - M)P_c e^{-(K_{h1}/K_{h2})} \tag{7.7}$$

Inserting Eq. (7.7) into Eq. (7.6) yields

$$LCC_{st} = \sum_{i=1}^{3} K_i + K_{h1} + K_f M + (G - M)K_f P_c e^{-(K_{h1}/K_{h2})} \tag{7.8}$$

Differentiating Eq. (7.8) with respect to K_{h1} results in

$$\frac{\partial(LCC_{st})}{\partial K_{h1}} = 1 - \frac{(G - M)K_f P_c e^{-(K_{h1}/K_{h2})}}{K_{h2}} \tag{7.9}$$

Setting the above result equal to zero yields

$$1 - \frac{(G - M)K_f P_c e^{-(K_{h1}/K_{h2})}}{K_{h2}} = 0 \tag{7.10}$$

Solving this equation yields

$$K_{h1}^* = K_{h2} \ln[(G - M)K_f P_c / K_{h2}] \tag{7.11}$$

where K_{h1}^* is the optimum value of K_{h1}.

Inserting Eq. (7.11) into Eq. (7.5) yields

$$p^* = K_{h2}/K_f(G - M) \tag{7.12}$$

where p^* is the optimum value of p.

7.5 HUMAN ERROR REDUCTION

This section discusses an approach and measures to reduce human errors.

An Approach for Reducing Human Errors

This is composed of the following four phases[28]:

1) Identification phase. This is concerned with the identification of probable errors. The human errors can be identified through routine surveillance of the failure or discrepancy reporting systems, periodic sampling of information flow (quality history, maintenance forms, etc.), monitoring field operation reports, and so on.

2) Investigation phase. This is concerned with an investigation to determine causal factors. In this phase, statements on event, reason, immediate cause, primary cause, source, root, and origination are sought.

3) Implementation phase. This is concerned with the implementation of preventive and corrective measures. The necessary corrective action may include items such as design changes, procedure changes, and changes in the organizational controls associated with personnel in the work area.

4) Verification phase. This is concerned with verification of the effectiveness of the changes made.

Design Practices for Human Error Reduction

This section presents a number of guidelines that will be useful to design engineers in reducing the human error in the product under consideration. These guidelines are[29]

1) Prevent connection errors as much as possible. These errors are an important problem in assembly or maintenance work.

2) Educate themselves on human factors.

3) Obtain information on human factors problems in similar systems.

4) Produce a design that takes into consideration the human operator in the operator's environment.

5) Make sure that the tasks in question are made interesting.

6) Remember that human beings are not very reliable monitors.

7.6 HUMAN RELIABILITY DATA BANKS

These banks play an important role in human reliability analysis. There have been a number of data banks developed in the last few years, some of which are discussed here.

Historical Background

The U.S. Department of Defense and Atomic Energy Commission (AEC) were probably the first to provide an impetus to develop formal data collection and analysis methods for quantifying human performance reliability.[30] Their efforts were associated with weapon delivery systems and nuclear weapons, respectively. The study of Shapero et al.[1] was probably one of the first to quantify system malfunctions due to human error. They discovered that human error was the cause for about 39% of the missile malfunctions.

The study conducted by Rook[4] in 1962 is also regarded as an important milestone in the history of human reliability data banks. His study was concerned with approximately 23,000 defects in the production of nuclear components. He found that approximately 82% of the defects were due to the human element.

In the same year, American Institute for Research (AIR) Data Store was developed by Munger et al.[16] It contained time and human performance reliability estimates. Ever since, the data banks such as the following have been developed:

1) Bunker-Ramo tables for predicting the operational performance of personnel.[31]

2) Operational Recording and Data System (OPREDS).[32]

3) Nuclear Plant Reliability Data System (NPRDS).[33]

4) Aerojet General human reliability in the performance of maintenance.[34]

Several other data banks are listed in Refs. 20 and 30.

Brief Descriptions of Selected Human Reliability Data Banks

This section briefly describes a number of selected human reliability data banks. A detailed description of several such data banks may be found in Ref. 30.

Data Store. It is normally referred to as American Institute for Research (AIR) Data Store. It was first published in 1962 by Munger et al.[16] and contains time and human performance reliability estimates. This data bank was developed so that the reliability estimates can be used with the index of electronic equipment operability (IEEO). This index had such objectives as determining the time and human performance reliability and the identification of specific design features associated with the degradation of human performance.

All in all, the following two things may be said about the Data Store: 1) it possesses important data relating human error probabilities to design features and 2) it is probably the most established data bank for use during the design process.

Operational Recording and Data System. This is concerned with collecting operational human performance data. More specifically, the Operational Recording and Data System (OPREDS) permits automatic monitoring of human operator actions. In the late 1960's and early 1970's, this method of data collection was used by the U.S. Navy Electronics Laboratory in San Diego. The OPREDS equipment was installed in the Navy Tactical Data System (NTDS) of many ships. The outputs from the OPREDS are obtained in the form of 30 bit parallel words and are recorded on magnetic tapes after their conversion to serial bit streams. The OPREDS data are restricted to switch-turning and button-pushing. In recent years, the OPREDS data have been integrated with an automated task inventory.[35]

Nuclear Plant Reliability Data System. This system is basically an operating-equipment-oriented system and is being used by the participating utilities. Subcommittee ANS-58.20 of the American Nuclear Society initially developed the Nuclear Plant Reliability Data System (NPRDS) operated by the Southwest Research Institute.[23] Since 1982, the sponsorship of the NPRDS has been transferred from the American Nuclear Society to the Institute for Nuclear Power Operations (INPO). In its present form, the NPRDS possesses only 83 entries of system failures where the "cause" can be attributed to human error. All in all it may be said that NPRDS has the potential to serve as a nuclear human reliability data bank provided its present difficulties are overcome.

Aerojet General Method. In the development of this method, the purpose was to determine the effectiveness of personnel during the maintenance and scheduled check-out activities. These activities were associated with the Titan II propulsion system. The method was developed by Irwin et al.[34] and depends to a certain degree on the use of the AIR Data Store. More specifically, in order to estimate human reliability, the method makes use of AIR Data Store, expert judgment, and task analysis. The Aerojet General human reliability method is composed of eight steps, which are described in detail in Ref. 30.

Aviation Safety Reporting System. This system was developed by the National Aeronautics and Space Administration (NASA) and contains civil aircraft accident/incident data. It is managed by Battelle Columbus Laboratories and is based upon voluntary reporting. The system receives about 400 human error reports each month and contains a total of about 30,000 reports. The system is described in detail in Ref. 30.

Bunker-Ramo Tables. These are used to determine the operational performance of personnel. These tables first appeared in 1967 as an appendix to Ref. 31. The probability estimates of human reliability are based upon empirical data. These data were obtained from 37 different experimental studies. The government reports, behavioral journals, abstracts of literature dating back to 1950, etc., formed data sources of such studies. The Bunker-Ramo tables are composed of four parts, namely, the activity to be carried out with specific display types, the parameters considered to obtain estimates for probability, the estimates for probability, and the appended notes (which discuss, e.g., the rationale for the data source). In comparison to the AIR Data Source, these tables represent only a slight advance over it.

U.S. Air Force Aircraft Accident Statistics

This section presents selected statistics associated with U.S. Air Force aircraft accidents.[30] Table 7.2 presents Air Force aircraft accident (selected) cause frequency for 1971. Similarly, Figs. 7.3 and 7.4 present U.S. Air Force major aircraft accidents related to human factors.

Table 7.2 Aircraft Accident Cause Frequency for 1971 (U.S. Air Force)

No.	Selected cause	Frequency	Year
1	Selected incorrect action	35	1971
2	Inattention	6	1971
3	Navigation error	6	1971
4	Instrument reading misinterpretation	9	1971
5	Delay in executing required action	42	1971
6	Error in judging distance or speed	26	1971
7	Selected incorrect action	38	1978–1979
8	Instrument reading misinterpretation	8	1978–1979
9	Delay in executing required action	54	1978–1979

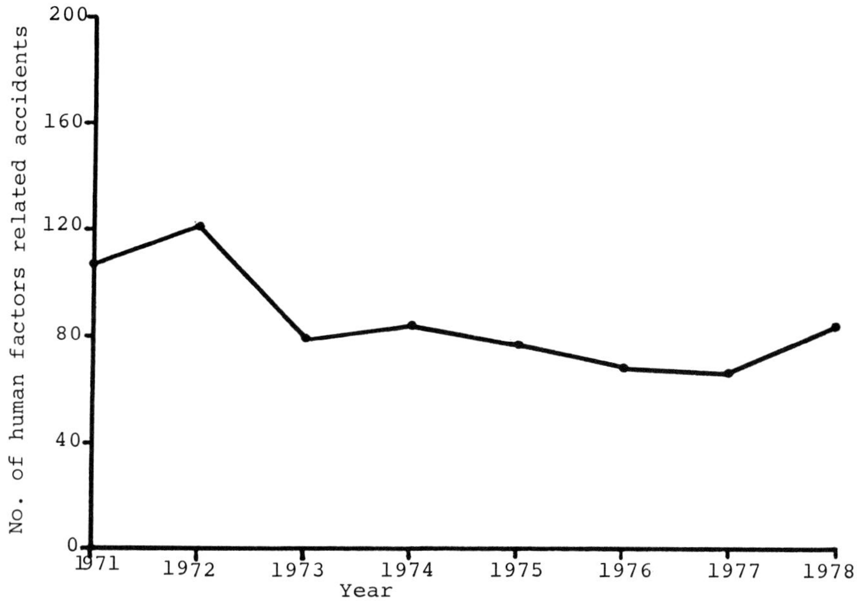

Fig. 7.3 Major aircraft accidents in the U.S. Air Force related to human factors.

Fig. 7.4 Percentage of total major aircraft accidents in the U.S. Air Force related to human factors.

Table 7.3 Human Error Rates for Selected Tasks

No.	Task	Human error rate (%)
1	General maintenance action	0.02–2.0
2	Mechanical linkage adjustment	1.67
3	Arithmetic tasks (simple)	10
4	Error correction (i.e., error will not be corrected)	0.2–20.0
5	Errors of commission (operating actions)	0.3
6	Nuts and bolts installation	0.06
7	Assembly errors	0.4
8	O-ring installation	6.67
9	Monitoring tasks (passive)	10

Human Error Rates for Selected Tasks

This section presents human error rates for selected tasks. The sources to obtain human error rates for various different types of tasks are listed in Ref. 20. Table 7.3 presents error rates for various different tasks.[36]

7.7 RELIABILITY EVALUATION OF MECHANICAL SYSTEMS WITH HUMAN ERROR

This section presents two Markov models representing mechanical systems with human error.

Mechanical System Type I: Model I

In this case, the system can fail due either to a hardware failure or a human error. The mechanical system transition diagram is shown in Fig. 7.5. The diagram shows that the system has two mutually exclusive failure modes, i.e., the system failure due to a human error (failure mode I) and the system failure due to a hardware failure (failure mode II).

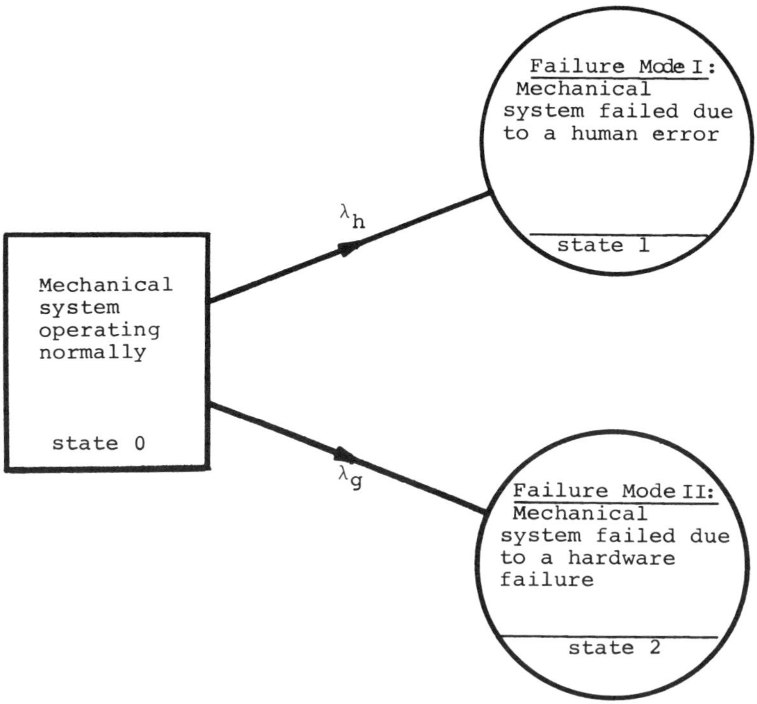

Fig. 7.5 Mechanical system transition diagram.

The Markov technique was used to develop probability equations for the three system states shown in Fig. 7.5. The following were assumed in developing the state probability equations: 1) system failures due to hardware failures or human errors can be separated, 2) a human error can cause the system failure, 3) the system human error rate is constant, and 4) the system hardware failure rate is constant. The following symbols are associated with this Markov model:

i = state of the system; $i = 0$ means the mechanical system operating normally, $i = 1$ means the mechanical system failed due to a human error, $i = 2$ means the mechanical system failed due to a hardware failure
$P_i(t)$ = probability that mechanical system is in state i at time t, for $i = 0,1,2$
λ_h = constant human error rate of the mechanical system
λ_g = constant hardware failure rate of the mechanical system
t = time

The following system of differential equations is associated with Fig. 7.5:

$$\frac{\mathrm{d}P_0(t)}{\mathrm{d}t} + (\lambda_h + \lambda_g)P_0(t) = 0 \tag{7.13}$$

$$\frac{\mathrm{d}P_1(t)}{\mathrm{d}t} - \lambda_h P_0(t) = 0 \tag{7.14}$$

$$\frac{\mathrm{d}P_2(t)}{\mathrm{d}t} - \lambda_g P_0(t) = 0 \tag{7.15}$$

At $t = 0$, $P_0(0) = 1$, $P_1(0) = 0$, and $P_2(0) = 0$. Solving the above differential equations results in

$$P_0(t) = e^{-(\lambda_h + \lambda_g)t} \tag{7.16}$$

$$P_1(t) = \frac{\lambda_h}{(\lambda_h + \lambda_g)}[1 - e^{-(\lambda_h + \lambda_g)t}] \tag{7.17}$$

$$P_2(t) = \frac{\lambda_g}{(\lambda_h + \lambda_g)}[1 - e^{-(\lambda_h + \lambda_g)t}] \tag{7.18}$$

From Eq. (7.16), the reliability $R_{\mathrm{ms}}(t)$ of the mechanical system is

$$R_{\mathrm{ms}}(t) = e^{-(\lambda_h + \lambda_g)t} \tag{7.19}$$

The system mean time to failure (MTTF) is given by

$$\mathrm{MTTF} = \int_0^{\infty} R_{\mathrm{ms}}(t)\,\mathrm{d}t = \frac{1}{\lambda_h + \lambda_g} \tag{7.20}$$

Mechanical System Type II: Model II

In this case, the mechanical system is composed of two independent and identical units in parallel.[37] Each unit can fail in two separate failure modes (i.e., due to human errors or hardware failures). At least one unit must operate normally for system success. In addition, both units may fail simultaneously due to common-cause failures. Some of the reasons for common-cause failures may be operations and maintenance errors, vibrations, temperature, dust, humidity, fire, flood, design deficiency, etc. As soon as any one of the operating units fails, the repair process begins. The active parallel system fails when both units are in the failed state. The system transition diagram is shown in Fig. 7.6.

The following assumptions were made to develop state probability equations by using the Markov approach: 1) failures are statistically independent, 2) the unit hardware failure and human error rates are constant, 3) the repaired unit is as good as new, 4) the failed system is never repaired, 5) each unit's human and hardware failures can be separated, 6) the common-cause failure rate is constant, 7) both units may fail simultaneously due to the occurrence of a common-cause failure, and 8) the system units are identical.

The following notation is associated with this Markov model:

i = ith state (shown in Fig. 7.6) of the system; $i = 0$ means the parallel system operating normally, i.e., both units operating normally, $i = 1$ means one unit failed due to human errors, while the other was operating normally, $i = 2$ means one unit failed due to hardware failures, while the other operating normally, $i = 3$ means both units failed, at least one due to human errors, $i = 4$ means both units failed, at least one due to hardware failures, $i = c$ means both units failed due to common-cause failures

$P_i(t)$ = probability that system is in state i at time t, for $i = 0,1,2,3,4,c$

$P_i(s)$ = Laplace transform of the probability that system is in state i, for $i = 0,1,2,3,4,c$

λ_c = constant common-cause failure rate of the system

λ_h = constant human error rate of the unit

λ = constant hardware failure rate of the unit

μ_1 = constant repair rate of the unit from state 1 to state 0

μ_2 = constant repair rate of the unit from state 2 to state 0

s = Laplace transform variable

The following system of differential equations is associated with Fig. 7.6:

$$P'_0(t) + (\lambda_c + 2\lambda + 2\lambda_h)P_0(t) = P_1(t) \cdot \mu_1 + P_2(t) \cdot \mu_2 \tag{7.21}$$

$$P'_c(t) = P_0(t) \cdot \lambda_c \tag{7.22}$$

$$P'_1(t) + (\lambda + \lambda_h + \mu_1)P_1(t) = P_0(t) \cdot 2\lambda_h \tag{7.23}$$

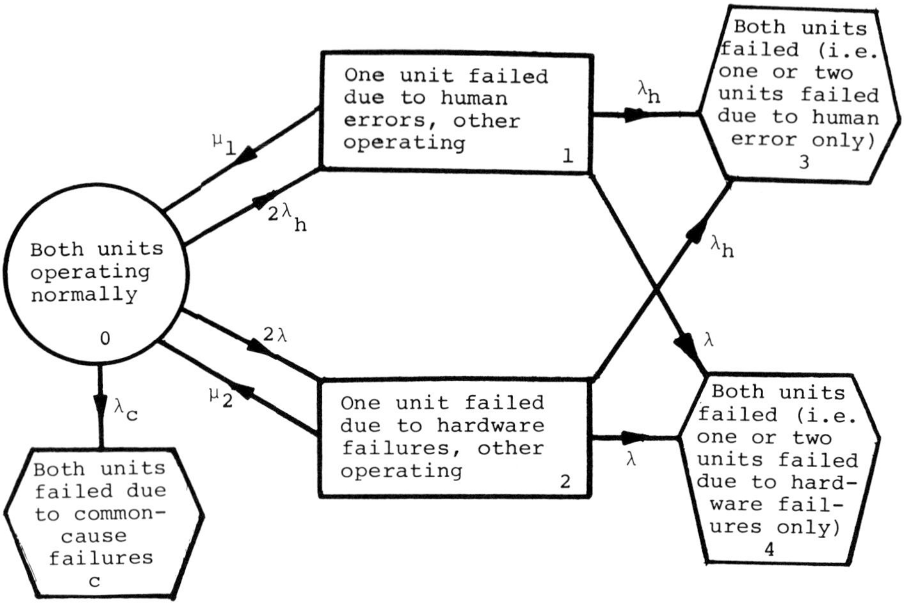

Fig. 7.6 System transition diagram.

$$P_2'(t) + (\lambda + \lambda_h + \mu_2)P_2(t) = P_0(t) \cdot 2\lambda \tag{7.24}$$

$$P_3'(t) = P_1(t) \cdot \lambda_h + P_2(t) \cdot \lambda_h \tag{7.25}$$

$$P_4'(t) = P_1(t) \cdot \lambda + P_2(t) \cdot \lambda \tag{7.26}$$

At $t = 0$, $P_0(0) = 1$,

$$P_1(0) = P_2(0) = P_3(0) = P_4(0) = P_c(0) = 0$$

The prime on $P(t)$ denotes the differentiation with respect to time t. Laplace transforms of the solution are

$$P_0(s) = \frac{EF}{E \cdot FG - 2\mu_1\lambda_h \cdot F - 2\mu_2\lambda E} \tag{7.27}$$

where

$$E \equiv (s + \lambda + \lambda_h + \mu_1), \qquad F \equiv (s + \lambda + \lambda_h + \mu_2), \qquad G \equiv (s + 2\lambda_h + 2\lambda + \lambda_c)$$

$$P_c(s) = \lambda_c \cdot P_0(s)/s \tag{7.28}$$

$$P_1(s) = \frac{2\lambda_h \cdot P_0(s)}{E} \tag{7.29}$$

$$P_2(s) = \frac{2\lambda}{F} P_0(s) \tag{7.30}$$

$$P_3(s) = \frac{1}{s}\left(\frac{2\lambda_h^2}{E} + \frac{2\lambda\lambda_h}{F}\right) \cdot P_0(s) \tag{7.31}$$

$$P_4(s) = \frac{1}{s}\left(\frac{2\lambda \cdot \lambda_h}{E} + \frac{2\lambda^2}{F}\right) \cdot P_0(s) \tag{7.32}$$

In order to obtain the time domain solution, Eqs. (7.27–7.32) have to be inverted.

The Laplace transform of the system reliability is

$$R(s) = P_0(s) + P_1(s) + P_2(s) \tag{7.33}$$

The system mean time to failure (MTTF) is given by

$$\text{MTTF} = \lim_{s \to 0} R(s) \tag{7.34}$$

Thus, with the aid of Eqs. (7.33) and (7.34), we get

$$\text{MTTF} = \left(\frac{H \cdot I}{J \cdot H \cdot I - 2\mu_1\lambda_h \cdot I - 2\mu_2\lambda H}\right)\left(1 + \frac{2\lambda_h}{H} + \frac{2\lambda}{I}\right) \tag{7.35}$$

where

$$H \equiv \lambda + \lambda_h + \mu_1, \qquad I \equiv \lambda + \lambda_h + \mu_2, \qquad J \equiv 2\lambda_h + 2\lambda + \lambda_c$$

References

[1]Shapero, A., Cooper, J. I., Rappaport, M., Schaeffer K. H., and Bates, C. J., "Human Engineering Testing and Malfunction Data Collection in Weapon System Programs," Wright Air Development Division, Wright-Patterson Air Force Base, OH, WADD TR 60-36, Feb. 1960.

[2]Joos, D. W., Sabri, Z. A., and Husseiny, A. A., "Analysis of Gross Error Rates in Operation of Commercial Nuclear Power Stations," *Nuclear Engineering and Design*, Vol. 52, 1979, pp. 265–300.

[3]Kinney, G. C., Spahn, M. J., and Amato, R. A., "The Human Element in Air Traffic Control: Observations and Analysis of Performance of Controllers and Supervisors in Providing Air Traffic Control Separation Services," METREK Div., MITRE Corp., Rept. MTR-7655, Dec. 1977.

[4]Rook, L. W., "Reduction of Human Error in Industrial Production," Sandia National Laboratories, Albuquerque, NM, Rept. SCTM 93-63 (14), June 1962.

[5]LeVan, W. I., "Analysis of the Human Error Problem in the Field," Bell Aerosystems Co., Buffalo, NY, Rept. 7-60-932004, June 1960.

[6]Christensen, J. M., Howard, J. M., and Stevens, B. S., "Field Experience in Maintenance," *Human Detection and Diagnosis of System Failures*, edited by J. Rasmussen and W. B. Rouse, Plenum Press, New York, 1981, pp. 111–113.

[7]Meister, D., "The Problem of Human-Initiated Failures," *Proceedings of the 8th National Symposium on Reliability and Quality Control*, 1962, pp. 234–239.

[8]Hettinger, C. W., "Analysis of Equipment Failures Caused by Human Errors," Convair-Astronautics, Rept. REL-R-005, April 15, 1958.

[9]Robinson, J. E., Deutsch, W. E., and Rogers, J. G., "The Field Maintenance Interface Between Human Engineering and Maintainability Engineering," *Human Factors*, Vol. 12, 1970, pp. 253–259.

[10]Juran, J. M., *Quality Control Handbook*, McGraw-Hill, New York, 1974.

[11]Ronan, W. W., "Training for Emergency Procedures in Multiengine Aircraft," American Institute for Research, Rept. AIR-153-53-FR-44, Pittsburgh, PA, 1953.

[12]Meister, D., "Human Reliability," *Human Factors Review*, edited by F. A. Muckler, Human Factors Society, 1984, pp. 13–53.

[13]Swain, A. D., "Overview and Status of Human Factors Reliability Analysis," *Proceedings of the 8th Annual Reliability and Maintainability Conference*, Gordon and Breach Publishers, Baltimore, MD, 1969, pp. 251–254.

[14]Williams, H. L., "Reliability Evaluation of the Human Component in Man-Machine Systems," *Electrical Manufacturing*, April 1958, pp. 78–82.

[15]Meister, D., "Human Engineering Survey of Difficulties in Reporting Failure and Consumption Data," Convair-Astronautics, Rept. ZX-7-035, Sept. 1958.

[16]Munger, S. J., Smith, R. W., and Payne, D., "An Index of Electronic Equipment Operability: Data Store," American Institute for Research, Pittsburgh, PA, Rept. AIR-C43-1/62 RP(1), 1962.

[17]Rook, L. W., "Motivation and Human Error," Sandia National Laboratories, Albuquerque, NM, Rept. SCTM-65-135, 1965.

[18]Swain, A. D., "A Method for Performing a Human Factors Reliability Analysis," Sandia National Laboratories, Albuquerque, NM, Rept. SCR-685, 1963.

[19]Meister, D., "Methods of Predicting Human Reliability in Man-Machine Systems," *Human Factors*, Vol. 6, 1964, pp. 621–646.

[20]Dhillon, B. S., *Human Reliability: With Human Factors*, Pergamon Press, New York, 1986, p. 239.

[21]Dhillon, B. S., "On Human Reliability—Bibliography," *Microelectronics and Reliability*, Vol. 20, 1980, pp. 371–373.

[22]Meister D., "The Problem of Human-Initiated Failures," *Proceedings of the 8th National Symposium on Reliability and Quality Control*, 1962, pp. 234–239.

[23]Lincoln, R. S., "Human Factors in the Attainment of Reliability," *IRE Transactions on Reliability*, Vol. 9, 1960, pp. 97–103.

[24]Thoft-Christensen, P. and Baker, M. J., *Structural Reliability Theory and Its Applications*, Springer-Verlag, Berlin, 1982, pp. 239–248.

[25]*Structural Failures in Buildings, Symposium*, Institution of Structural Engineers, London, April 1980.

[26]"Rationalization of Safety and Serviceability Factors in Structural Codes," Construction Industry Research and Information Association, London, Rept. CIRIA-63, July 1977.

[27]Nowak, A. S., "Effect of Human Error on Structural Safety," *Journal of the American Concrete Institute*, Vol. 76, 1979, pp. 959–972.

[28]Peters, G. A., "Human Error: Analysis and Control," *Journal of the American Society of Safety Engineers*, Vol. XI, 1966, pp. 9–15.

[29]Swain, A. D., "Human Factors in Design of Reliable Systems," *Proceedings of the Tenth Symposium on Reliability and Quality Control*, 1964, pp. 250–259.

[30]Topmiller, D. A., Eckel, J. S., and Kozinsky, E. J., "Human Reliability Data Bank for Nuclear Power Plant Operations: A Review of Existing Human Reliability Data Banks," U.S. Nuclear Regulatory Commission, Rept. NUREG/CR 2744/1 of 2, 1982.

[31]Hornyak, S. J., "Effectiveness of Display Subsystems Measurement and Prediction Techniques," Rome Air Development Center, Griffiss Air Force Base, NY, RADC Rept. TR-67-292, Sept. 1967.

[32]Urmston, R., "Operational Performance Recording and Evaluation Data System (OPREDS)," Navy Electronics Laboratory Center, San Diego, CA, Descriptive Brochures, Code 3400, Nov. 1976.

[33]"Reporting Procedures Manual for the Nuclear Plant Reliability Data System (NPRDS)," Southwest Research Institute, San Antonio, TX, Dec. 1980.

[34]Irwin, I. A., Levitz, J. J., and Freed, A. M., "Human Reliability in the Performance of Maintenance," Bunker-Ramo Corp., Rept. BRC 317-BSD-TDR-63-218, 1963.

[35]Osga, G. A., "Guidelines for Development, Use, and Validation of a Human Performance Data Bank for NTDS Combat Operations," Systems Exploration, Inc., San Diego, CA, Contract Nov-00123-80-D0263 (Naval Ocean Systems Center), March 1981.

[36]"Reliability and Maintainability Data for Industrial Plants," A. P. Harris and Associates, Ottawa, Canada, Rept. TD-84-3, 1984.

[37]Dhillon, B. S., "Stochastic Models for Evaluating Probability of System Failure Due to Human Error," *Microelectronics and Reliability*, Vol. 24, 1984, pp. 921–924.

Exercises

7.1 Define the following two terms: human error and human reliability.

7.2 Write a short essay on the history of the human reliability discipline.

7.3 What is a gross error?

7.4 List at least five categories of human errors.

7.5 List at least eight existing human reliability data banks.

7.6 A mechanical system can fail due either to a hardware failure or to a human error. The hardware failure and human error rates of the system are 0.0006 failures/h and 0.0001 errors/h, respectively. Calculate the probability that the system will be in its failed state due to a human error at time $t = 50$ h. Assume that the system starts operating at time $t = 0$.

7.7 Assume that in Fig. 7.6 $\mu_1 = \mu_2 = 0$. Develop an expression for the system mean time to failure.

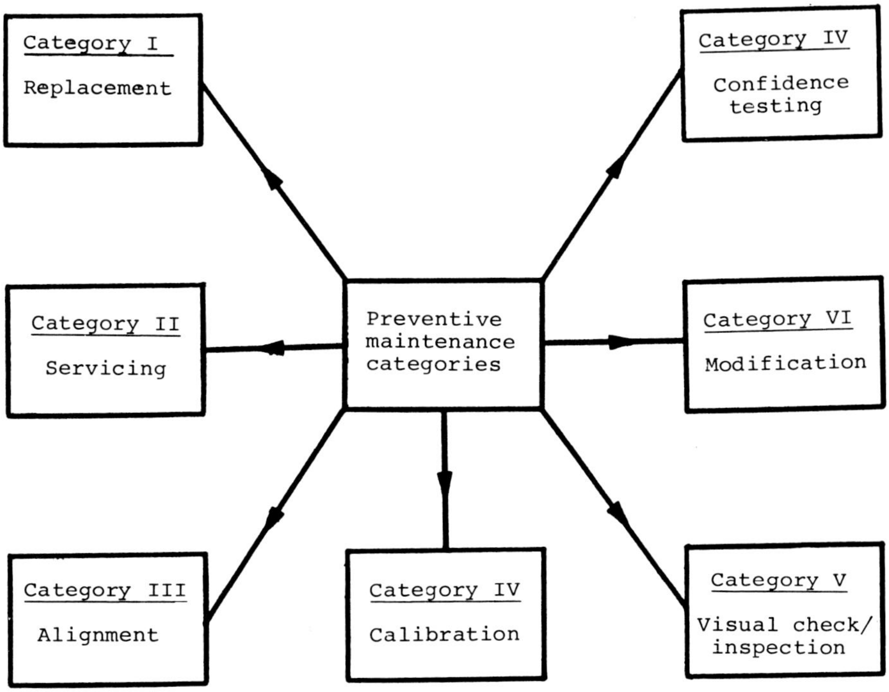

Fig. 8.1 Categories of preventive maintenance.

optimum frequency of inspections for the items under consideration. When establishing an item's frequency of inspection, attention must be paid to factors such as[3] safety requirements; age of item; criticality of the item's service; susceptibility to damage, wear and loss of adjustment; condition of item; value of item; and time in operation.

Equipment for Inclusion in Preventive Maintenance Program. There are various types of equipment included in the preventive maintenance program. According to a survey of the maintenance practices in 542 plants in the United States, the following percentages of equipment were covered by preventive maintenance: 65.9% production machinery, 58.9% materials-handling equipment, 50.5% plant services, 78.4% motors, 51.8% process equipment, 61.9% buildings, 48.8% lighting, 63.6% controls.

According to Ref. 3, items such as the following were covered in the preventive maintenance program of a process plant: pumps, fire-protection equipment, plant buildings, safety equipment (e.g., pressure-relief valves, flame arrestors, emergency-relief equipment), auxiliary equipment, heat exchangers, storage tanks, main boilers, furnaces, compressed-air pipelines, compressors, motors, electric generators, instruments, and distribution systems for water and steam.

Guidelines for Improving a Preventive Maintenance Program. A preventive maintenance program has to be refined on a continuous basis in order to be effective. The following guidelines are examples of the item that should be considered in developing a preventive maintenance program:

1) Make use of statistical and modern diagnostic tools.
2) Make every effort to avoid overmaintenance.
3) Design for low-cost maintenance in new equipment.
4) Obtain accurate estimates of costs.
5) Make use of industrial engineering methods.
6) Examine inspection frequencies.
7) Make use of repair codes.
8) Make available specific craft instructions and keep them updated.

Advantages of Performing Preventive Maintenance. The preventive maintenance has provided its users with many returns, some of which are reductions in the cost of maintenance, large-scale repairs, costs of repairs, unit cost of manufacture, production downtime, the need for standby equipment, number of product rejects, and overtime for maintenance workers. It has also led to improvements in industrial relations, spare-parts control, and safety for concerned workers.

8.3 IDENTIFICATION OF PROBLEMS WITH PUMPS AND GEAR TEETH

As pumps and gears are the concerns of the many mechanical maintenance personnel, this section identifies problems with steam, rotary, centrifugal, and reciprocating pumps and gear teeth. Each of these items is discussed below.[4,5]

Steam Pumps

The typical symptoms of steam pump troubles are vibrations, erratic operation, failure to develop rated pressure, short strokes, and capacity loss after starting. The possible causes of each of these symptoms are as follows[4]:

1) Vibrations. In this case, the cause can be misalignment, too tight packing of liquid piston, foundation not rigid, or lack in piping support.

2) Erratic operation. The possible cause for this trouble could be vapors in the liquid, a cavitation, leak in the suction line, or too high suction lift.

3) Failure to develop rated pressure. There are four possible causes for this trouble, e.g., piston rings in the steam end are worn out, steam pressure is low, piston rings at the liquid end are binding, or the exhaust pressure is too high.

4) Short strokes. In this case, the possible cause for the trouble can be too tight packing of liquid piston, too much steam cushioning, or steam valves out of adjustment.

5) Capacity loss after starting. There are three possible causes for the capacity loss: Vapors in liquid, too high suction lift, and leak in the suction line.

Rotary Pumps

There are a number of symptoms of the rotary pump troubles, some of which are noise, consumption of too much power, rapid wear, failure to discharge, and loss of suction after start. The possible causes for each of these symptoms of trouble are listed in Ref. 4.

Centrifugal Pumps

There are many symptoms that indicate trouble in centrifugal pumps. Some of them are overheating, failure to deliver water, insufficient pressure, excessive leakage in the stuffing box, short life of bearings or packing, vibrations, delivery of insufficient capacity, and noise. Possible causes for each of these are listed in Ref. 4.

Reciprocating Pumps

Reciprocating pumps are also prone to various kinds of trouble and they exhibit many symptoms, such as water in the crankcase, pitted seats, overheating of the power end, oil leaks in the crankcase, rapid plunger wear, prime loss, and noise (liquid end or power end). A list of the possible causes for these symptoms can be found in Ref. 4.

Gear Teeth

The wear and failure of gear teeth are important problems in mechanical reliability. Most of these problems may be classified under the following nine headings[4]:

1) Cracking. This is due to residual stresses or incorrect operating conditions. In hardenable material, the residual stresses are induced because of ineffective manufacturing control.

2) Fatigue. Fatigue breakage is due to a large number of revolutions or cycles (repetitions of a load).

3) Welding. In general the welding of gear tooth surfaces is caused by poor lubrication and overloading.

4) Plastic yielding. This is a deformation of the gear tooth surfaces.

5) Wear. This includes normal wear, abrasive wear, overload wear, and scratching.

6) Overload. Overload breakage occurs from an unexpected shock overload. An example of a possible cause is a jamming of the connected machinery.

7) Surface fatigue. This includes spalling, initial pitting, and destructive pitting and can be described simply as the formation of cavities in the teeth surfaces.

8) Quenching cracks. These occur because of tooling marks in mating gears, incorrect heat treatment of gears, or excessively sharp tooth fillets.

9) Miscellaneous. This includes burning, corrosive wear, grinding checks, and interference.[5]

8.4 MAINTENANCE- AND MAINTAINABILITY-RELATED FORMULAS

This section presents a number of formulas related to maintenance and maintainability.[6,7]

Maintenance Support Indicator

This is an important measure of maintenance manpower requirements. The measure is generally known as the maintenance support index. Both the repairability and the reliability of a certain item are reflected by this index. The measure is expressed as follows:

$$M_{si} = M_{am}(1000)/T \tag{8.1}$$

where M_{si} is the maintenance support index, M_{am} the number of active maintenance man-hours, and T the system operating hours.

Man-Hour Indicator

Another important index providing information about maintenance manpower needs is defined as

$$M_r = M_h/T_{ar} \tag{8.2}$$

where M_r is the man-hour ratio, M_h the total number of man-hours, and T_{ar} the total active repair time.

Mean Corrective Maintenance Labor-Hours

This is useful information in maintainability work and is expressed by

$$H_{cm} = \frac{\left(\sum \mathrm{FR}_j\right)(H_j)}{\sum \mathrm{FR}_j} \tag{8.3}$$

where H_{cm} is the average corrective maintenance labor-hours, H_j the mean labor-hours required to accomplish repair of the item j, and FR_j the constant failure rate (failures/h) of the item j.

Similarly, one can obtain mean preventive maintenance labor-hours.

Average Preventive and Corrective Maintenance Downtimes

Information on the average values of preventive and corrective maintenance downtimes play an important role in maintainability analysis. The mean value of the preventive maintenance downtime can be calculated from

$$T_{mp} = \sum_{1}^{k} T_{ap} \Big/ k \tag{8.4}$$

where T_{mp} is the average preventive maintenance downtime, k the total number of preventive maintenance tasks, and T_{ap} the active maintenance downtime for one preventive maintenance task.

Similarly, the mean value of the corrective maintenance downtime can be computed from

$$T_{mc} = \sum_{1}^{n} T_{ac} \Big/ n \tag{8.5}$$

where T_{mc} is the average corrective maintenance downtime, n the total number of corrective maintenance tasks, and T_{ac} the active maintenance downtime for one corrective maintenance task.

Average Downtime

This is given by

$$T_{md} = \frac{\theta(1 - \mathrm{AV}_I)}{\mathrm{AV}_I} \tag{8.6}$$

where θ is the mean time between failures, T_{md} the average downtime, and AV_I the inherent (intrinsic) availability (steady state). This availability does not include logistic and administrative downtime.

Average Time between Maintenance

This is defined by

$$\mathrm{ATBM} = \left(\frac{1}{T_c} + \frac{1}{T_p}\right)^{-1} \tag{8.7}$$

where ATBM is the average time between maintenance (corrective and preventive), T_c the average corrective maintenance time interval, and T_p the average preventive maintenance time interval.

Mean Time to Repair

This is given by

$$\mathrm{MTTR} = \left(\sum_{j=1}^{m} \mathrm{FR}_j T_j \right) \Big/ \sum_{j=1}^{m} \mathrm{FR}_j \tag{8.8}$$

where MTTR is the mean time to repair, FR_j the constant failure rate of the jth repairable part, m the total number of repairable parts, and T_j the necessary time to repair equipment because of the failure of the jth part.

Maintainability Function

This is used to predict the probability that the repair will be completed in time t. The maintainability function is expressed as

$$M(t) = \int_0^t f(t)\,\mathrm{d}t = 1 - e^{-\int_0^t \mu(t)\,\mathrm{d}t} \tag{8.9}$$

where $M(t)$ is the probability that repair will be completed in time t provided it began at time $t = 0$, $f(t)$ the probability density function of the repair times (examples of the repair time probability density functions are lognormal, exponential, Weibull, and gamma), t the repair time variable, and $\mu(t)$ the instantaneous repair rate.

Other related formulas are as follows:

The instantaneous repair rate is defined by

$$\mu(t) = \frac{f(t)}{1 - M(t)} \tag{8.10}$$

where $\mu(t)$ is the instantaneous repair rate.

The mean time to repair is given by

$$\mathrm{MTTR} = \int_0^{\infty} t f(t)\,\mathrm{d}t = \int_0^{\infty} [1 - M(t)]\,\mathrm{d}t \tag{8.11}$$

where MTTR is the mean time to repair.

For a given instantaneous repair time, the probability density function can be obtained from

$$f(t) = \mu(t) \exp\left[-\int_0^t \mu(t)\,\mathrm{d}t \right] \tag{8.12}$$

Maintainability Index

This index was developed by the Naval Electronics Laboratory in San Diego for the use in the U.S. Navy. Basically, the index is a measure of the operational time of the system/equipment and is defined as

$$I_m = \left(1 - \frac{D_{Tc} + D_{Tp}}{T_{ct}}\right) \times (100) \tag{8.13}$$

where I_m is the maintainability index, $T_{ct} = 2184$ h, the length of anticipated continuous cruise period under combat conditions (e.g., 91 days $\times$ 24 h = 2184 h), D_{Tp} the total scheduled or preventive maintenance time (downtime) in hours during 91 days, and D_{Tc} the total corrective maintenance time (downtime) in hours during 91 days.

8.5 ITEM (EQUIPMENT) AVAILABILITIES

There are several different types of steady-state availabilities one can calculate for a piece of equipment. These are presented in this section.

Item Inherent Availability

The inherent or intrinsic availability can be defined in three different ways as

$$\mathrm{AV}_I = \frac{\mathrm{OT}}{\mathrm{OT} + \mathrm{AT}} \tag{8.14}$$

or

$$\mathrm{AV}_I = \frac{\mathrm{MTBF}}{\mathrm{MTTR} + \mathrm{MTBF}} \tag{8.15}$$

or

$$\mathrm{AV}_I = \frac{\mathrm{MTCM}}{\mathrm{MTCM} + \mathrm{AADT}} \tag{8.16}$$

where AV_I is the item's (equipment) inherent or intrinsic availability, OT the operating time, AT the active downtime, MTBF the mean time between failures, MTTR the mean time to repair, MTCM the mean time to corrective maintenance, and AADT the average active downtime.

Item Achieved Availability

This is expressed as

$$\mathrm{AV}_A = \frac{\mathrm{ATBM}}{\mathrm{ATBM} + \mathrm{AADT}} \tag{8.17}$$

where AV_A is the item's (equipment) achieved availability, ATBM the average time between maintenance (corrective and preventive), and AADT the average active downtime (corrective and preventive maintenance).

Item Use Availability

This is defined as

$$AV_{use} = \frac{OT + OFT}{TDT + OT + OFT} \tag{8.18}$$

where AV_{use} is the item's (equipment) use availability, OT the operating time, OFT the off time, and TDT the total downtime.

Item Operational Availability

This is given by

$$AV_{op} = \frac{\text{operating time}}{\text{downtime} + \text{operating time}} \tag{8.19}$$

or

$$AV_{op} = \frac{OT}{OT + DT} \tag{8.20}$$

where OT is the operating time, DT the downtime, and AV_{op} the item's (equipment) operational availability.

In Eq. (8.19), one should note that total values for the operating time and downtime are required.

8.6 MATHEMATICAL MODELS FOR EQUIPMENT REPLACEMENT OR MAINTENANCE

This section presents a number of mathematical models concerned with various aspects of equipment replacement and maintenance.

Model I

This model estimates the economic life of a piece of equipment—or, more specifically, the time limit beyond which it is not advisable to perform repairs. The economic life L_e of the equipment is expressed as[8,9]

$$L_e = \left[\frac{2(K - S_v)}{K_R}\right]^{\frac{1}{2}} \tag{8.21}$$

where K_R is the yearly increase in repair cost, K the initial cost of equipment (installed), and S_v the equipment scrap value.

Example 8.1

A piece of equipment costs \$40,000 (installed) and its expected scrap value is \$1000. The estimated yearly increase in repair cost is \$200. Compute the time limit beyond which repairs are not advisable.

Substituting the given data into Eq. (8.21) results in

$$L_e = \left[\frac{2(40{,}000 - 1000)}{200}\right]^{\frac{1}{2}}$$
$$= 19.75$$

Thus, the economic repair life of the equipment is 19.75 years.

Model II

This model is used to estimate mean number of spares. The average number of spares is given by

$$S_A = \frac{T_0 k}{\text{MTTF}_A} \tag{8.22}$$

where S_A is the average number of spares, k the total number of identical items (equipment), T_0 the estimated cumulative operational time of each item (equipment), and MTTF_A the mean time to failure of each item (equipment).

Model III

This is another model concerned with determining the required number of spares. The following equation is used to estimate number of spares for each item in use:

$$S_N = (\text{FR})t_m + z[(\text{FR})t_m]^{\frac{1}{2}} \tag{8.23}$$

where t_m is the length of the mission time in hours, FR the failure rate per hour of the item under consideration, S_N the number of spares, and z is associated with the cumulative normal distribution function. Its value depends on a given confidence level for "no stock out." The value of z is obtained from the standardized cumulative normal distribution function table for a specified confidence level.

The standardized cumulative normal distribution function is given by

$$F(z) = \frac{1}{\sqrt{2\pi}} \int_{-\infty}^{z} e^{-x^2/2}\, dx \qquad (8.24)$$

The tabulated values of this function can be found in several books on reliability.

Example 8.2

Assume that the failure rate of a pump is 0.0007 failures/h. For a mission time of 3000 h and a 0.9773 confidence level for "no stock out," determine the required number of spares.

For the given confidence level of 0.9773, the value of z from the table of normal distribution function is 2. Substituting this value and the other given data into Eq. (8.23) yields

$$S_n = (0.0007)(3000) + 2[(0.0007)(3000)]^{\frac{1}{2}}$$

$$\simeq 5 \text{ pumps}$$

Thus, five spare pumps are required to satisfy the 0.9773 confidence level for "no stock out."

Model IV

This is another model used to determine the quantity of spare parts. The following equation was proposed for this model[10]:

$$F(k) = e^{-\lambda T} \sum_{j=0}^{k} (\lambda T)^j / j! \qquad (8.25)$$

where $F(k)$ is the probability of having k or less failures in specified time T and λ the item's constant failure rate.

Model V

This model is concerned with computing the optimum number of inspections M^* per facility per unit time. This quantity is important to a decision maker because the inspection is often disruptive; however, such inspection generally reduces the downtime because it results in fewer breakdowns. In this model, the total downtime is minimized to obtain the optimum number of inspections.

The total downtime D_T per unit time is expressed as[11]

$$D_T = MD_I + c \cdot D_B \cdot M^{-1} \qquad (8.26)$$

where c is a constant for specific facility, D_I the downtime per inspection for a facility in question, and D_B the downtime per breakdown for a facility in question.

Differentiating Eq. (8.26) with respect to M yields

$$\frac{\mathrm{d}D_T}{\mathrm{d}M} = D_I - \frac{cD_B}{M^2} \tag{8.27}$$

Equating Eq. (8.27) to zero yields

$$D_I - \frac{cD_B}{M^2} = 0 \tag{8.28}$$

Thus, from Eq. (8.28), we get

$$M^* = (cD_B/D_I)^{\frac{1}{2}} \tag{8.29}$$

Substituting Eq. (8.29) into Eq. (8.26) yields

$$D_T^* = 2(D_I \cdot cD_B)^{\frac{1}{2}} \tag{8.30}$$

where D_T^* is the minimum total downtime.

Example 8.3

For a certain mechanical equipment, the following data have been obtained from the past experience: $c = 3$, $D_B = 0.15$ months, and $D_I = 0.05$ months. Calculate the optimum number of inspections per month and the minimum total downtime.

Substituting these given data into Eqs. (8.29) and (8.30) yields

$$M^* = \left[\frac{3(0.15)}{0.05}\right]^{\frac{1}{2}} = 3 \text{ inspections/month}$$

and

$$D_T^* = 2[(0.05)(3)(0.15)]^{\frac{1}{2}} = 0.3 \text{ months}$$

Model VI

This model is concerned with finding the optimum time between replacements. In this model, the average annual total cost TK is minimized with respect to the time between replacements (i.e., the life of the equipment in years). The average total cost has three components: mean maintenance, mean operating, and mean investment costs. Thus, the mean total cost is

$$TK = \left(\sum_{i=1}^{2} k_i\right) + \frac{c_I}{t} + \frac{(t-1)}{2}(i_{oc} + i_{mc}) \tag{8.31}$$

where k_1 is the operational cost for the first year, k_2 the maintenance cost for the first year, t the life of the equipment in years, c_I the cost of investment, i_{oc} the amount by which operational cost increases per year, and i_{mc} the amount by which maintenance cost increases per year.

Differentiating Eq. (8.31) with respect to t results in

$$\frac{\mathrm{d}TK}{\mathrm{d}t} = \frac{1}{2}(i_{oc} + i_{mc}) - \frac{c_I}{t^2} \tag{8.32}$$

Setting Eq. (8.32) equal to zero and rearranging yields

$$t^* = \left(\frac{2c_I}{i_{oc} + i_{mc}}\right)^{\frac{1}{2}} \tag{8.33}$$

where t^* is the optimum replacement interval.

Using Eq. (8.33) in Eq. (8.31) yields

$$TK^* = \sum_{i=1}^{2} K_i - \left(\frac{i_{oc} + i_{mc}}{2}\right) + [2c_I(i_{oc} + i_{mc})]^{\frac{1}{2}} \tag{8.34}$$

where TK^* is the minimum average annual total cost.

Example 8.4

The following data are associated with a mechanical equipment: $c_I = \$10{,}000$, $i_{oc} = \$400$, and $i_{mc} = \$200$. Calculate the value of the optimum replacement interval.

Using these given data in Eq. (8.33) yields

$$t^* = \left[\frac{2(10{,}000)}{400 + 200}\right]^{\frac{1}{2}} = 5.8 \text{ years}$$

Thus, the value of the mechanical equipment optimum replacement interval is 5.8 years.

Model VII

This is another model used to determine the equipment optimum replacement period. In this case, the mean annual cost incurred by the equipment is minimized. The total running cost $C_E(T)$ for equipment used for T years is defined by

$$C_E(T) = \int_0^T K_0(t)\,\mathrm{d}t \tag{8.35}$$

where $K_0(t)$ is the equipment operating cost at time t.

The total cost $TC_E(T)$ associated with the equipment in T years is given by

$$TC_E(T) = K_a + \int_0^T K_0(t)\,\mathrm{d}t - V_s \tag{8.36}$$

where K_a is the equipment acquisition cost and V_s the scrap value of the equipment.

Dividing Eq. (8.36) by T results in

$$AC_E(T) = \left[K_a + \int_0^T K_0(t)\,\mathrm{d}t - V_s\right] \Big/ T \tag{8.37}$$

where $AC_E(T)$ is the annual average cost of the equipment.

In order to find minimum value of $AC_E(T)$, we differentiate Eq. (8.37) with respect to T and equate the resulting expression equal to zero, as

$$\frac{\mathrm{d}AC_E(T)}{\mathrm{d}T} = -\frac{K_a}{T^2} - \frac{1}{T^2}\int_0^T K_0(t)\,\mathrm{d}t + \frac{K_0(t)}{T} + \frac{V_s}{T^2} = 0 \tag{8.38}$$

The value of T, for which $AC_E(T)$ is minimum, can be obtained from Eq. (8.38).

Model VIII

This model is concerned with finding the minimum total parallel system cost K_{PT} due to maintenance, failures, and operation per unit time. The parallel system is composed of N independent and identical items (equipment). At least one item (equipment) must operate normally for the system's success. Thus, the total parallel system cost is given by

$$K_{PT} = \frac{K_{pd}T_D + K_{om}T_U N}{T_D + T_U} \tag{8.39}$$

where T_U is the operating time of the system, T_D the downtime of the system, K_{pd} the penalty cost because of downtime (per unit time), and K_{om} the operation and maintenance cost (per unit time) for each item (equipment).

The N item system availability AV_s and unavailability UV_s are given by

$$\mathrm{AV}_s = 1 - (\mathrm{UV})^N \tag{8.40}$$

and

$$\mathrm{UV}_s = (\mathrm{UV})^N \tag{8.41}$$

where UV is the item (equipment) unavailability. Also, AV_s and UV_s can be expressed as follows:

$$\mathrm{AV}_s = \frac{T_U}{T_U + T_D} \tag{8.42}$$

and

$$\mathrm{UV}_s = \frac{T_D}{T_U + T_D} \tag{8.43}$$

The parallel system downtime is given by

$$T_D = T_U \cdot \mathrm{UV}_s/\mathrm{AV}_s \tag{8.44}$$

Substituting Eqs. (8.42) and (8.43) into Eq. (8.39) results in

$$K_{PT} = K_{pd} \cdot \mathrm{UV}_s + K_{om} \cdot \mathrm{AV}_s N \tag{8.45}$$

After substituting for AV_s and UV_s from Eqs. (8.40) and (8.41), respectively, into Eq. (8.45), we get

$$\begin{aligned} K_{PT} &= K_{pd} \cdot (\mathrm{UV})^N + K_{om} N[1 - (\mathrm{UV})^N] \\ &= K_{om} N + (K_{pd} - K_{om} N)(\mathrm{UV})^N \end{aligned} \tag{8.46}$$

In order to find the minimum value of K_{PT}, differentiate Eq. (8.46) with respect to N, set the resulting expression equal to zero, and then solve for N. Substitute the value of N back into Eq. (8.46) to obtain the minimum value of K_{PT}.

Model IX

This is another very similar model concerned with the parallel system. In this case, the parallel system is composed of N identical machines (equipment) and its output is fed into the next stage of the production process.[12] At least one machine must operate normally for system success. In this case, we are concerned with minimizing the total cost K_{TS} (per unit time) involved in system operation and downtime losses with respect to N, or more specifically with determining the optimal value of N.

Using queuing theory, the average proportion of unit time that the parallel system is unavailable for service is given by

$$U_s = \left(\frac{\lambda}{\lambda + \mu}\right)^N \tag{8.47}$$

where λ is the constant failure rate of a machine and μ the constant repair rate of a machine.

The total cost K_{TS} is expressed as

$$K_{TS} = U_s \cdot K_{DT} + NK_0 \tag{8.48}$$

where K_{DT} is the downtime cost per unit time and K_0 the single machine's operational cost per unit time.

Substituting Eq. (8.47) into Eq. (8.48) yields

$$K_{TS} = \left(\frac{\lambda}{\lambda + \mu}\right)^N K_{DT} + NK_0 \tag{8.49}$$

Differentiating the above equation with respect to N results in

$$\frac{\partial K_{TS}}{\partial N} = \left(\frac{\lambda}{\lambda + \mu}\right)^N K_{DT} \ell n \left(\frac{\lambda}{\lambda + \mu}\right) + K_0 \tag{8.50}$$

Setting the above equation equal to zero yields

$$K_{DT} \cdot \left(\frac{\lambda}{\lambda + \mu}\right)^N \ell n \left(\frac{\lambda}{\lambda + \mu}\right) + K_0 = 0 \tag{8.51}$$

Solving Eq. (8.51) for N gives

$$N^* = \ell n[-K_0/(K_{DT} \ell n U)] / \ell n U \tag{8.52}$$

where $U = \lambda/(\lambda + \mu)$ and N^* is the optimal number of machines to be used in the parallel configuration.

Inserting Eq. (8.52) back into Eq. (8.49) yields

$$K_{TS}^* = U^{N^*} K_{DT} + N^* K_0 \tag{8.53}$$

where K_{TS}^* is the minimum value of K_{TS}.

Example 8.5

From the past experience with a mechanical system used for production purposes, we obtained the following data: $K_{DT} = \$4000$, $K_0 = \$200$,

$\lambda = 10$ failures/unit time, and $\mu = 20$ repairs/unit time. Calculate the value of N^* by using Eq. (8.52). Utilizing the given data in Eq. (8.52) yields

$$U = \frac{10}{20 + 10} = 0.3333$$

$$N^* = \ell n[-200/(4000)\ell n(0.3333)/\ell n(0.3333)]$$

$$= 2.813 \simeq 3 \text{ machines}$$

In order to obtain the minimum value of K_{TS}, the parallel system must have three machines.

Model X

This model is concerned with determining point and asymptotic mission availabilities when two items (equipment) are tested simultaneously. The identical and independent items form either a series or a parallel configuration. The testing is performed at intervals of time T_I.

The point availabilities for two items tested simultaneously at time $t = 0$, $T_I, 2T_I, 3T_I, 4T_I, \ldots$, are

$$\mathrm{AV_a}(t) = e^{-\lambda_I t}, \qquad 0 \leqslant t < T_I \tag{8.54}$$

and

$$\mathrm{AV_b}(t) = e^{-\lambda_I t}, \qquad 0 \leqslant t < T_I \tag{8.55}$$

where $\mathrm{AV_a}(t)$ is the point availability of item a, $\mathrm{AV_b}(t)$ the point availability of item b, and λ_I the item's constant failure rate.

Both the items under consideration may be arranged in either a series or a parallel configuration.

For the series system, the point availability is given by

$$\mathrm{AV_{ss}}(t) = [\mathrm{AV_a}(t)][\mathrm{AV_b}(t)]$$

$$= e^{-2\lambda_I t}, \qquad 0 \leqslant t < T_I \tag{8.56}$$

Similarly, for the parallel system, the point availability is

$$\mathrm{AV_{ps}}(t) = 1 - [1 - \mathrm{AV_a}(t)][1 - \mathrm{AV_b}(t)]$$

$$= 2e^{-\lambda_I t} - e^{-2\lambda_I t}, \qquad 0 \leqslant t < T_I \tag{8.57}$$

The asymptotic mission availability for the series system is

$$\mathrm{AV}_{ss}(\infty) = \mathrm{AV}_{ss}(T_I) = \frac{1}{T_I}\int_0^{T_I} e^{-2\lambda_I t}\,\mathrm{d}t$$

$$= \frac{1}{2\lambda_I T_I} - \frac{e^{-2\lambda_I T_I}}{2\lambda_I T_I} \tag{8.58}$$

Similarly, the asymptotic mission availability for the parallel system is given by

$$\mathrm{AV}_{ps}(\infty) = \mathrm{AV}_{ps}(T_I) = \frac{1}{T_I}\int_0^{T_I} \left(2e^{-\lambda_I t} - e^{-2\lambda_I t}\right)\mathrm{d}t$$

$$= \frac{3}{2\lambda_I T_I} - \frac{2e^{-\lambda_I T_I}}{\lambda_I T_I} + \frac{e^{-2\lambda_I T_I}}{2\lambda_I T_I} \tag{8.59}$$

Model XI

This model is concerned with determining the reliability and mean time to failure of a system subject to periodic maintenance. The model is based on the assumptions[13] that periodic preventive maintenance is carried out on the system after every X hours, beginning at time zero, and that a failed component is replaced with a new and statistically identical one.

For periodic maintenance, the time interval of x hours is expressed as

$$x = kX + T; \qquad k = 0,1,2,3,4,\ldots; \qquad 0 \leqslant T < X \tag{8.60}$$

The reliability of a redundant system subject to periodic maintenance after every X hours is denoted by $R_X(x)$. For $T = 0$ and $k = 1$,

$$R_X(x = X) = R(X) \tag{8.61}$$

For $T = 0$ and $k = 2$,

$$R_X(x = 2X) = [R(X)]^2 \tag{8.62}$$

In this case, the system under consideration must function the first X hours without failure. In addition, it must operate another X hours without failure after the replacement of any failed components.

For $0 < T < X$, another T hours of operation without a failure is necessary. Thus,

$$R_X(x = 2X + T) = [R(X)]^2 R(T) \tag{8.63}$$

The above reliability equation is generalized as

$$R_X(x = kX + T) = [R(X)]^k[R(T)]$$

$$\text{for } k = 0,1,2,3,4,...; \qquad 0 \leqslant T < X \tag{8.64}$$

The mean time to failure MTTF_{PM} of a redundant system with periodic maintenance is given by

$$\text{MTTF}_{\text{PM}} = \int_0^{\infty} R_X(x)\, \text{d}x \tag{8.65}$$

To evaluate this equation, we write the integral over the range $0 < x < \infty$ as the sum of integrals over time intervals of length X as

$$\text{MTTF}_{\text{PM}} = \sum_{k=0}^{\infty} \int_{kX}^{(k+1)X} R_X(x)\, \text{d}x \tag{8.66}$$

In the above expression, the integral of Eq. (8.65) is divided into time intervals of length X. For $x = kX + T$, inserting Eq. (8.64) into Eq. (8.66) yields

$$\text{MTTF}_{\text{PM}} = \sum_{k=0}^{\infty} \int_0^X [R(X)]^k R(T)\, \text{d}T \tag{8.67}$$

In the above expression for $x = kX + T$, $\text{d}x = \text{d}t$ and the limits become 0 and X.

Rearranging Eq. (8.67), we get

$$\text{MTTF}_{\text{PM}} = \sum_{k=0}^{\infty} [R(X)]^k \int_0^X R(T)\, \text{d}T \tag{8.68}$$

The sum of the series given in Eq. (8.68) is

$$\sum_{k=0}^{\infty} [R(X)]^k = \frac{1}{1 - R(X)} \tag{8.69}$$

Inserting Eq. (8.69) into Eq. (8.68) yields

$$\text{MTTF} = \frac{\int_0^X R(T)\, \text{d}T}{1 - R(X)} \tag{8.70}$$

Example 8.6

Three identical and independent pumps form an active-parallel configuration. At least one pump is required for the system's success. The failure times of each pump are described by the exponential distribution with mean life of 100 h. Assuming that periodic maintenance is performed after every 100 h, compute the redundant system mean time to failure. In addition, calculate the system mean time to failure without periodic maintenance.

The three-pump parallel system reliability is given by

$$R(x) = 3e^{-\lambda x} - 3e^{-2\lambda x} + e^{-3\lambda x} \tag{8.71}$$

where x is the time and λ the constant failure rate of a pump (i.e., $\lambda = 1/\text{mean life} = 1/100 = 0.01$ failures/h).

Inserting Eq. (8.71) and these given data into Eq. (8.70) yields

$$\text{MTTF}_{\text{PM}} = \frac{\int_0^{100} [3e^{-(0.01)T} - 3e^{-2(0.01)T} + e^{-3(0.01)T}]\,\mathrm{d}T}{1 - [3e^{-(0.01)(100)} - 3e^{-2(0.01)(100)} + e^{-3(0.01)(100)}]}$$

$$= \frac{91.61}{0.2526} = 362.67 \text{ h}$$

The mean time to failure (MTTF) without periodic maintenance (i.e., $X = \infty$) is

$$\text{MTTF} = \int_0^{\infty} [3e^{-(0.01)x} - 3e^{-2(0.01)x} + e^{-3(0.01)x}]\,\mathrm{d}x$$

$$= \frac{11}{6(0.01)} = 183.333 \text{ h}$$

Thus, periodic maintenance helped to improve the system's mean time to failure from 183.333 to 362.67 h.

References

[1]Gabrielson, O. N., "Frequency of Maintenance," *Proceedings of Fifth Reliability and Maintainability Conference*, AIAA, New York, July 1966, pp. 428–433.

[2]Wild, R., *Essentials of Production and Operations Management*, Holt, Rinehart and Winston, London, 1985, pp. 356–368.

[3]Wyder, C. G., "Preventive Maintenance," *Maintenance Engineering Handbook*, edited by L. C. Morrow, McGraw-Hill, New York, 1966, pp. 1.94–1.125.

[4]Karassik, I. J., "Installation, Operation, and Maintenance," *Pump Handbook*, edited by I. J. Karassik, W. C. Krutzsch, W. H. Fraser, and J. P. Messina, McGraw-Hill, New York, 1985, pp. 12.1–12.19.

[5]Ball, R. C., "Gear Drives and Speed Reducers," *Maintenance Engineering Handbook*, edited by L. C. Morrow, McGraw-Hill, New York, 1966, pp. 8.162–8.181.

[6]Von-Alven, W. H., *Reliability Engineering*, Prentice-Hall, Englewood Cliffs, NJ, 1964, pp. 160–161, 345–450.

[7]Blanchard, B. S. and Fabrycky, W. J., *Systems Engineering and Analysis*, Prentice-Hall, Englewood Cliffs, NJ, 1981, pp. 370–419.

[8]Varnum, E. C. and Bassett, B. B., "Machine and Tool Replacement Practices," *Manufacturing Planning and Estimating Handbook*, edited by F. W. Wilson and P. D. Harvey, McGraw-Hill, New York, 1963, pp. 18.1–18.22.

[9]Eidmann, F. L., *Economic Control of Engineering and Manufacturing*, McGraw-Hill, New York, 1931.

[10]Ebel, G. and Lang. A., "Reliability Approach to the Spare Parts Problem," *Proceedings of Ninth National Symposium on Reliability and Quality Control*, IEEE, New York, 1963, pp. 85–92.

[11]Wild, R., *Essentials of Production and Operations Management*, Holt, Rinehart and Winston, London, 1985, pp. 356–368.

[12]Jardine, A. K. S., *Maintenance, Replacement and Reliability*, Pitman Publishing, London, 1973, pp. 181–183.

[13]Rosenheim, D. E., "Analysis of Reliability Improvement Through Redundancy," *Proceedings of the New York University Conference on Reliability Theory*, New York University, New York, June 1958, pp. 119–142.

Exercises

8.1 What are the benefits of performing preventive maintenance?

8.2 What is the difference between preventive maintenance and corrective maintenance?

8.3 What are the reasons for creating the demand for preventive maintenance?

8.4 Discuss at least five different categories of preventive maintenance.

8.5 What are the symptoms of steam pump problems?

8.6 The repair times of a centrifugal pump are described by the lognormal probability density function. Obtain separately expressions for the maintainability function and mean time to repair.

8.7 A machine costs \$90,000 (installed) and its estimated scrap value is \$6000. The projected annual increase in repair costs is \$300. Calculate the time limit beyond which repairs are not justified.

8.8 Assume that the failure times of a ball bearing are exponentially distributed with mean value of 200 h. For a mission period of 4500 h and a 0.9885 confidence level (for no stock out), determine the required number of spares.

8.9 Two identical and independent pumps form an active-parallel configuration. The constant failure rate of a pump is 0.002 failures/h. Periodic maintenance is performed after every 200 h. Calculate the system mean time to failure with and without periodic maintenance. Comment on the results.

9
FAILURE DATA COLLECTION AND ANALYSIS

This chapter is concerned with failure data collection and analysis. Five basic functions of the failure reporting, analysis, and corrective measure system are presented. The usage of failure data is discussed. A number of guidelines for designing a failure reporting and documentation system are listed, along with information to be considered in the design of a failure data collection form. Sources of failure data collection during the equipment life cycle are identified. Ten reliability related tests are listed, as well as the typical deficiencies encountered in failure data sources. A number of sources of data existing in various countries are presented. Failure rates for selected mechanical items are given, along with failure modes for hydraulic valves, gages, pneumatic valves, fans, clutches, bearings, gears, heat exchangers, seals, brakes, filters, and hydraulic pumps.

Formulas to calculate the mean time to failure and variance are given. A hazard rate plot is developed and a graphical technique to determine the statistical distributions of failure times is presented. The maximum-likelihood method is demonstrated for the normal distribution. The Kolmogorov-Smirnov goodness-of-fit test is described, along with one solved example.

9.1 INTRODUCTION

Product failure data provide invaluable information to reliability engineers, design engineers, management, etc. These data are the final proof of the success or failure of the effort expended during the design and manufacture of a product operating under the designed conditions. There are many failure data banks currently in use in several countries. Even though their end purpose may be the same, the size, type, contents, and capability of these data banks vary considerably from one installation to another.

Failure data are regarded as the backbone of reliability studies. Without effective knowledge of an item's failure data, its associated studies (reliability) will be incomplete. Therefore, it may be said the existence of effective failure data banks is very important and must be given a top priority. These data banks can exist only if all the proper procedures are taken to collect the required data.

Generally, it may be said that the failure experience is underutilized and frequently left out in important decisions regarding the product in question. There are various factors for such underutilization, some of which are incomplete failure analyses, poorly documented failure case histories, and poorly organized failure experiences for general use.[1]

Any collected failure data must be analyzed properly in order to take full advantage of their value. Otherwise, these data are of little use. There are several statistical techniques that can be used to perform failure data analysis. Several different aspects of failure data collection and analysis are addressed in subsequent sections of this chapter.

9.2 A SYSTEM FOR FAILURE REPORTING, ANALYSIS, AND CORRECTIVE MEASURE

A failure reporting, analysis, and corrective measure system is vitally important for a company irrespective of whether it is industrial, space, consumer, or military oriented. Such a system helps to assure better reliability of a product. There are five basic functions associated with this system,[2] as shown in Fig. 9.1: failure data recording, transmitting the recorded data to the analysis group, failure data analysis, corrective measure, and follow-up.

9.3 REQUIREMENTS FOR COLLECTING AND REPORTING FAILURE DATA AND USING FAILURE DATA

There are certain requirements concerned with data collection and reporting, the three most important of which are shown in Fig. 9.2.

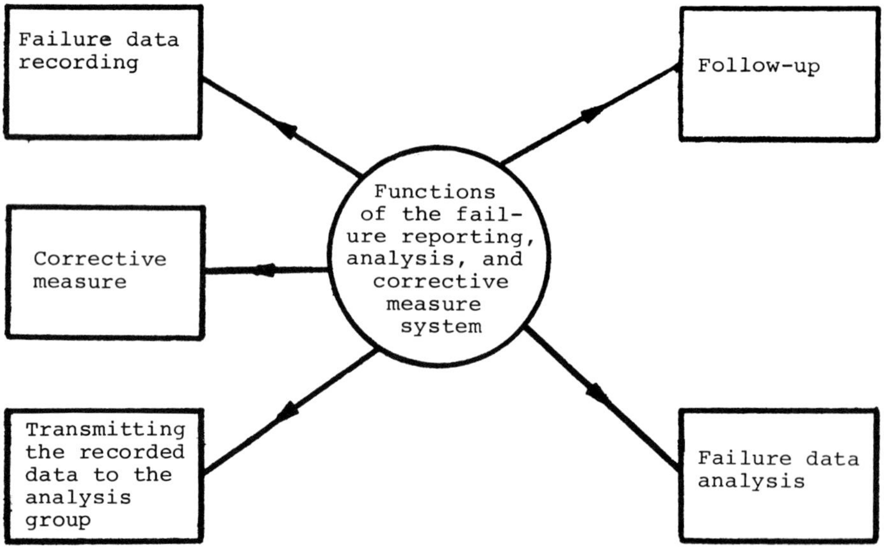

Fig. 9.1 Functions of the failure reporting, analysis, and corrective measure system.

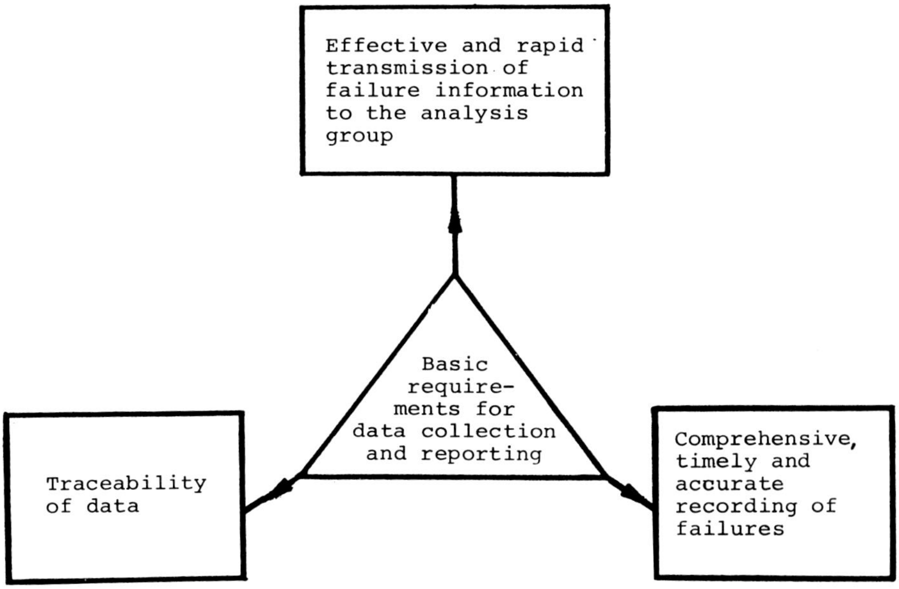

Fig. 9.2 Three basic requirements for failure data collection and reporting.

Failure data become useful in many ways and can be used for different purposes. Some of these purposes are[3,4]

1) To calculate hazard rate (failure rate) of an item.
2) To make decisions regarding the introduction of a redundant item(s).
3) To perform tradeoff studies (cost reliability).
4) To conduct the item's replacement studies.
5) To perform the preventive maintenance studies of the item.
6) To conduct effective design reviews.
7) To determine the maintenance needs of a new item(s).
8) To perform life cycle cost studies.
9) To predict the reliability and availability of the item.
10) To recommend design changes for improving the item's reliability.

9.4 GUIDELINES FOR DESIGNING A FAILURE REPORTING AND DOCUMENTATION SYSTEM

In order to have an effective failure reporting and documentation system, there are several factors that require attention during the design stage, some of which are[5]: 1) simplicity and clarity of the failure reporting form, 2) quickness in performing an analysis of the recorded failure data, 3) involvement of the users, 4) feasibility of prerecording any static information on the data form, 5) elimination of the requirements for memorizing codes,

6) feedback of the analyzed data to concerned people, 7) documentation of the information on failures, 8) cost and efficiency, 9) flexibility, and 10) clarity of the terms and definitions used.

9.5 FAILURE DATA FORMS

Forms are used to collect failure data. It may be said that such forms are the backbone of the failure data banks. Thus, care must be given in designing them. The forms must be designed so that minimum effort is required to collect the necessary data. Generally, the failure data collection form is tailored to satisfy the needs of a particular program or organization. Information such as given below should be considered for inclusion during the design of the form[2,6]:

1) Description of the item in question.
2) Location of the hardware.
3) Form number.
4) Serial number of the failed item.
5) Name of the manufacturer.
6) Failure description.
7) Date the failure occurred.
8) Date the failure was detected.
9) Date the failure was corrected.
10) Form completion date.
11) Serial number of the replaced part.
12) Manufacturer of the replaced part.
13) Time taken for repair.
14) Operating hours from the previous failure.
15) Cost of the repair.
16) Names of the repairmen.
17) Signatures.

9.6 DATA COLLECTION SOURCES

There are several sources from which data can be collected. Some of these are maintenance records, inspection records, test log sheets, welding and x-ray reports, market research reports, customers' complaints or praises, quality control departments, and field failure reports. However, in the equipment life cycle, there are eight identifiable data sources[7]:

1) Past experience with similar or identical material.
2) Inspection data produced by quality control and manufacturing groups.
3) Failure reporting systems developed and used by customers.
4) Reports generated by the repair facility.
5) Data obtained during the breadboard or the development of the item (system).
6) Tests: field demonstration, environmental qualification, and field installation.
7) Factory acceptance testing of equipment, modules, and assemblies.
8) Warranty claims.

There are various types of reliability-related tests. Their output of success or failure is of significant interest in reliability evaluation. Some of the reliability-related tests cover[2] reliability and maintainability demonstration, system life, burn-in, qualification, destruction, environmental, longevity, production verification, screening, nonstandard parts or components.

Data Collection from Warranty Claims

The warranty claims are an important source for obtaining product failure-related information and, therefore, can be utilized for reliability

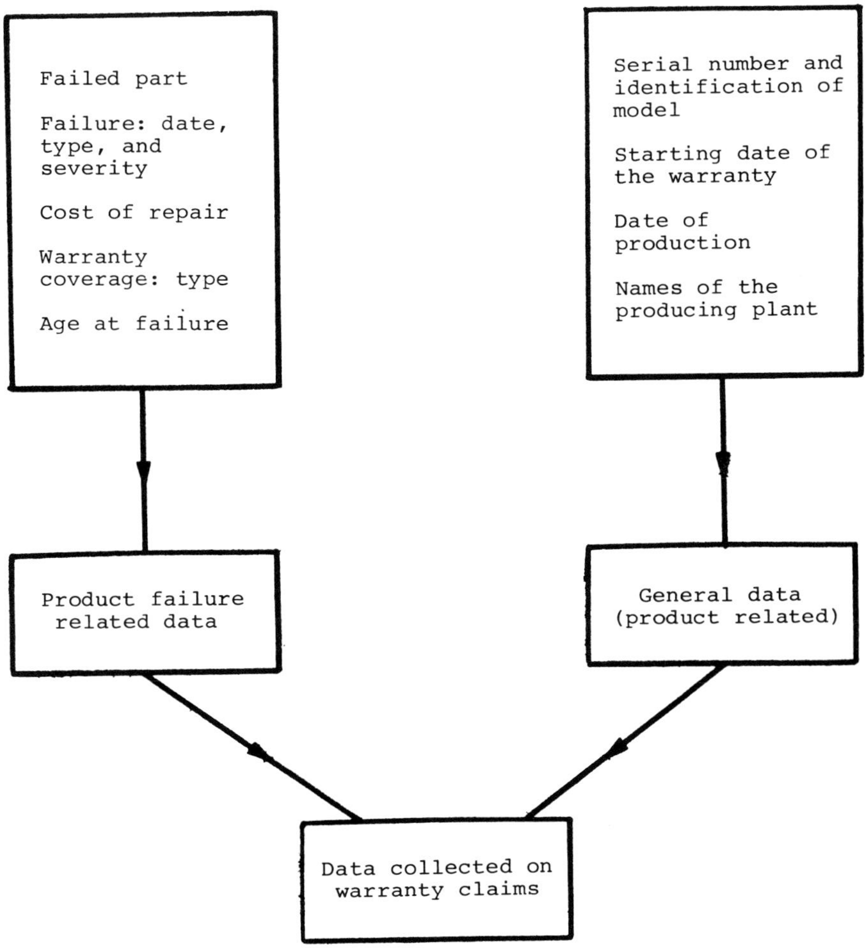

Fig. 9.3 Type of data collected on warranty claims.

monitoring and anlaysis purposes. Generally, information such as shown in Fig. 9.3 is collected on warranty claims.[8] As shown, it can be divided into two main categories: general product-related data and product failure-related data.

Typical Deficiencies in Failure Data Sources

Various deficiencies may exist in the existing failure data sources. Some of them[9] are nonstandardized data reporting, poor descriptions of the item in question, time-consuming data feedback, lack of accuracy in the repair times required for many parts, inconsistent and vague definitions of terms on the reporting forms, difficulty in pinpointing the cause(s) of the failure, government security or company proprietary classification of the data, and the lack of information about the relationship between the part's failure and its operating milieu. In order to develop effective failure data, the data sources must be free of such shortcomings.

9.7 FAILURE DATA BANKS

There are may established sources (banks) from which various types of failure data can be obtained. Some of these are presented in Table 9.1. In addition, a number of sources of human reliability data may be found in Ref. 10. The sources presented in Table 9.1 are biased toward mechanical item failure data. One should note that the type of information listed is subject to change.

9.8 FAILURE RATES FOR SELECTED MECHANICAL ITEMS

Over the last 10 or so years, failure data on many mechanical items have been collected and analyzed. Tables 9.2–9.5 list the failure rates for selected mechanical items.[11–14] The relation between the failure rate and the item's environment is indicated in Tables 9.3–9.5.

9.9 FAILURE MODES OF SELECTED MECHANICAL ITEMS

Mechanical items fail in several distinct failure modes. This section presents failure modes of the following items[13]: hydraulic and pneumatic valves, gages, fans, clutches, bearings, gears, heat exchangers, seals, brakes, filters, and hydraulic pumps.

In hydraulic valves, leaking is the most important failure mode and typically accounts for 57% of the failures. Other failure modes associated with these valves are stuck open (9%), improper flow (1%), out of specification (6%), stuck closed (9%), cracked/fractured (4%), out of adjustment (4%), and other (10%). The pneumatic valves can fail in six failure modes: leaking (60%), unstable (3%), stuck open (8%), out of specification (8%), stuck closed (7%), and other (14%).

Table 9.1 Sources of Failure Data

No.	Author or contact person	Source title or type of data	Published in or developed by
1	M. J. Rossi	Nonelectronic parts reliability data, Rept. NPRD-3, 1985	Reliability Analysis Center, Rome Air Development Center (RADC), Griffiss Air Force Base, NY 13441-5700
2	R. E. Schafer, J. E. Angus, J. M. Finkelstein M. Yerasi, D. W. Fulton	RADC Non-Electronic Reliability Notebook, Rept. RADC-TR-85-194, 1985	Reliability Analysis Center, Rome Air Development Center (RADC), Griffiss Air Force Base, NY 13441-5700
3	—	Data on general equipment	Military Electronics Laboratory (FTL/FOA), FACK, 10450 Stockholm 80, Sweden
4	A. E. Green	Safety Systems Reliability, 1983	John Wiley & Sons, Chichester, UK
5	—	Government industry data exchange program (GIDEP)	GIDEP Operations Center, U.S. Dept. of the Navy, Naval Weapons Station, Seal Beach, Corona Annex, Corona, CA 91720
6	C. H. Levinson	*Systems Reliability Analysis Handbook*, 1964, pp. 18.37–18.54	McGraw-Hill Book Co., 1221 Avenue of the Americas, New York, NY 10019
7	R. J. Sadlon	Analysis Techniques for Mechanical Reliability, Rept. NPS-1, 1985, p. 95	Reliability Analysis Center, Rome Air Development Center (RADC), Griffiss Air Force Base, NY 13441–5700
8	R. G. Arno	Nonelectronic Parts Reliability Data, Rept. NPPD-2, 1981	Reliability Analysis Center, Rome Air Development Center (RADC), Griffiss Air Force Base, NY 13441-5700

(Table continued on next page.)

Table 9.1 (cont.) Sources of Failure Data

No.	Author or contact person	Source title or type of data	Published in or developed by
9	—	SYREL: reliability data bank	Systems Reliability Service, Safety and Reliability Directorate, UKAEA, Wigshaw Lane, Culcheth, Warrington, Lancashire, WA3 4NE, England
10	—	Data on general equipment	Det Norske Veritas, P.O. Box 6060, Etterstad, Oslo 6, Norway
11	—	Reactor Safety Study: An Assessment of Accident Risks in U.S. Commercial Nuclear Power Plants, Rept. Wash-1400 (NUREG 75/014), Appendix 3 & 4, 1975.	U.S. Nuclear Regulatory Commission[a]
12	—	IEEE Nuclear Reliability Data Manual, IEEE Std. 500-1977	Institute of Electrical and Electronics Engineers (avaliable from John Wiley & Sons, 605 Third Ave., New York, NY 10017)
13	—	Data on nuclear powerplant equipment	ASEA-ATOM, Box 53, 72104, Vasteras 1, Sweden
14	H. P. Bloch, F. K. Geitner	*Practical Machinery Management for Process Plants: Machinery Failure Analysis and Troubleshooting* 1983, pp. 628–630	Gulf Publishing Co., Houston, TX

(Table continued on next page.)

Table 9.1 (cont.) Sources of Failure Data

No.	Author or contact person	Source title or type of data	Published in or developed by
15	D. Hauck	Failure Rates of Mechanical Components for Nuclear Reactors: A Literature Survey, AECL Rept. CRNL-739 (SP-R-10)	Chalk River Nuclear Laboratories (CRNL), Atomic Energy of Canada Ltd., Chalk River, Ontario, Canada
16	—	Data related to automobiles	Dept. 56470 (Reliability Engineering), Car Div., AB Volvo, Torslanda, Goteborg, 40508, Sweden
17	—	Data related to space systems	Product Assurance Div., European Space Research and Technology Center, European Space Research Organization, Domeinweg, Noordwijk, the Netherlands
18	—	(Useful to obtain information on available data sources)	Defense Technical Information Center, Defense Logistics Agency, Cameron Station, Alexandria, VA 22314
19	—	Data related to space systems	Parts Reliability Information Center (PRINCE), NASA, George C. Marshall Space Flight Center, Reliability Office, Astrionics, Laboratory, Huntsville, AL 35812
20	—	Data on equipment used in steel works	Corporate Engineering Laboratory, 140 Battersea Park Road, London SW11 4LZ, UK

(Table continued on next page.)

Table 9.1 (cont.) Sources of Failure Data

No.	Author or contact person	Source title or type of data	Published in or developed by
21	—	Data on mechanical components	SNIA Viscosa Via Borgonuova 14, 20121 Milano, Italy
22	—	Prototype development (military)	Stephen Howe, (Consultants) Ltd., Kings Avenue House, New Malden, Surrey KT3 4DY, England, UK
23	A. P. Harris	Reliability and Maintainability Data for Industrial Plants, Rept. TD-84-3, 1984	A. P. Harris & Associates, Box 6130, Station J, Ottawa, Ontario K2A 1T2, Canada
24	W. H. Dickinson	Report on Reliability of Electric Equipment in Industrial Plants	*IEEE Transactions on Industrial Applications*, July 1962, pp. 132–151
25	C. H. Karr, G. L. Wagner	R & M—Today's Heating and Cooling versus Solar Energy	*Proceedings of the Annual Reliability and Maintainability Symposium*, 1976, pp. 491–499 (available from IEEE, 345 W. 47th St., New York, NY 10017)
26	—	Data on equipment used in electric power generation	Reliability Engineering Group, Generation Development and Construction Div., Barnwood, Gloucester GL4 7RS, England, UK
27	—	ESA electronic components data bank	Space Documentation Service, European Space Agency, Via Galileo Galilei, 00044 Frascati, Italy

(Table continued on next page.)

Table 9.1 (cont.) Sources of Failure Data

No.	Author or contact person	Source title or type of data	Published in or developed by
28	—	Satellite and rocket experiments—data catalog	NASA Space Science Data Center, Goddard Space Flight Center, Greenbelt, MD 20771
29	—	—	Mechanical Properties Data Center (MPDC), P.O. Box 8128, Columbus, OH 43201
30	—	Failure Rates and Failure Modes of Small Rotary Electrical Devices, Rept. AD 267315, Nov. 1961	ASTIA[a]
31	—	Data on trucks and vans	Commanding General, Attn: DRSTA-QRA, U.S. Army Automotive-Tank Command, Warren, MI 48090
32	E. Bassett	General Part Failure Rates for Spacecraft Application, GIDEP, Rept. E0410857, Aug. 12, 1974	Report prepared by TRW Systems Group, GIDEP Operations Center, U.S. Dept. of the Navy, Naval Weapons Station, Seal Beach, Corona Annex, Corona, CA 91720
33	—	Data on components used in nuclear powerplants	Process Engineering, Euratom Joint Research Center, 21020 Centro Euratom di ISPRA, Varese, Italy
34	—	Failure Assessment Handbook. Rept. RD-101, Feb. 1, 1963	Astro Reliability Corp., 14827 Ventura Blvd., Sherman Oaks, CA

(Table continued on next page.)

Table 9.1 (cont.) Sources of Failure Data

No.	Author or contact person	Source title or type of data	Published in or developed by
35	—	Component Parts Failure Data Compendium	Engineering Dept., Electronic Industries Association, 11 W. 42nd St., New York, NY 10036
36	G. Guth	Development of Nonelectronic Part Cycle Failure Rates, Rept. 17D/A-050 678, Dec. 1977	Martin Marietta Corp., Orlando, FL[a]
37	—	Reliability Analysis Data for Systems and Component Design Engineers, Rept. TRA-873-74, Sept. 1, 1961	Missile and Space Vehicle Dept., General Electric Co.[a]
38	M. Anderson	Nuclear Reliability Assurance Data Source Guide, Rept. ORNL-EWG-TM-2, Nov. 1976	Union Carbide Corp., Oak Ridge, TN (available as GIDEP Rept. F010-2254, GIDEP Operations Center, U.S. Dept. of the Navy, Naval Weapons Station, Seal Beach, Corona Annex, Corona, CA 91720)
39	—	Data on equipment used in steel works	ITALSIDER, Via Corsica 4, 16128 Genova, Italy
40	—	Failure Rates—Reliability Engineering Data Series	Avco Corp., Research and Advanced Development Div., 201 Lowell Street, Wilmington, MA
41	—	Information service in mechanical engineering	Data Courier, Inc., Dept. DM, 620 South Fifth St., Louisville, KY 40202

(Table continued on next page.)

Table 9.1 (cont.) Sources of Failure Data

No.	Author or contact person	Source title or type of data	Published in or developed by
42	—	Information on metals	Defense Metals Information Center, Battelle Memorial Institute, 505 King Ave., Columbus, OH 43201
43	—	Data on space equipment	Centre National d'Etudes Special (CNES), c/o Matra, 37 Av. Louis Breguet, 78140 Velizy, France
44	—	Data on equipment used in fossil fuel powerplants	National Electricity Generating Board (ENEL), Bastioni di Porta Volta 10, 20121 Milano, Italy
45	—	Data on equipment used in electric power generation	Equipment Reliability Information System (ERIS), Canadian Electrical Association, Montreal, Quebec, Canada
46	J. P. Drago, R. J. Borkowski, D. H. Pike, F. F. Goldberg	The In-plant Reliability Data Base (IPRD) for Nuclear Power Plant Components: Data Collection and Methodology Report, Rept. NUREG/CR-2641 (ORNL/TM 8271)	Oak Ridge National Laboratories (ORNL), Oak Ridge, TN 37830
47	B. S. Dhillon	*Human Reliability: With Human Factors*, 1986 (a number of human reliability data sources are listed in this text)	Pergamon Press, 395 Saw Mill River Rd., Elmsford, NY 10523

(Table continued on next page.)

Table 9.1 (cont.) Sources of Failure Data

No.	Author or contact person	Source title or type of data	Published in or developed by
48	S. O. Nilsson	Reliability Data on Automotive Components	*Proceedings of the Annual Reliability and Maintainability Symposium*, 1975, pp. 276–279 (available from IEEE, 345 W. 47th St., New York, NY 10017)
49	—	Insurance facts (disaster impact data)	Insurance Information Institute, 110 William St., New York NY, 10017
50	—	Data on main steam piping (ships)	NAVSHIPS report, Maintenance Support Office, Naval Ship Systems Command, U.S. Dept. of the Navy, Arlington, VA
51	—	Data on piping failure	NASA Literature Search, Information Tisco Inc., NASA Scientific and Technical Information Facility, College Park, MD
52	A. E. Green, A. J. Bourne	*Reliability Technology*, 1972, pp. 567–568	John Wiley & Sons, Chichester, UK
53	—	Data on pipeline leaks	Office of the Pipeline Safety, U.S. Dept. of Transportation (DOT), Office of the Secretary, Washington, D.C.
54	—	Data on piping failure (GEAP)	Atomic Power Dept., General Electric Co., San Jose, CA

(Table continued on next page.)

Table 9.1 (cont.) Sources of Failure Data

No.	Author or contact person	Source title or type of data	Published in or developed by
55	T. Anderson, A. Misund	Pipe Reliability: An Investigation of Pipeline Failure Characteristics and Analysis of Pipeline Failure Rates for Submarine and Cross-Country Pipelines	*Journal of Petroleum Technology*, April 1983, pp. 709–717
56	G. L. Wells	*Safety in Process Plant Design*, 1980, p. 50	John Wiley & Sons, 605 Third Ave., New York, NY 10017
57	—	Reliability Prediction of Electronic Equipment	MIL-HDBK-217, Dept. of Defense, Rome Air Development Center, Griffiss Air Force Base, New York

[a]Available from National Technical Information Service, U.S. Department of Commerce, 5285 Port Royal Rd., Springfield, VA 22151.

The failure modes are given below for various mechanical items:

1) Gages: leaking (39%), broken (26%), out of adjustment (14%), intermittent (4%), binding (9%), and other (8%).

2) Fans: no movement (11%), noisy (25%), intermittent (7%), out of adjustment (14%), and other (43%).

3) Clutches: displaced (17%), worn out (66%), and jammed (17%).

4) Bearings: worn out (54%), binding (12%), scored (5%), sticking (5%), loss of lubrication (9%), and other (15%).

5) Gears: broken (8%), worn out (54%), binding (8%), jammed (8%), stripped (12%), and other (10%).

6) Heat exchangers: leaking (70%), improper flow (8%), broken (4%), out of specification (4%), cracked/fractured (9%), and other (5%).

7) Seals: cracked/fractured (5%), leaking (6%), deteriorated (28%), ripped/cut/torn (51%), and other (10%).

8) Brakes: scored (12%), worn out (58%), leaking (27%), overheated (1%), and other (2%).

9) Filters: leaking (73%), cracked/fractured (18%), out of tolerance (3%), broken (3%), improper flow (1%), and other (2%).

10) Hydraulic pumps: noisy (6%), leaking (61%), intermittent (3%), cracked/fractured (9%), out of specification (4%), improper flow (7%), and other (10%).

Table 9.2 Failure Rates for Selected Mechanical Items

No.	Mechanical item	Failure rate/10^6 h	No.	Mechanical Item	Failure rate/10^6 h
1	Pipes	0.2	25	Mechanical joints	0.2
2	Bellows	5.0	26	Pistons	1.0
3	Ball valves	0.5	27	Hinges	0.02–5.5
4	Nuts or bolts	0.02	28	Pulleys, idler (roller)	0.025–0.16
5	Conveyor belts (light load)	3.6–19.0	29	Pumps, all types	65.74
6	Pipe joints	0.5	30	Valves, all types	11.33
7	Motors, general	10.0	31	Gaskets	0.5
8	Heavily stressed shafts	0.2	32	Relief valves	0.5–10.0
9	Conveyor belts (heavy load)	20.0–140.0	33	Cylinders	0.1
10	Brake assemblies (mechanical)	2.8–7.6	34	Solenoid valves	0.66–30.0
11	Hair springs	1.0		Bearings	
	Gears		35	Roller	0.139–7.31
12	Helical	1.0	36	Ball (heavy duty)	20.0
13	Spurs	10.0	37	Sleeve	0.008–7.3
14	Ducts	1.0	38	Ball (light duty)	10.0
15	Brake assemblies	0.8–114.0	39	Filters, blockage	1.0
	(hydraulic, pneumatic)		40	Diesel engines	30.3–1733.0
16	Compressors	0.84–198.0	41	Rotating seals	4.4
	Seals		42	Gages, dial type	0.5–21.5
17	O-ring	0.2	43	Clamps, cable	2.0
18	Sliding	3.0	44	Cups, grease	0.091
19	Heat exchangers	6.11–244.3	45	Motors, stepper	5.0
20	Pivots	1.0	46	Tachometers	5.0
21	Washers	0.5	47	Rivets	0.01–1.5
22	Flanges	5.69	48	Welded fittings	0.01–0.5
23	Knobs	0.09	49	Inset locks	4.0–27.0
24	Shims	0.003	50	Shackles	0.05

Table 9.3 Failure Rates for Selected Mechanical Items (Use Environment: Airborne)

No.	Mechanical item	Failure rate/10^6 h	Data source: Military	Data source: Commercial
1	Instrument counter	346.985	x	
	Fans			
2	Centrifugal	89.200	x	
3	Centrifugal	21.739		x
4	Axial	211.557	x	
5	Axial	5.510		x
6	Power transmitter, mechanical	10.0		x
7	Gimbal	7.829	x	
	Pumps			
8	Coolant	657.251	x	
9	Engine driven	18.519		x
10	Electric motor driven	6.889		x
11	Electric motor driven	142.857	x	
12	Thermostat	233.746	x	
13	Pump, hydraulic	573.711	x	
	Bearings			
14	Roller	0.863	x	
15	Ball	5.133	x	
16	Ball	1.372		x
17	Seal, O-ring	2.387	x	
18	Shock absorber, mount	6.309	x	
19	Tank, storage	45.404	x	
	Valves			
20	Relief	26.796	x	
21	Solenoid	28.128	x	
22	Solenoid	29.827		x
23	Gate	32.448	x	
24	Check	30.277	x	
25	Manual activated	32.895	x	
26	Servo	38.730	x	

Table 9.4 Failure Rates for Selected Mechanical Items (Use Environment: Ground, Mobile)

No.	Mechanical item	Failure rate/10^6 h	No.	Mechanical item	Failure rate/10^6 h
1	Gearbox (reduction)	18.755		Washers	
	Gears		17	Flat	0.165
2	Spur	5.95	18	Lock	0.116
3	Helical	19.725	19	Spring	4.302
4	Bevel	2.941	20	Tank (nonpressurized)	529.100
	Filters			Bearings	
5	Liquid	14.944	21	Ball	6.040
6	Gas (air)	3.242	22	Roller	8.323
7	Duct (general)	4.267	23	Sleeve	4.661
	Couplings		24	Bellows (general)	13.317
8	Rigid	1.908	25	V-belt	16.812
9	Flexible	9.987	26	Axle (general)	9.539
10	Clutch (friction)	38.155	27	Antenna (radar)	114.812
11	Compressor (general)	33.624	28	Crankshaft (general)	33.292
12	Actuator (mechanical)	33.624		Blowers and fans	
13	Shock absorbers (resilient)	19.078	29	Centrifugal	4.84
			30	Axial	15.608
14	Slip rings (general)	263.158	31	Gear (antirotation)	1578.948
	Valves		32	Gaskets and seals (dynamic)	4.643
15	Hydraulic	19.608			
16	Pneumatic	16.537			

Table 9.5 Failure Rates for Selected Mechanical Items (Use Environment: Ground Fixed)

No.	Mechanical Item	Failure rate/10^6 h	No.	Mechanical Item	Failure rate/10^6 h
1	Duct (general)	2.902		Pumps	
2	Clutch (friction)	5.318	20	Pneumatic	9.420
3	Bushings (general)	6.114	21	Vacuum-ring seal type	11.111
	Bearings			Instruments	
4	Sleeve	4.94	22	Indicator	4.12
5	Ball	2.237	23	Ammeter	8.000
6	Brake (electromechanical)	16.00	24	Gear (spur)	3.152
7	Fittings (threaded)	2.4	25	Gaskets and seals (static)	3.004
8	Gaskets and seals (dynamic)	3.295		Filters	
	Gears		26	Liquid	6.000
9	Helical	5.000	27	Gas-air	10.635
10	Bevel	1.333	28	Slip rings (general)	0.667
11	Hypoid	5.000	29	Solenoids (rotary)	33.906
	Keyboards		30	Sprocket (general)	5.693
12	General	6.909	31	Solenoids (rotary)	33.906
13	Mechanical	4.666	32	Tank (pressurized)	8.000
14	Knob (general)	2.081		Valves	
	Pumps		33	Solenoid operated	11.111
15	Vacuum	10.610	34	Pneumatic	4.000
16	Vacuum-lobe type	244.444	35	Hydraulic	6.000
17	Pulley (gear belt)	5.287	36	Washer (lock)	0.586
18	Metal tubing (general)	0.150	37	Thermocouple (general)	1.333
19	Lamp holder (general)	4.540	38	Spring (torsion)	14.296
			39	Washer (flat)	0.614
			40	Air conditioner (comfort)	711.111

9.10 FAILURE DATA ANALYSIS

In order to utilize the collected failure data effectively, they have to be analyzed. There are several methods and procedures one can use for such a purpose. This section presents some of them.

Mean Time to Failure

For practical purposes, this is expressed as

$$\text{MTTF} = \sum_{j}^{m} T_j \Big/ m \tag{9.1}$$

where MTTF is the mean time to failure, m the total number of identical (mechanical) items, and T_j the time to failure of (mechanical) item j for $j = 1,2,...,m$.

Example 9.1

Assume that in life testing of 12 identical bearings, the time to failure, in hours, were 1500, 1800, 1000, 2100, 2500, 2000, 1300, 1450, 1840, 1350, 1960, and 1750. Calculate bearing mean time to failure by using Eq. (9.1).

Substituting the above data into Eq. (9.1), we get

$$\text{MTTF} = \frac{(1500 + 1800 + 1000 + 2100 + \ldots + 1750)}{12}$$

$$= 1712.5 \text{ h}$$

Thus, the bearing mean time to failure is 1712.5 h.

Variance

The variance σ^2 of a data set is expressed as

$$\sigma^2 = \left[\sum_{j=1}^{m} (T_j - T_{am})^2 \right] \Big/ m \tag{9.2}$$

where T_{am} is the arithmetic mean (i.e., MTTF).

Example 9.2

Using Eq. (9.2), compute the variance of the failure data set given in Example 9.1. The result is

$$\sigma^2 = \frac{(1500 - 1712.5)^2 + (1800 - 1712.5)^2}{12}$$

$$+ \frac{(1000 - 1712.5)^2 + \ldots + (1750 - 1712.5)^2}{12}$$

$$= 155{,}235.417$$

Hazard Rate Plot

A failure data plot can be used to estimate the hazard rates of mechanical items. For the time interval $T_k < T \leqslant T_k + \Delta T_k$, the data hazard rate is defined as

$$\lambda_d(T) = \frac{f}{T_L \cdot S} = \frac{[M(T_k) - M(T_k + \Delta T_k)]}{(\Delta T_k) \cdot M(T_k)} \tag{9.3}$$

where T is the time, $\lambda_d(T)$ the data hazard rate, f the number of failures in the given interval of time, T_L the length of the time interval, S the survivors at the start of a given interval of time, ΔT_k the time interval, and $M(T_k)$ the survivors at time T_k.

Example 9.3

Failure data for specific identical mechanical components were obtained as shown in Table 9.6. Develop a hazard rate plot by using Eq. (9.3).

Using Eq. (9.3), the hazard rate values given in Table 9.7 were calculated. A hazard rate plot for the data of Table 9.7 is given in Fig. 9.4. This plot clearly shows that the hazard rate for the mechanical components under study is constant, with estimated value of 0.45×10^{-2} failures/h. This means that the component failure times are exponentially distributed. The dotted line in Fig. 9.4 indicates the estimated (constant) failure rate of mechanical components.

A Graphical Technique to Determine a Failure Times Statistical Distribution

This technique is concerned with determining whether a hypothesized statistical distribution is a reasonable model to describe the failure data in question. The method is demonstrated for exponential and normal distributions.[15]

Table 9.6 Failure Times for 15 Mechanical Items

No.	Failure times, h	No.	Failure times, h
1	15	9	192
2	30	10	230
3	45	11	275
4	65	12	330
5	85	13	405
6	106	14	517
7	131	15	742
8	160		

Table 9.7 Calculated values for data hazard rate

Time interval	Hazard rate per hour, $\lambda_d(T)$
0–15	$1/(15-0)(15) = 0.444 \times 10^{-2}$
15–30	$1/(30-15)(14) = 0.476 \times 10^{-2}$
30–45	$1/(45-30)(13) = 0.513 \times 10^{-2}$
45–65	$1/(60-45)(12) = 0.417 \times 10^{-2}$
65–85	$1/(85-65)(11) = 0.455 \times 10^{-2}$
85–106	$1/(106-85)(10) = 0.476 \times 10^{-2}$
106–131	$1/(131-106)(9) = 0.444 \times 10^{-2}$
131–160	$1/(160-131)(8) = 0.431 \times 10^{-2}$
160–192	$1/(192-160)(7) = 0.446 \times 10^{-2}$
192–230	$1/(230-192)(6) = 0.439 \times 10^{-2}$
230–275	$1/(275-230)(5) = 0.444 \times 10^{-2}$
275–330	$1/(330-275)(4) = 0.455 \times 10^{-2}$
330–405	$1/(405-330)(3) = 0.444 \times 10^{-2}$
405–517	$1/(517-405)(2) = 0.446 \times 10^{-2}$
517–742	$1/(742-517)(1) = 0.444 \times 10^{-2}$

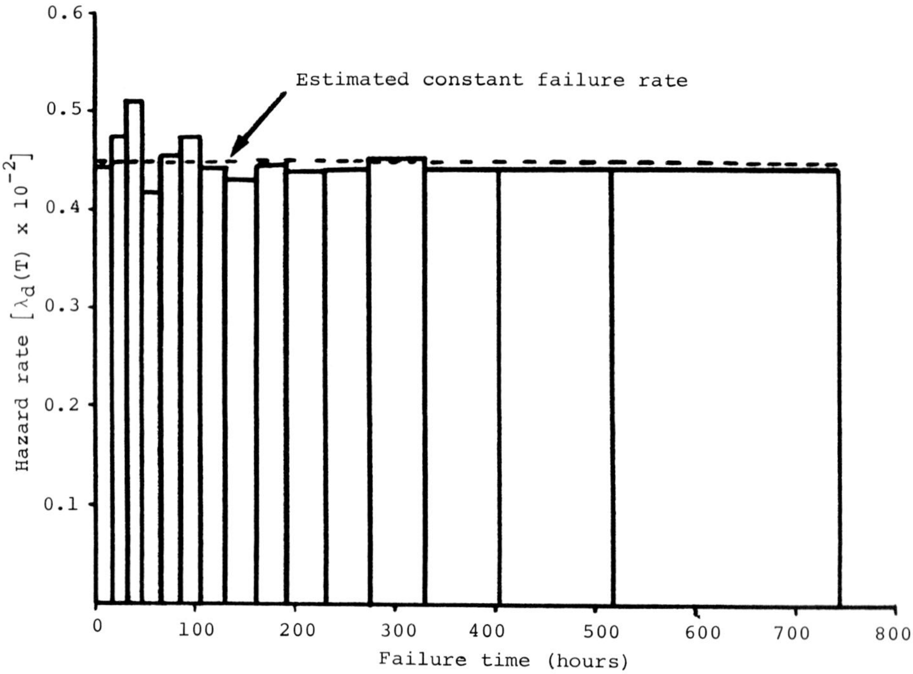

Fig. 9.4 Hazard rate plot.

Exponential Distribution. In this case, semilog graph paper can be used to determine the validity of the distribution for a sample of size k failure times. The paper has arithmetic (horizontal) and logarithmic (vertical) scales. The failure times in the sample are represented by $t_1,t_2,t_3,...,t_k$ (i.e., $t_1 \leqslant t_2 \leqslant t_3 \leqslant \ldots \leqslant t_k$). t_k denotes the highest failure time. The following steps are associated with the graphical technique:

1) Calculate the value of $(k + 1)/(k + 1 - j)$, for $j = 1,2,3,4,...,k$. j denotes the failure number. The computed values correspond to failure times $t_1,t_2,t_3,...,t_k$.

2) Plot the points t_j and $(k + 1)/(k + 1 - j)$ on the semilog graph paper [i.e., $(k + 1)/(k + 1 - j)$ on the vertical scale and t_j on the horizontal scale].

3) Draw a best-fit straight line through these points (i.e., points of step 2) and the origin.

4) Observe if the straight line really represents the trend of the plotted data points. If so, consider the exponential distribution. It means that the assumption of the exponential distribution is reasonable.

Normal Distribution. In this case, normal probability graph paper is used to find out whether a given sample of failure times follow the normal distribution or not. However, before using this distribution in reliability work the reader should consult section on normal distribution in Chapter 2. The probability paper has probability (vertical) and arithmetic (horizontal) scales. It is assumed that the sample is composed of k number of failure times $t_1,t_2,t_3,...,t_k$ (i.e., $t_1 \leqslant t_2 \leqslant t_3 \leqslant ... \leqslant t_k$). t_k denotes the highest failure time. The graphical approach is composed of the following steps:

1) Calculate the value of $100j/(k + 1)$, for $j = 1,2,3,...,k$. j denotes the failure number. The computed values correspond to failure times $t_1,t_2,t_3,...,t_k$.

2) Plot the points $[t_j,(100j)/(k + 1)]$ on normal probability graph paper [i.e., t_j on the horizontal scale and $100j/(k + 1)$ on the vertical scale].

3) Draw a best fit straight line through the plotted points of step 2.

4) Observe if the straight line really represents the trend of the plotted data points; if so, consider the normal distribution.

Example 9.4

Assume that a sample containing 15 identical gears was life tested. Their failure times were 100, 150, 250, 300, 400, 450, 500, 650, 750, 800, 900, 1050, 1200, 1300, and 1550 h. Tabulate values for coordinates of exponential and normal distributions.

For $k = 15$, coordinates for exponential and normal distributions are presented in Table 9.8.

Maximum-Likelihood Technique

This is a powerful method used to estimate parameters of statistical distributions representing failure times. In relation to other techniques, this method yields better estimates and is easy to apply. This technique calls for

Table 9.8 Coordinates for Exponential and Normal Distributions

j	$\frac{k+1}{k+1-j}$ (exponential)	$\frac{100j}{k+1}$ (normal)	t_j, h
1	1.07	6.25	100
2	1.14	12.50	150
3	1.23	18.75	250
4	1.33	25.00	300
5	1.46	31.25	400
6	1.60	37.50	450
7	1.78	43.75	500
8	2.00	50.00	650
9	2.29	56.25	750
10	2.67	62.50	800
11	3.20	68.75	900
12	4	75.00	1050
13	5.33	81.25	1200
14	8.00	87.50	1300
15	16	93.75	1550

the formulation of a joint density function known as likelihood function L. The function depends on the items' failure times t_i, for $i = 1,2,3,k$, and the density function parameters $\lambda_1,\lambda_2,\lambda_3,\ldots$. In order to maximize L, we let

$$\frac{\partial L}{\partial \lambda_i} = 0, \qquad \text{for } i = 1,2,3,\ldots \tag{9.4}$$

Generally, it is considered easier to maximize L after taking its natural logarithms (i.e., ℓnL). In this case, Eq. (9.4) becomes

$$\frac{\partial \ell nL}{\partial \lambda_i} = 0, \qquad \text{for } i = 1,2,3,\ldots \tag{9.5}$$

The likelihood function L of the sample is expressed as

$$L(t_1,t_2,t_3,\ldots,t_k;\lambda) = \prod_{j=1}^{k} f(t_j;\lambda) \tag{9.6}$$

where $f(t_j;\lambda)$ is the density function with parameter λ, t_j the jth time to failure, for $j = 1,2,3,\ldots k$, k the number of data values in the sample, and λ the distribution parameter to be estimated.

Example 9.5

Assume that the failure times of certain ball bearings are normally distributed. Obtain expressions for the distribution parameters by using the maximum-likelihood technique. The normal probability density function is defined by

$$f(t;\lambda,\sigma^2) = \frac{1}{\sigma(2\pi)^{\frac{1}{2}}} e^{-(t-\lambda)^2/2\sigma^2}$$

$$\text{for } \sigma^2 > 0, \qquad -\infty < \lambda < \infty, \qquad -\infty < t < \infty \tag{9.7}$$

where t is the time, σ the distribution parameter (standard deviation), and λ the distribution parameter (mean).

Using Eq. (9.7) in Eq. (9.6), we get

$$L(.) = \left[\frac{1}{\sigma(2\pi)^{\frac{1}{2}}}\right]^k \prod_{j=1}^{k} e^{-(t_j-\lambda)^2/2\sigma^2}$$

$$= \left[\frac{1}{\sigma(2\pi)^{\frac{1}{2}}}\right]^k \left[e^{-(1/2\sigma^2)\sum_{j=1}^{k}(t_j-\lambda)^2}\right] \tag{9.8}$$

Taking the natural logarithms of Eq. (9.8) yields

$$\ell n L(.) = \ell n L = k \ell n \left[\frac{1}{\sigma(2\pi)^{\frac{1}{2}}}\right] - \frac{1}{2\sigma^2}\sum_{j=1}^{k}(t_j-\lambda)^2 \tag{9.9}$$

After differentiating with respect to λ and setting the resulting expression equal to zero, Eq. (9.9) yields

$$\frac{\partial L}{\partial \lambda} = -\frac{1}{2\sigma^2}\sum_{j=1}^{k} 2(t_j-\lambda)(-1) = 0$$

$$= \frac{1}{\sigma^2}\left(\sum_{j=1}^{k} t_j - k\lambda\right) = 0 \tag{9.10}$$

Therefore,

$$\hat{\lambda} = \left(\sum_{j=1}^{k} t_j\right)\Big/ k \tag{9.11}$$

where $\hat{\lambda}$ is the maximum-likelihood estimation of λ.

Similarly, from Eq. (9.9), we get

$$\frac{\partial L}{\partial \sigma} = \frac{1}{\sigma}\left[-k + \frac{1}{\sigma^2}\sum_{j=1}^{k}(t_j-\lambda)^2\right] = 0$$

Therefore,

$$\hat{\sigma}^2 = \left[\sum_{j=1}^{k} (t_j - \hat{\lambda})^2\right] \bigg/ k \tag{9.12}$$

It is to be noted[16] that $\hat{\lambda}$ is unbiased and $\hat{\sigma}^2$ is biased, but $[k/(k-1)]\hat{\sigma}^2$ is unbiased. Thus,

$$\hat{\sigma}^2_{ub} = \frac{k}{(k-1)} \cdot \hat{\sigma}^2 = \frac{1}{(k-1)} \sum_{j=1}^{k} (t_j - \hat{\lambda})^2 \tag{9.13}$$

where $\hat{\sigma}^2_{ub}$ is the unbiased estimate of σ^2. For $k \geqslant 30$, the difference between $\hat{\sigma}^2$ and $\hat{\sigma}^2_{ub}$ is rather small.

Kolmogorov-Smirnov Test

The Kolmogorov-Smirnov goodness of fit test can be used to determine the underlying statistical distribution of failure times associated with mechanical items. More specifically, the objective of this test is to find out whether the assumptions made about statistical distributions of the failure times are reasonable or not. The test is quite useful for small or large samples, but is restricted to continuous distributions only. The following steps are associated with the test[17]:

1) Assume a statistical distribution for a given failure data sample or, more specifically the specific distribution to be tested.

2) Estimate the values of the assumed distribution's parameters using the given sample data.

3) Determine the value of the level of significance known as θ. The level of significance is the risk of not accepting the assumed distribution if it happened to be the real representative distribution of the failure data in question.

4) Calculate probability $F(t_j) = P(T \leqslant t_j)$ using the assumed distribution and parameters. The symbol t_j stands for the jth sample observation and $F(t_j)$ for the probability distribution function evaluated at t_j.

5) Calculate the value of $\hat{F}(t_j)$ for the jth ordered sample observation (use the given sampled data) by using the following relationship:

$$\hat{F}(t_j) = \frac{j}{m}, \qquad j = 1,2,3,\ldots,m \tag{9.14}$$

where m is the number of observations in the sample.

The value $\hat{F}(t_j)$ is the portion of the observations in the sample that are less than or equal to t_j. Also, shift j to $(j-1)$ and compute

$$\hat{F}(t_{j-1}) = \frac{j-1}{m}, \qquad j = 1,2,3,\ldots,m \tag{9.15}$$

6) Compute the statistic:

$$D = \max_j \{|F(t_j) - \hat{F}(t_j)|, |F(t_j) - \hat{F}(t_{j-1})|\} \tag{9.16}$$

7) Obtain the value of D_θ (for practical purposes) from Table 9.9[15–17] for a given value of the level of significance θ and the size of the sample.

8) Compare the value of D with the value of D_θ. If D_θ is less than D, reject the distribution in question.

Example 9.6

Assume that 10 identical mechanical items were life tested and their failure times were 10, 25, 30, 40, 50, 65, 75, 80, 90, and 100 h. From past experience on similar items, it is assumed that the items' failure times follow the normal distribution. Test this assumption by using the Kolmogorov-Smirnov method. The value of level of significance θ is 0.10.

Using the given failure data, we get the following values for the mean and standard deviation:

$$\mu = \left(\sum_{j=1}^{m} t_j \right) \bigg/ m$$

$$= \frac{10 + 25 + 30 + 40 + \ldots + 100}{10} = 56.50 \text{ h}$$

Table 9.9 Tabulated Critical Values D_θ

No. of data points m in a sample	Level of significance θ: $\theta = 0.05$	$\theta = 0.10$	$\theta = 0.20$
4	0.62	0.56	0.49
5	0.57	0.51	0.45
6	0.52	0.47	0.41
7	0.49	0.44	0.38
8	0.46	0.41	0.36
9	0.43	0.39	0.34
10	0.41	0.37	0.32
11	0.39	0.35	0.31
12	0.38	0.34	0.30
13	0.36	0.33	0.28
14	0.35	0.31	0.27
15	0.34	0.30	0.27
20	0.29	0.26	0.23
30	0.24	0.22	0.19
40	0.21	0.19	0.17
45	0.20	0.18	0.16
Over 50	$(1.36)m^{-\frac{1}{2}}$	$(1.22)m^{-\frac{1}{2}}$	$(1.07)m^{-\frac{1}{2}}$

where μ is the sample mean, m the sample size, t_j the time to failure of the jth item, for $j = 1,2,3,...,m$, and

$$\sigma = \left[\frac{\sum_{j=1}^{m} (t_j - \mu)^2}{(m-1)} \right]^{\frac{1}{2}}$$

$$= \left[\frac{(10-56.5)^2 + (25-56.5)^2 + (30-56.5)^2 + \ldots + (100-56.50)^2}{(10-1)} \right]^{\frac{1}{2}}$$

$$= 30.1$$

With the aid of cumulative normal distribution function (table) and the above calculated values for μ and σ, we obtained the values for $F(t_j)$ as shown in Table 9.10.

Using the sampled data, we obtained the values for $\hat{F}(t_j)$ and $\hat{F}(t_{j-1})$ as shown in Table 9.11. Using the calculated values of Tables 9.10 and 9.11, the absolute values of $[F(t_j) - \hat{F}(t_j)]$, and $[F(t_j) - \hat{F}(t_{j-1})]$ were listed in Table 9.12. From Table 9.12 it is to be noted that the maximum absolute difference is 0.13 (i.e., $D = 0.13$).

Table 9.10 Calculated Values for $F(t_j)$

j	t_j	$(t_j - \mu)/o$	$F(t_j)$
1	10	$(10 - 56.50)/30.1 = -1.55$	0.06
2	25	$(25 - 56.50)/30.1 = -1.05$	0.15
3	30	$(30 - 56.50)/30.1 = -0.88$	0.19
4	40	$(40 - 56.50)/30.1 = -0.55$	0.29
5	50	$(50 - 56.50)/30.1 = -0.22$	0.41
6	65	$(65 - 56.50)/30.1 = 0.28$	0.61
7	75	$(75 - 56.50)/30.1 = 0.62$	0.73
8	80	$(80 - 56.50)/30.1 = 0.78$	0.78
9	90	$(90 - 56.50)/30.1 = 1.11$	0.87
10	100	$(100 - 56.50)/30.1 = 1.45$	0.93

Table 9.11 Calculated Values for $\hat{F}(t_j)$ and $\hat{F}(t_{j-1})$

j	t_j	$\hat{F}(t_j)$	$\hat{F}(t_{j-1})$	j	t_j	$\hat{F}(t_j)$	$F(t_{j-1})$
1	10	$1/10 = 0.1$	$(1-1)/10 = 0$	6	65	$6/10 = 0.6$	$(6-1)/10 = 0.5$
2	25	$2/10 = 0.2$	$(2-1)/10 = 0.1$	7	75	$7/10 = 0.7$	$(7-1)/10 = 0.6$
3	30	$3/10 = 0.3$	$(3-1)/10 = 0.2$	8	80	$8/10 = 0.8$	$(8-1)/10 = 0.7$
4	40	$4/10 = 0.4$	$(4-1)/10 = 0.3$	9	90	$9/10 = 0.9$	$(9-1)/10 = 0.8$
5	50	$5/10 = 0.5$	$(5-1)/10 = 0.4$	10	100	$10/10 = 1$	$(10-1)/10 = 0.9$

Table 9.12 Absolute values for $[F(t_j) - \hat{F}(t_j)]$ and $[F(t_j) - \hat{F}(t_{j-1})]$

j	t_j	$\lvert F(t_j) - \hat{F}(t_j)\rvert$	$\lvert F(t_j) - \hat{F}(t_{j-1})\rvert$	j	t_j	$\lvert F(t_j) - \hat{F}(t_j)\rvert$	$\lvert F(t_j) - \hat{F}(t_{j-1})\rvert$
1	10	0.04	0.06	6	65	0.01	0.11
2	25	0.05	0.05	7	75	0.03	0.13 max
3	30	0.11	1.01	8	80	0.02	0.08
4	40	0.11	0.01	9	90	0.03	0.07
5	50	0.09	0.01	10	100	0.07	0.03

Table 9.13 Failure Times for 20 Mechanical Components

Component	Failure times, h	Component	Failure times, h	Component	Failure times, h
1	178	8	350	15	240
2	30	9	200	16	360
3	500	10	150	17	450
4	400	11	600	18	240
5	60	12	225	19	500
6	80	13	800	20	180
7	90	14	180		

For given $\theta = 0.10$ and sample size ($m = 10$), the value of D_θ (i.e., $D_{0.10}$) from Table 9.9 is 0.37 ($D_{0.10} = 0.37$). In this case, $D_{0.10} > D$. Therefore, one can see no reason to reject the normal distribution to represent the failure times of the mechanical items under consideration. However, before using this distribution in reliability work the reader should consult section on normal distribution in Chapter 2.

References

[1]Collins, J. A., "Failure Experience Feedback in Design Improvement," *Failure Prevention and Reliability*, edited by S. B. Bennett, A. L. Ross, and P. Z. Zemanick, ASME, New York, 1978, pp. 245–249.

[2]"A Reliability Guide to Failure Reporting, Analysis, and Corrective Action Systems," Committee on Reliability Reporting, American Society for Quality Control (ASQC), Milwaukee, 1977.

[3]Kletz, T., "The Uses, Availability and Pitfalls of Data on Reliability," *Process Technology*, Vol. 18, 1973, pp. 111–113.

[4]Mitchell, R. L. and Rutter, R. R., "A Study of Automotive Reliability and Associated Cost of Maintenance in the USA," SAE Paper 780277, Jan. 1978.

[5]Parascos, E. T., "A New Approach to the Establishment and Maintenance of Equipment Failure Rate Data Bases," *Failure Prevention and Reliability*, edited by S. B. Bennett, A. L. Ross, and P. Z. Zemanick, ASME, New York, 1977, pp. 263–268.

[6]Dhillon, B. S., *Systems Reliability, Maintainability and Management*, Petrocelli Books, New York, 1983.

[7]Hahn, R. F., "Data Collection Techniques," *Proceedings of the Annual Reliability and Maintainability Symposium*, IEEE, New York, 1972, pp. 38–43.

[8]Schneider, R. W. and Tibbetts, D. R., "Assessing Product Reliability from Warranty Data," SAE Paper 831200, 1983.

[9]Duphily, R. J. and Long, R. L., "Enhancement of Electric Power Plant Reliability Data Systems," Paper presented at Fourth Annual Reliability Engineering Conference for the Electric Power Industry, June 1977 (available from Holmes and Narver, Inc., Anaheim, CA).

[10]Dhillon, B. S., *Human Reliability: with Human Factors*, Pergamon Press, New York, 1986.

[11]"Reliability and Maintainability Data for Industrial Plants," A. P. Harris & Associates, Rept. TD-84-3, Ottawa, Canada, 1984.

[12]Green, A. E. and Bourne, A. J., *Reliability Technology*, Wiley, Chichester, England, 1972, pp. 567–570.

[13]Rossi, M. J., "Nonelectronic Parts Reliability Data," Reliability Analysis Center, Rome Air Development Center, Griffiss Air Force Base, NY, Rept. NPRD-3, 1985.

[14]Schafer, R. E., Angus, J. E., Finkelstein, J. M., Yerasi, M., and Fulton, D. W., "RADC Nonelectronic Reliability Notebook," Reliability Analysis Center, Rome Air Development Center, Griffiss Air Force Base, NY, Rept. RADC-TR-85-194, 1985.

[15]"Quality Assurance, Reliability Handbook," U.S. Army Material Command, Washington DC, AMCP 702–3, Oct. 1968, pp. F4–F7.

[16]Lloyd, D. K. and Lipow, M., *Reliability: Management, Methods, and Mathematics*, Prentice-Hall, Englewood Cliffs, NJ, 1962, pp. 170–171.

[17]Massey, F., "The Kolmogorov-Smirnov Test for Goodness of Fit," *Journal of the American Statistical Association*, Vol. 46, 1951, p. 70.

Exercises

9.1 List the functions of the failure reporting, analysis, and corrective measure system.

9.2 List the applications of the collected failure data.

9.3 What are the guidelines to be followed in the design of a failure reporting and documentation system?

9.4 Discuss the type of information to be considered for inclusion in the failure data collection form.

9.5 List eight data collection sources during the equipment life cycle.

9.6 What are the typical deficiencies encountered in failure data sources?

9.7 List the possible failure modes of the following mechanical items: gearbox, clutch, brake, fan, and seal.

9.8 Failure data for certain identical mechanical items are as given in Table 9.13. Develop a plot for hazard rate by using Eq. (9.3).

9.9 Assume that the failure times of relief valves are described by the following probability density function:

$$f(t) = \lambda e^{-\lambda t} \tag{9.15}$$

where t is the time and λ the distribution parameter.

Develop an expression to estimate λ by using the maximum-likelihood technique.

9.10 Assume that 12 identical mechanical devices were life tested and their failure times were 80, 90, 100, 115, 130, 140, 150, 165, 185, 195, 208, and 225 h. From past experience on similar items, it was assumed that their failure times are Weibull distributed. Test this assumption by using the Kolmogorov-Smirnov method.

10
TRANSIT SYSTEM RELIABILITY MODELING

This chapter presents a number of mathematical models concerned with transit systems. Service availability and unavailability are defined. Three mathematical models representing a transit system with exponentially distributed failure, repair, and towing times are presented. For all three models, steady-state probability, availability, reliability, mean time to failure, and the variance of the time to failure equations are developed. A mathematical model representing a transit system with nonexponentially distributed repair times is analyzed. The Laplace transforms of the state probability equations are developed. Steady-state probability solutions are obtained for Erlangian distributed repair times. A general expression for the transit system's steady-state availability is established when the repair distribution is Erlangian. Reliability and availability analyses of vehicles operating in alternating environments are presented. Steady-state probability, availability, and mean time to failure equations are developed.

10.1 INTRODUCTION

From a number of decades, reliability engineering principles have been applied in aerospace and military fields quite successfully. The success in these two areas has encouraged the application of reliability engineering principles in other fields as well. The transit system is one of them. In modern times, reliable and on-time transportation is very important to both the transit operator and the passengers. Lower reliability, in general, leads to decreased equipment availability and increased unscheduled maintenance. The incidence of system failures cannot be decreased with more vehicles; in addition, an unreliable transit system requires extremely good maintenance. Therefore, in order to provide effective transportation, a reliable transit system is a must. That is why reliability has become one of the vital ingredients in the design, development, and operational phases of a transit system.

Bibliography of literature on transit system reliability is given in Ref. 1. This chapter presents a number of mathematical models concerned with transit systems. In this chapter the term transit system is used loosely. However, it is self-explanatory wherever used.

10.2 SERVICE AVAILABILITY AND UNAVAILABILITY

According to Ref. 2, the service availability and unavailability in transit systems may be expressed as

$$s_A = \frac{T_{PH}}{T_{PH} + T_{PHD}} \tag{10.1}$$

$$s_U = \frac{T_{PHD}}{T_{PH} + T_{PHD}} \tag{10.2}$$

where s_A is the service availability, T_{PH} the number of annual passenger-hours of travel on the transit system, T_{PHD} the number of annual passenger-hours of delay because of failures, and s_U the service unavailability.

Dividing the top and bottom of the right-hand side of Eq. (10.1) with T_{PH}, we get

$$s_A = \frac{1}{1 + (T_{PHD}/T_{PH})} \tag{10.3}$$

For values of (T_{PHD}/T_{PH}) much less than 0.1, Eq. (10.3) yields

$$s_A \simeq 1 - (T_{PHD}/T_{PH}) \tag{10.4}$$

The number of annual passenger-hours of travel on the transit system is given by

$$T_{PH} = (PT) \cdot (T_m) \tag{10.5}$$

where PT is the person-trips per year and T_m the mean trip time.

10.3 RELIABILITY AND AVAILABILITY MODELING OF A VEHICLE

This section presents a number of mathematical models concerned with a vehicle.

Reliability and Availability Analysis of a Transit System with Exponentially Distributed Failure, Repair, and Towing Times

This section presents three Markov models[3] as given below.

Model I. This model deals with a three-state transit system. In this case, the vehicle can be in any one of the three states, namely: vehicle operating normally in the field, vehicle failed in the field, or failed vehicle in the workshop for repair. The failed vehicle is towed or by alternative means taken to the repair workshop from the field. The repaired vehicle is put back

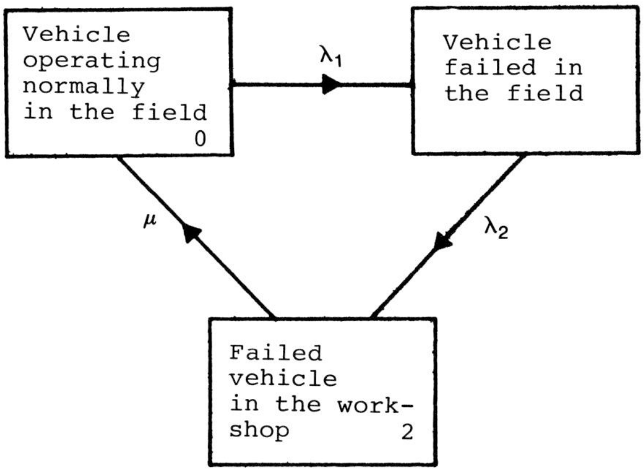

Fig. 10.1 State space diagram for model I.

into its normal operational state. The state space diagram of the model is shown in Fig. 10.1. The numerals in the boxes denote the system states. The following assumptions are associated with this Markov model: 1) the failure, repair, and towing rates are constant and 2) the repaired vehicle is as good as "new." The symbols used in describing the model are

λ_1 = failure rate of the vehicle
λ_2 = towing rate of the vehicle from state 1 to state 2 (it also includes the rate of taking the vehicle to the workshop by alternative means, if used)
μ = repair rate of the vehicle from state 2 to state 0
s = Laplace transform variable
$P_i(t)$ = probability that the transit system is in state i at time t, for $i = 0,1,2$
t = time

The system of differential equations associated with Fig. 10.1 is

$$\frac{dP_0(t)}{dt} + \lambda_1 P_0(t) = P_2(t) \cdot \mu \tag{10.6}$$

$$\frac{dP_1(t)}{dt} + \lambda_2 P_1(t) = P_0(t) \cdot \lambda_1 \tag{10.7}$$

$$\frac{dP_2(t)}{dt} + \mu P_2(t) = P_1(t) \cdot \lambda_2 \tag{10.8}$$

At time $t = 0$, $P_0(0) = 1$ and $P_1(0) = P_2(0) = 0$.

By setting the derivatives of Eqs. (10.6–10.8) equal to zero and utilizing the relationship, $P_0 + P_1 + P_2 = 1$, leads to the following steady-state probabilities:

$$P_0 = \frac{1}{1 + \rho_1 + \rho_2} \tag{10.9}$$

where $\rho_1 = \lambda_1/\lambda_2$ and $\rho_2 = \lambda_1/\mu$.

$$P_1 = \rho_1 \cdot P_0 \tag{10.10}$$

$$P_2 = \rho_2 \cdot P_0 \tag{10.11}$$

The steady-state availability of the transit system is given by

$$\mathrm{AV}_{ss} = (1 + \rho_1 + \rho_2)^{-1} \tag{10.12}$$

The plots of Eq. (10.12) are shown in Fig. 10.2 for the specified values of ρ_1 and ρ_2.

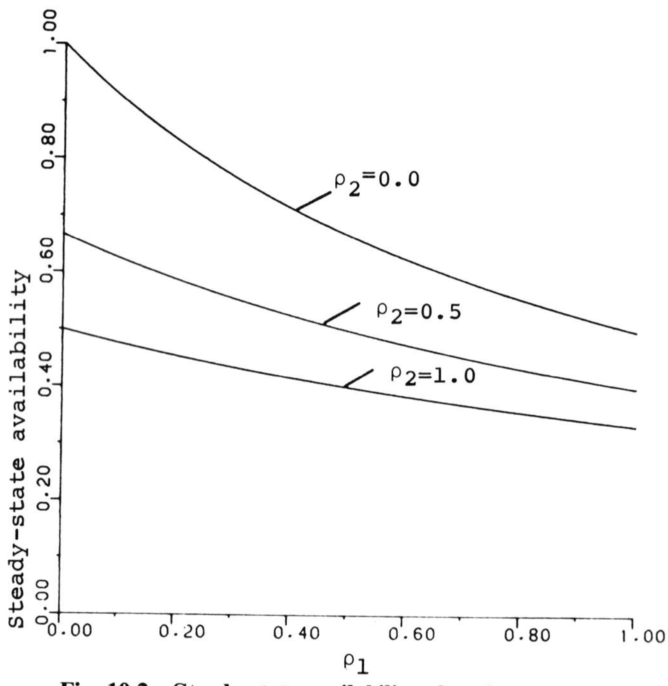

Fig. 10.2 Steady-state availability plots for model I.

These plots show the effect of ρ_1 and ρ_2 on transit system steady-state availability. From these plots, it is observed that the steady-state availability of the transit system decreases with the corresponding increasing values of ρ_1 and ρ_2. Setting $\mu = 0$ in Eqs. (10.6–10.8) and solving for $P_0(s)$ with the aid of Laplace transforms yields

$$R(s) = P_0(s) = \frac{1}{s + \lambda_1} \tag{10.13}$$

where $R(s)$ is the Laplace transform of the transit system reliability function and $P_0(s)$ the Laplace transform of the probability that the system is in state 0.

Thus, the reliability of the transit system is given by

$$R(t) = e^{-\lambda_1 t} \tag{10.14}$$

The transit system mean time to failure (MTTF) is given by

$$\text{MTTF} = \lim_{s \to 0} R(s) = 1/\lambda_1 \tag{10.15}$$

The plot of Eq. (10.15) is shown in Fig. 10.3 for the specified values of λ_1. The figure shows the effect of varying λ_1 on transit system MTTF.

The variance of the time to failure is

$$\sigma^2 = -2 \lim_{s \to 0} R'(s) - [\lim_{s \to 0} R(s)]^2 = \frac{1}{\lambda_1^2} \tag{10.16}$$

The prime denote differentiation with respect to s.

Example 10.1

The constant failure rate of a vehicle is 0.004 failures/h. Calculate the vehicle's reliability for a 10 h mission and its mean time to failure (MTTF).

Substituting the given data into Eqs. (10.14) and (10.15), we get

$$R(10) = e^{-(0.004)(10)} = 0.9608$$

and

$$\text{MTTF} = 1/0.004 = 250 \text{ h}$$

Thus, the reliability and mean time to failure (MTTF) of the vehicle are 0.9608 and 250 h, respectively.

Model II. This model is concerned with a four-state transit system. In this case, the vehicle can be in any one of the four states, namely, vehicle

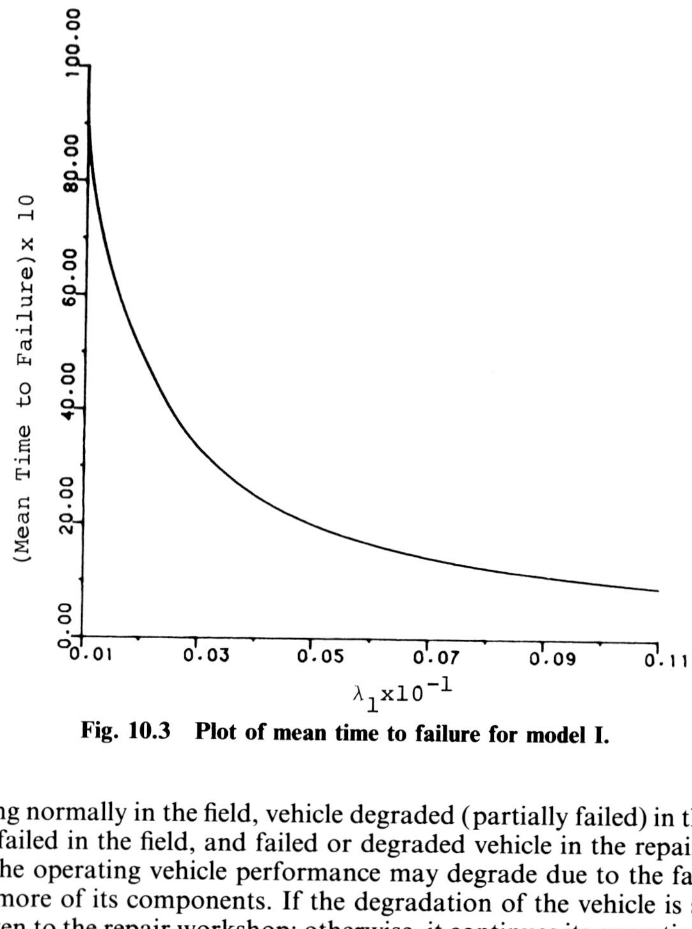

Fig. 10.3 Plot of mean time to failure for model I.

operating normally in the field, vehicle degraded (partially failed) in the field, vehicle failed in the field, and failed or degraded vehicle in the repair workshop. The operating vehicle performance may degrade due to the failure of one or more of its components. If the degradation of the vehicle is serious, it is driven to the repair workshop; otherwise, it continues its operation in the field. The vehicle may fail either from its normal operating state or from its degraded state. The failed vehicle is towed or by alternative means taken to the repair workshop. The fully repaired vehicle is put back into its normal operational state. On the other hand, the partially repaired vehicle can be put back into its degraded state operation. The system transition diagram is shown in Fig. 10.4. The numerals in boxes denote the system state numbers.

The following assumptions are associated with this Markov model: 1) failure, repair, towing, and other transition rates are constant and 2) the repaired vehicle is as good as "new."

The symbols used to develop this model are

λ_1 = degradation rate of the vehicle from state 0 to state 1
λ_2 = failure rate of the vehicle from state 1 to state 2

λ_3 = failure rate of the vehicle from state 0 to state 2
λ_4 = transition rate of the vehicle from state 1 to state 3
λ_5 = towing rate of the vehicle from state 2 to state 3 (it includes the rate of taking the vehicle to the workshop by alternative means, if used)
μ_1 = repair rate of the vehicle from state 3 to state 1
μ_2 = repair rate of the vehicle from state 3 to state 0
s = Laplace transform variable
t = time
$P_i(t)$ = probability that the transit system is in state i at time t, for $i = 0,1,2,3$

The system of differential equations associated with Fig. 10.4 is:

$$\frac{dP_0(t)}{dt} + (\lambda_1 + \lambda_3)P_0(t) = P_3(t) \cdot \mu_2 \tag{10.17}$$

$$\frac{dP_1(t)}{dt} + (\lambda_2 + \lambda_4)P_1(t) = P_0(t) \cdot \lambda_1 + P_3(t) \cdot \mu_1 \tag{10.18}$$

$$\frac{dP_2(t)}{dt} + \lambda_5 P_2(t) = P_0(t) \cdot \lambda_3 + P_1(t) \cdot \lambda_2 \tag{10.19}$$

$$\frac{dP_3(t)}{dt} + (\mu_1 + \mu_2)P_3(t) = P_1(t) \cdot \lambda_4 + P_2(t) \cdot \lambda_5 \tag{10.20}$$

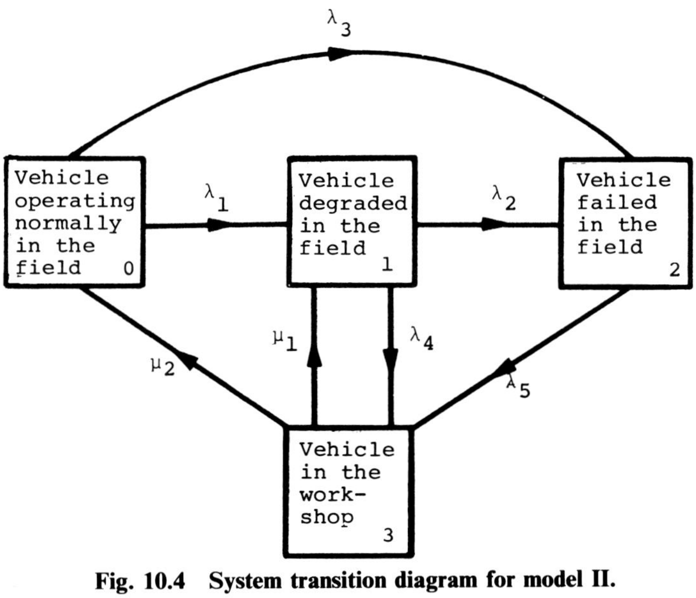

Fig. 10.4 System transition diagram for model II.

At time $t = 0$, $P_0(0) = 1$ and all other initial condition probabilities are equal to zero.

By setting the derivatives of Eqs. (10.17–10.20) equal to zero and utilizing the relationship, $\Sigma_{i=0}^{3} P_i = 1$, leads to the following results:

$$P = (1 + A_2 + A_3 + A_4)^{-1} \tag{10.21}$$

where

$$A_1 = \frac{\mu_1}{\mu_1 + \mu_2}, \qquad A_2 = \frac{\lambda_1 + \lambda_3 A}{(\lambda_2 + \lambda_4)(1 - A_1)}$$

$$A_3 = \frac{\lambda_3 + \lambda_2 A_2}{\lambda_5}, \qquad A_4 = \frac{\lambda_4 A_2 + \lambda_5 A_3}{\mu_1 + \mu_2}$$

$$P_1 = A_2 \cdot P_0 \tag{10.22}$$

$$P_2 = A_3 \cdot P_0 \tag{10.23}$$

$$P_3 = A_4 \cdot P_0 \tag{10.24}$$

The plots of Eqs. (10.21–10.24) are shown in Fig. 10.5. These plots show the impact of increasing λ_3 on transit system steady-state probabilities.

The transit system steady-state availability is

$$\mathrm{AV}_{ss} = P_0 + P_1 \tag{10.25}$$

By setting $\mu_1 = \mu_2 = 0$ in Eqs. (10.17–10.20), taking the Laplace transforms, and solving for $P_0(s)$ and $P_1(s)$, we get

$$R(s) = P_0(s) + P_1(s) \tag{10.26}$$

where $R(s)$ is the Laplace transform of the transit system reliability function and

$$P_0(s) = \frac{1}{s + \lambda_1 + \lambda_3} \tag{10.27}$$

$$P_1(s) = \frac{\lambda_1}{(s + \lambda_1 + \lambda_3)(s + \lambda_2 + \lambda_4)} \tag{10.28}$$

where $P_0(s)$ is the Laplace transform of the probability that the system is in state 0 and $P_1(s)$ the Laplace transform of the probability that the system is in state 1.

Thus, the reliability of the transit system is

$$R(t) = (1 + B)e^{-(\lambda_1 + \lambda_3)t} - Be^{-(\lambda_2 + \lambda_4)t} \tag{10.29}$$

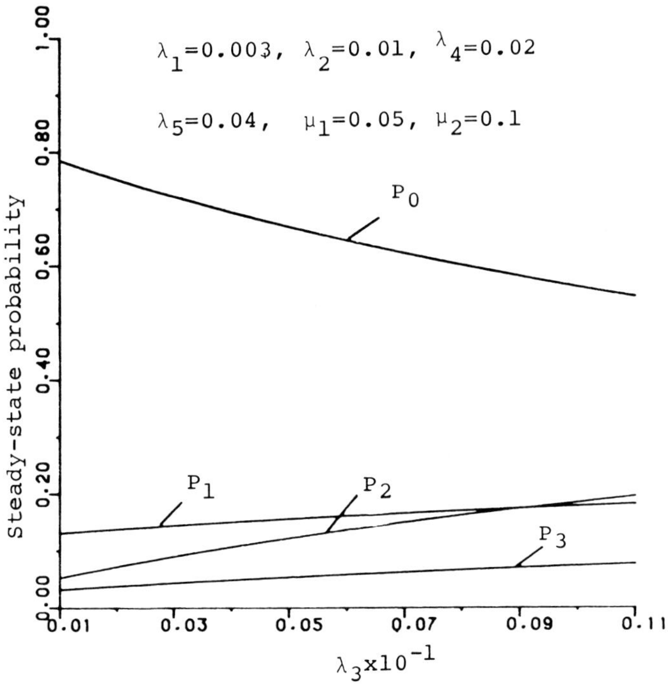

Fig. 10.5 Steady-state probability plots for model II.

where

$$B = \frac{\lambda_1}{\lambda_2 + \lambda_4 - \lambda_1 - \lambda_3} \tag{10.30}$$

The transit system mean time to failure (MTTF) and the variance of time to failure are

$$\text{MTTF} = \lim_{s \to 0} R(s) = \frac{\lambda_1 + \lambda_2 + \lambda_4}{(\lambda_1 + \lambda_3)(\lambda_2 + \lambda_4)} \tag{10.31}$$

and

$$\begin{aligned} \sigma^2 &= -2 \lim_{s \to 0} R'(s) - [\lim_{s \to 0} R(s)]^2 \\ &= \frac{(\lambda_2 + \lambda_4)^2 + \lambda_1(\lambda_1 + 2\lambda_3)}{(\lambda_1 + \lambda_3)^2(\lambda_2 + \lambda_4)^2} \end{aligned} \tag{10.32}$$

where the prime denotes the differentiation with respect to s.

Example 10.2

Assume that we have the following data: $\lambda_1 = 0.005$ failures/h, $\lambda_2 = 0.003$ failures/h, $\lambda_3 = 0.002$ failures/h, and $\lambda_4 = 0.001$/h. Calculate the transit system mean time to failure using Eq. (10.31)

Substituting the above data into Eq. (10.31) results in

$$\text{MTTF} = \frac{\lambda_1 + \lambda_2 + \lambda_4}{(\lambda_1 + \lambda_3)(\lambda_2 + \lambda_4)}$$

$$= \frac{(0.005) + (0.003) + (0.001)}{(0.005 + 0.002)(0.003 + 0.001)} = 321.43 \text{ h}$$

Thus, the mean time to failure of the transit system is 321.43 h.

Model III. This model is also concerned with a four-state transit system. The system transition diagram is shown in Fig. 10.6. The numerals in circles denote states of the system. The failed vehicle is towed or by alternative means taken to the repair workshop. The fully repaired vehicle is put back to its normal operation.

The following assumptions are associated with this Markov model: 1) failure, repair, and towing rates are constant and 2) the repaired vehicle is as good as "new." The symbols used to develop this model are

λ_1 = fail-safe failure rate of the vehicle
λ_2 = fail-accident failure rate of the vehicle
λ_3 = towing rate of the vehicle from state 1 to state 3 (also includes the rate of taking the vehicle to workshop by alternative means, if used)
λ_4 = towing rate of the vehicle from state 2 to state 3 (also includes the rate of taking the vehicle to workshop by alternative means, if used)
μ = repair rate of the vehicle from state 3 to state 0
s = Laplace transform variable
t = time
$P_i(t)$ = probability that the transit system is in state i at time t, for $i = 0,1,2,3$

The system of differential equations associated with Fig. 10.6 is

$$\frac{dP_0(t)}{dt} + (\lambda_1 + \lambda_2)P_0(t) = P_3(t) \cdot \mu \qquad (10.33)$$

$$\frac{dP_1(t)}{dt} + \lambda_3 P_1(t) = P_0(t) \cdot \lambda_1 \qquad (10.34)$$

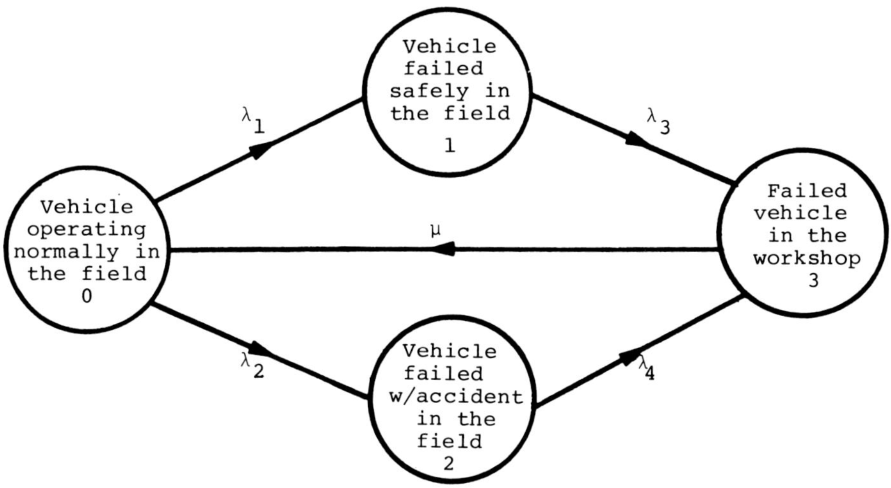

Fig. 10.6 State space diagram for model III.

$$\frac{dP_2(t)}{dt} + \lambda_4 P_2(t) = P_0(t) \cdot \lambda_2 \tag{10.35}$$

$$\frac{dP_3(t)}{dt} + \mu P_3(t) = P_1(t) \cdot \lambda_3 + P_2(t) \cdot \lambda_4 \tag{10.36}$$

At time $t = 0$, $P_0(0) = 1$ and $P_1(0) = P_2(0) = P_3(0) = 0$.

By setting the derivatives of Eqs. (10.33–10.36) equal to zero and utilizing the relationship, $\Sigma_{i=0}^{3} P_i = 1$, leads to the following probability equations:

$$P_0 = \left(1 + \frac{\lambda_1}{\lambda_3} + \frac{\lambda_2}{\lambda_4} + \frac{\lambda_1 + \lambda_2}{\mu}\right)^{-1} \tag{10.37}$$

$$P_1 = \frac{\lambda_1}{\lambda_3} \cdot P_0 \tag{10.38}$$

$$P_2 = \frac{\lambda_2}{\lambda_4} \cdot P_0 \tag{10.39}$$

$$P_3 = \frac{\lambda_1 + \lambda_2}{\mu} \cdot P_0 \tag{10.40}$$

The steady-state availability of the transit system is given by

$$AV_{ss} = P_0 \tag{10.41}$$

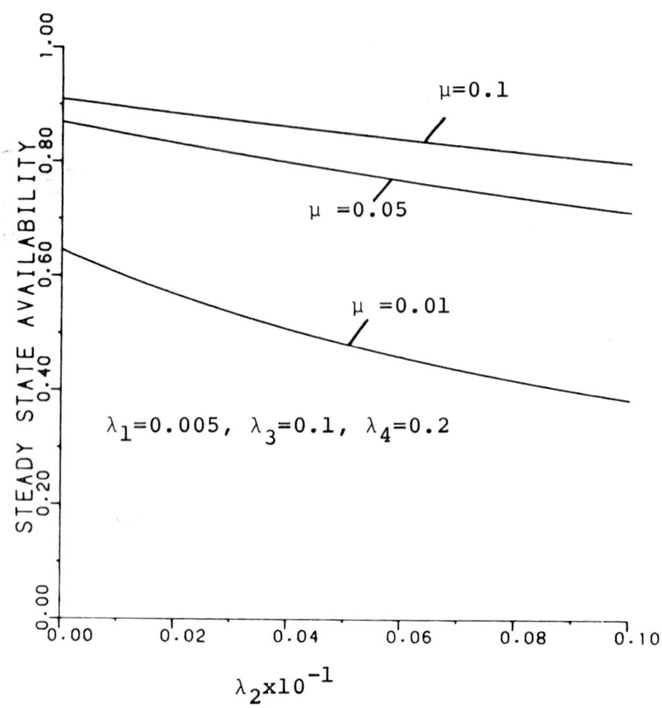

Fig. 10.7 Steady-state availability plots for model III.

The plots of the above equation are given in Fig. 10.7. The steady-state probability of the vehicle failing safely in the field is

$$P_s = P_1 \tag{10.42}$$

On the other hand, the steady-state probability of the vehicle failing with accident in the field is

$$P_A = P_2 \tag{10.43}$$

By setting $\mu = 0$ in Eqs. (10.33–10.36), taking Laplace transforms, and solving for $P_0(s)$ we get

$$R(s) = P_0(s) = \frac{1}{s + \lambda_1 + \lambda_2} \tag{10.44}$$

where $R(s)$ is the Laplace transform of the transit system reliability function and $P_0(s)$ the Laplace transform of the probability that the transit system is in state 0.

Thus, the reliability of the transit system is

$$R(t) = e^{-(\lambda_1 + \lambda_2)t} \tag{10.45}$$

The mean time to failure (MTTF) of the transit system is given by

$$\text{MTTF} = \lim_{s \to 0} R(s) = \frac{1}{\lambda_1 + \lambda_2} \tag{10.46}$$

The variance of the time to failure is given by

$$\sigma^2 = -2 \lim_{s \to 0} R'(s) - [\lim_{s \to 0} R(s)]^2 = \frac{1}{(\lambda_1 + \lambda_2)^2} \tag{10.47}$$

where the prime denotes differentiation with respect to s.

Example 10.3

A vehicle may fail in two mutually exclusive failure modes as shown in Fig. 10.6. Calculate vehicle's steady-state availability using Eq. (10.41) for the following given data: $\lambda_1 = 0.005$ failures/h, $\lambda_2 = 0.002$ failures/h, $\lambda_3 = 0.1$/h, $\lambda_4 = 0.2$/h, and $\mu = 0.01$ repairs/h. Using these data in Eq. (10.41) results in

$$\begin{aligned} AV_{ss} = P_0 &= \left[1 + \frac{\lambda_1}{\lambda_3} + \frac{\lambda_2}{\lambda_4} + \frac{(\lambda_1 + \lambda_2)}{\mu}\right]^{-1} \\ &= \left[1 + \frac{(0.005)}{(0.1)} + \frac{(0.002)}{(0.2)} + \frac{(0.005) + (0.002)}{(0.01)}\right]^{-1} \\ &= 0.5682 \end{aligned}$$

Availability Modeling of a Transit System with Nonexponentially Distributed Repair Times

General Model. This model[4] is basically the same as model I given in the previous section, but with the exception that the repair rate is time dependent (i.e., nonexponentially distributed repair times) instead of being constant (i.e., exponentially distributed repair times) as in the case of model I. The system transition diagram is shown in Fig. 10.8. The numerals in the circles denote the system states.

The following assumptions are associated with this model: 1) the vehicle failure and towing rates are constant, 2) vehicle repair times are distributed arbitrarily, and 3) the repaired vehicle is as good as "new."

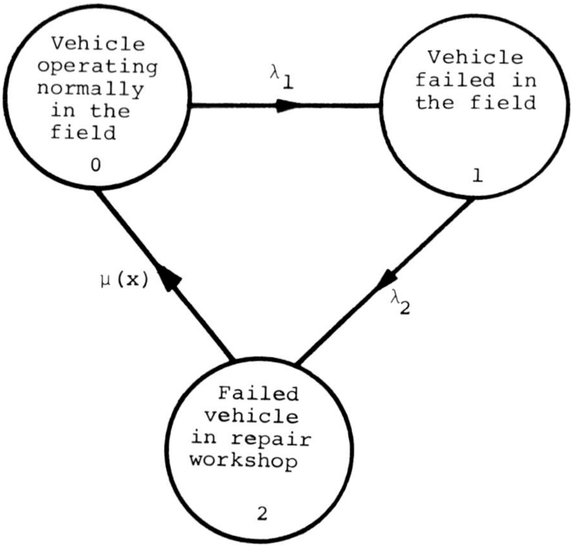

Fig. 10.8 System transition diagram for the general model.

The symbols used to develop this model are

t = time
λ_1 = constant failure rate of the vehicle
λ_2 = constant towing rate of the vehicle
$\mu(x),q(x)$ = repair rate and probability density function of the repair time when the transit system is in state 2 and has an elapsed repair time of x
$P_i(t)$ = probability that the transit system is in state i at time t, for $i = 0,1,2$
$P_2(x,t)$ = probability density (with respect to repair time) that the failed transit system is in state 2 and has an elapsed repair time of x
s = Laplace transform variable
AV(s) = Laplace transform of the availability of the transit system
AV_{ss} = steady-state availability of the transit system

The following integrodifferential equations[5] (and the associated boundary initial condition) are associated with Fig. 10.8:

$$\frac{dP_0(t)}{dt} + \lambda_1 P_0(t) = \int_0^\infty P_2(x,t)\mu(x)\,dx \tag{10.48}$$

$$\frac{dP_1(t)}{dt} + \lambda_2 P_1(t) = \lambda_1 P_0(t) \tag{10.49}$$

$$\left[\frac{\partial}{\partial t} + \frac{\partial}{\partial x} + \mu(x)\right] P_2(x,t) = 0 \tag{10.50}$$

$$P_2(0,t) = \lambda_2 P_1(t) \tag{10.51}$$

At time $t = 0$, $P_0(0) = 1$ and all other initial condition probabilities are equal to zero.

Solving Eqs. (10.48–10.51) with the aid of Laplace transforms yields

$$P_0(s) = (s + \lambda_2)/[(s + \lambda_1)(s + \lambda_2) - \lambda_1\lambda_2 q(s)] \tag{10.52}$$

where

$$q(s) = \int_0^\infty \exp(-sx)\mu(x) \exp\left[-\int_0^x \mu(x)\, dx\right] dx$$

$$P_1(s) = \lambda_1/[(s + \lambda_1)(s + \lambda_2) - \lambda_1\lambda_2 q(s)] \tag{10.53}$$

$$P_2(s) = \lambda_1\lambda_2[1 - q(s)]/s[(s + \lambda_1)(s + \lambda_2) - \lambda_1\lambda_2 q(s)] \tag{10.54}$$

The Laplace transform of the transit system availability is

$$\mathrm{AV}(s) = P_0(s) \tag{10.55}$$

The steady-state availability of the transit system is given by

$$\mathrm{AV}_{ss} = \lim_{s \to 0} s\mathrm{AV}(s) \tag{10.56}$$

The special-case models given below are associated with this model when the repair time distribution is described by

$$q(x) = \frac{\mu(\mu x)^{n-1} \exp(-\mu x)}{(n-1)!} \tag{10.57}$$

where n and μ are Erlangian parameters (n is a positive integer) and $q(x)$ the probability density function.

Special-Case Model A. For $n = 1$ and $q(x) = \mu \exp(-\mu x)$, we get

$$P_0(s) = [s^2 + (\lambda_2 + \mu)s + \lambda_2\mu]/D(s) \tag{10.58}$$

where

$$D(s) = s[s^2 + (\lambda_1 + \lambda_2 + \mu)s + \lambda_2\mu + \lambda_1(\lambda_2 + \mu)]$$

$$P_1(s) = \lambda_1(s + \mu)/D(s) \tag{10.59}$$

$$P_2(s) = \lambda_1\lambda_2/D(s) \tag{10.60}$$

The Laplace transform of the availability of the transit system is

$$\mathrm{AV}(s) = P_0(s) \tag{10.61}$$

Using Eqs. (10.58–10.60), the transit system steady-state probabilities are

$$P_0 = \frac{\lambda_2\mu}{\lambda_2\mu + \lambda_1(\lambda_2 + \mu)} \tag{10.62}$$

$$P_1 = \frac{\lambda_1\mu}{\lambda_2\mu + \lambda_1(\lambda_2 + \mu)} \tag{10.63}$$

$$P_2 = \frac{\lambda_1\lambda_2}{\lambda_2\mu + \lambda_1(\lambda_2 + \mu)} \tag{10.64}$$

Thus, the steady-state availability of the transit system is

$$\mathrm{AV}_{ss} = P_0 = \frac{\lambda_2\mu}{\lambda_1\lambda_2 + \mu(\lambda_1 + \lambda_2)} \tag{10.65}$$

Special-Case Model B. For $n = 2$ and $q(x) = \mu^2 x \exp(-\mu x)$ we get

$$P_0(s) = \frac{\sum_{i=1}^{4} a_i s^{4-i}}{s \sum_{i=1}^{4} b_i s^{4-i}} \tag{10.66}$$

where

$$a_1 = 1, \qquad a_2 = \lambda_2 + 2\mu, \qquad a_3 = \mu(\mu + 2\lambda_2), \qquad a_4 = \mu^2\lambda_2$$

$$b_1 = 1, \qquad b_2 = \lambda_1 + \lambda_2 + 2\mu, \qquad b_3 = \lambda_1\lambda_2 + 2\mu(\lambda_1 + \lambda_2) + \mu^2$$

$$b_4 = 2\mu\lambda_1\lambda_2 + \mu^2(\lambda_1 + \lambda_2)$$

$$P_1(s) = \frac{\lambda_1(s + \mu)^2}{s \sum_{i=1}^{4} b_i s^{4-i}} \tag{10.67}$$

$$P_2(s) = \frac{\lambda_1\lambda_2(s + 2\mu)}{s \sum_{i=1}^{4} b_i s^{4-i}} \tag{10.68}$$

The Laplace transform of the transit system availability is

$$\mathrm{AV}(s) = P_0(s) \tag{10.69}$$

Using Eqs. (10.66–10.68), the transit system steady-state probabilities are

$$P_0 = a_4/b_4 \tag{10.70}$$

$$P_1 = \lambda_1\mu^2/b_4 \tag{10.71}$$

$$P_2 = 2\lambda_1\lambda_2\mu/b_4 \tag{10.72}$$

Thus, the steady-state availability of the transit system is

$$\mathrm{AV}_{ss} = P_0 = \frac{\lambda_2\mu^2}{2\mu\lambda_1\lambda_2 + \mu^2(\lambda_1 + \lambda_2)} \tag{10.73}$$

Special-Case Model C. For $n = 3$ and $q(x) = \mu^3 x^2 \exp(-\mu x)/2!$, we get

$$P_0(s) = \frac{\sum_{i=1}^{5} c_i s^{5-i}}{s \sum_{j=1}^{5} d_j s^{5-j}} \tag{10.74}$$

where

$$c_1 = 1, \quad c_2 = 3\mu + \lambda_2, \quad c_3 = 3\mu(\mu + \lambda_2), \quad c_4 = \mu^2(\mu + 3\lambda_2), \quad c_5 = \lambda_2\mu^3$$

$$d_1 = 1, \quad d_2 = \lambda_1 + \lambda_2 + 3\mu, \quad d_3 = 3\mu(\lambda_1 + \lambda_2 + \mu) + \lambda_1\lambda_2$$

$$d_4 = 3\lambda_1\lambda_2\mu + 3\mu^2(\lambda_1 + \lambda_2) + \mu^3, \quad d_5 = 3\mu^2\lambda_1\lambda_2 + \mu^3(\lambda_1 + \lambda_2)$$

$$P_1(s) = \frac{(\lambda_1 s^3 + 3\lambda_1\mu s^2 + 3\lambda_1\mu^2 s + \lambda_1\mu^3)}{s \sum_{j=1}^{5} d_j s^{5-j}} \tag{10.75}$$

$$P_2(s) = \frac{\lambda_1\lambda_2(s^2 + 3\mu s + 3\mu^2)}{s \sum_{j=1}^{5} d_j s^{5-j}} \tag{10.76}$$

The Laplace transform of the availability of the transit system is given by

$$\mathrm{AV}(s) = P_0(s) \tag{10.77}$$

Using Eqs. (10.74–10.76), the transit system steady-state probabilities are

$$P_0 = c_5/d_5 \tag{10.78}$$

$$P_1 = \lambda_1\mu^3/d_5 \tag{10.79}$$

$$P_2 = 3\lambda_1\lambda_2\mu^2/d_5 \tag{10.80}$$

The transit system steady-state availability is given by

$$AV_{ss} = P_0 = \frac{\lambda_2 \mu_3}{3\mu^2 \lambda_1 \lambda_2 + \mu^3(\lambda_1 + \lambda_2)} \tag{10.81}$$

Special-Case Model D. For $n = 4$ and $q(x) = \mu^4 x^3 \exp(-\mu x)/3!$, we get

$$P_0(s) = \frac{\sum_{i=1}^{6} e_i s^{6-i}}{s \sum_{j=1}^{6} f_j s^{6-j}} \tag{10.82}$$

where

$$e_1 = 1, \quad e_2 = 4\mu + \lambda_2, \quad e_3 = \mu(6\mu + 4\lambda_2), \quad e_4 = \mu^2(4\mu + 6\lambda_2)$$

$$e_5 = \mu^3(\mu + 4\lambda_2), \quad e_6 = \lambda_2 \mu^4, \quad f_1 = 1, \quad f_2 = (\lambda_1 + \lambda_2 + 4\mu)$$

$$f_3 = (\lambda_1 \lambda_2 + 4\mu(\lambda_1 + \lambda_2) + 6\mu^2)$$

$$f_4 = (4\mu^3 + 6\mu^2(\lambda_1 + \lambda_2) + 4\mu\lambda_1\lambda_2)$$

$$f_5 = (\mu^4 + 4\mu^3(\lambda_1 + \lambda_2) + 6\mu^2\lambda_1\lambda_2)$$

$$f_6 = (\mu^4(\lambda_1 + \lambda_2) + 4\mu^3\lambda_1\lambda_2)$$

$$P_1(s) = \frac{\lambda_1(s^4 + 4\mu s^3 + 6\mu^2 s^2 + 4\mu^3 s + \mu^4)}{s \sum_{j=1}^{6} f_j s^{6-j}} \tag{10.83}$$

$$P_2(s) = \frac{\lambda_1 \lambda_2(s^3 + 4\mu s^2 + 6\mu^2 s + 4\mu^3)}{s \sum_{j=1}^{6} f_j s^{6-j}} \tag{10.84}$$

The Laplace transform of the availability of the transit system is

$$AV(s) = P_0(s) \tag{10.85}$$

The transit system steady-state probabilities are obtained using Eqs. (10.82–10.84) and are

$$P_0 = e_6/f_6 \tag{10.86}$$

$$P_1 = \lambda_1 \mu^4/f_6 \tag{10.87}$$

$$P_2 = 4\mu^3 \lambda_1 \lambda_2/f_6 \tag{10.88}$$

The steady-state availability of the transit system is

$$AV_{ss} = P_0 = \frac{\mu^4 \lambda_2}{4\mu^3 \lambda_1 \lambda_2 + \mu^4(\lambda_1 + \lambda_2)} \tag{10.89}$$

For all of these special-case models, the time-dependent state probabilities of the transit system can be obtained by inverting the Laplace transforms of the state probabilities. Using Eqs. (10.57), (10.65), (10.73), (10.81), and (10.89), the following general expression for the transit system steady-state availability is established:

$$AV_{ss} = \frac{\lambda_2 \mu^n}{n\mu^{n-1}\lambda_1 \lambda_2 + \mu^n(\lambda_1 + \lambda_2)} \tag{10.90}$$

The plots of the above equation are shown in Fig. 10.9 for different values of n.

The plots in Fig. 10.9 show the impact of varying vehicle towing rate λ_2 on the steady-state availability of the transit system for various specific values of n.

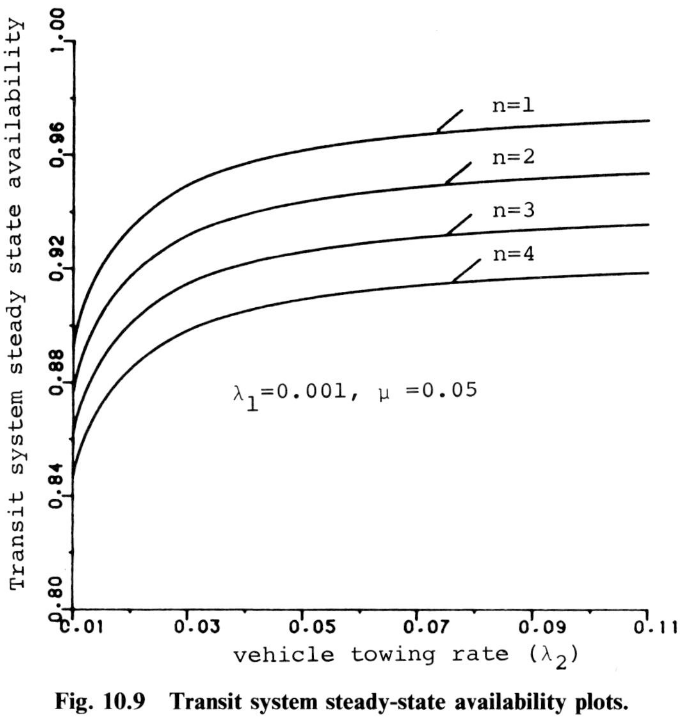

Fig. 10.9 Transit system steady-state availability plots.

10.4 RELIABILITY AND AVAILABILITY ANALYSIS OF VEHICLES IN CHANGING ENVIRONMENTS

This section presents two Markov models[6] concerned with a vehicle operating in alternating environments (i.e., normal and abnormal).

Model A

This Markov model represents a vehicle operating in alternating environments (i.e., normal and abnormal). In either situation, the vehicle can fail due to an accident or a hardware failure. The failed vehicle is repaired back to its normal operation mode. The system state space diagram is shown in Fig. 10.10. The numerals and symbols used in the diagram are defined below.

The symbols associated with the model are as follows. i is the ith state of the vehicle: $i = 0$ (vehicle operating in normal environment), $i = 1$ (vehicle operating in abnormal environment), $i = 2$ (vehicle failed due to an accident), $i = 3$ (vehicle failed due to a hardware failure). $P_i(t)$ is the probability that vehicle is in state i at time t, for $i = 0,1,2,3$. λ_i is the ith constant rate of the vehicle: $i = 1$ (state 1 to state 2), $i = 2$ (state 0 to state 3), $i = 3$ (state 0 to state 2), $i = 4$ (state 1 to state 3). α is the environment changeover constant rate from state 1 to state 0. β is the environment changeover constant rate from state 0 to state 1. μ_i is the ith constant repair rate of the vehicle: $i = 1$ (state 3 to state 0), $i = 2$ (state 2 to state 0), $i = 3$ (state 2 to state 1), $i = 4$ (state 3 to state 1). s is the Laplace transform variable.

The system of differential equations associated with Figure 10.10 is

$$\frac{dP_0(t)}{dt} + (\beta + \lambda_2 + \lambda_3)P_0(t) = \alpha P_1(t) + P_2(t) \cdot \mu_2 + P_3(t) \cdot \mu_1 \qquad (10.91)$$

$$\frac{dP_1(t)}{dt} + (\alpha + \lambda_1 + \lambda_4)P_1(t) = P_0(t) \cdot \beta + \mu_3 \cdot P_2(t) + \mu_4 P_3(t) \qquad (10.92)$$

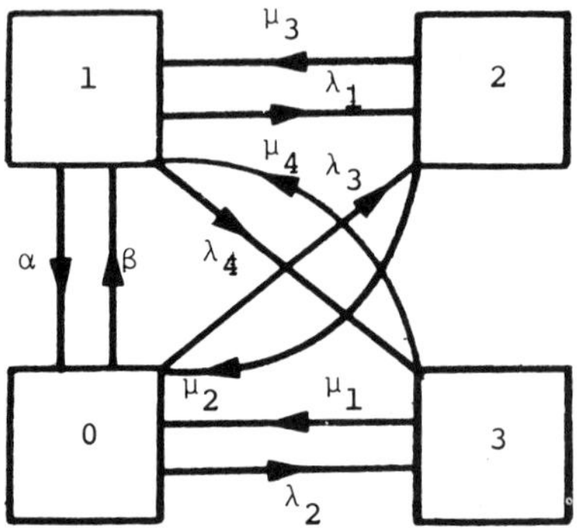

Fig. 10.10 State space diagram for model A.

$$\frac{dP_2(t)}{dt} + (\mu_3 + \mu_2)P_2(t) = P_1(t) \cdot \lambda_1 + P_0(t) \cdot \lambda_3 \tag{10.93}$$

$$\frac{dP_3(t)}{dt} + (\mu_1 + \mu_4)P_3(t) = P_0(t) \cdot \lambda_2 + P_1(t) \cdot \lambda_4 \tag{10.94}$$

At time $t = 0$, $P_0(0) = 1$ and $P_1(0) = P_2(0) = P_3(0) = 0$.

Setting the derivatives of Eqs. (10.91–10.94) equal to zero, $\mu_3 = \mu_4 = 0$ and using the relationship $\Sigma_{i=0}^{3} P_i = 1$ leads to the following steady-state probability solutions:

$$P_0 = \frac{\mu_1\mu_2 \cdot A_1}{A_1\mu_1\mu_2 + \beta\mu_1\mu_2 + \lambda_1\beta \cdot \mu_1 + \lambda_3\mu_1 \cdot A_1 + \lambda_2 \cdot \mu_2 \cdot A_1 + \lambda_4 \cdot \beta\mu_2} \tag{10.95}$$

$$A_1 \equiv \alpha + \lambda_1 + \lambda_4 \tag{10.96}$$

$$P_1 = P_0 \cdot \beta / A_1 \tag{10.97}$$

$$P_2 = P_0 \left(\frac{\lambda_1 \cdot \beta}{\mu_2 \cdot A_1} + \frac{\lambda_3}{\mu_2} \right) \tag{10.98}$$

$$P_3 = P_0 \left(\frac{\lambda_2}{\mu_1} + \frac{\lambda_4 \cdot \beta}{A_1 \cdot \mu_1} \right) \tag{10.99}$$

Setting $\mu_1 = \mu_2 = 0$ in Eqs. (10.91–10.94), taking the Laplace transforms, and solving for $P_0(s)$ and $P_1(s)$ leads to

$$\begin{aligned} \text{MTTF}_v &= \lim_{s \to 0} R(s) = \lim_{s \to 0} [P_0(s) + P_1(s)] \\ &= \left(\frac{B}{A_1} + 1 \right) A_1 \cdot [(\beta + \lambda_2 + \lambda_3) \cdot A_1 \cdot - \alpha\beta]^{-1} \end{aligned} \tag{10.100}$$

where $R(s)$ is the Laplace transform of the vehicle reliability function, MTTF_v the vehicle mean time to failure, $P_0(s)$ the Laplace transform of the probability that the vehicle is in state 0, and $P_1(s)$ the Laplace transform of the probability that the vehicle is in state 1.

Model B

This Markov model also represents a vehicle operating in an alternating environment. The vehicle can fail operating in either a normal or abnormal environment. The failed vehicle is repaired back to both its operating states. Essentially, this model is the special case of the earlier model A. The state space diagram of the Markov model is shown in Fig. 10.11. The numerals and symbols used in the diagram are defined below.

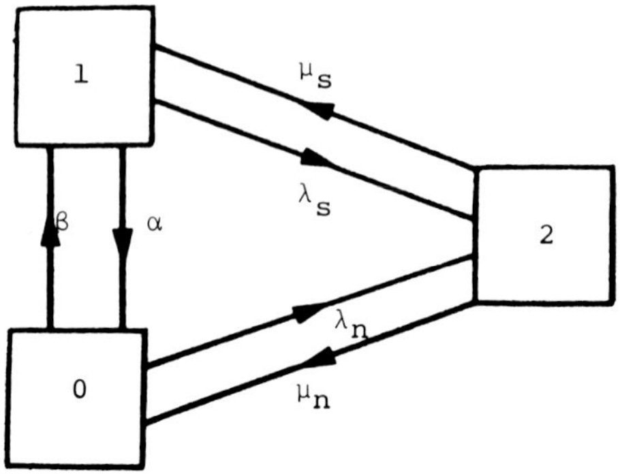

Fig. 10.11 State space diagram for model B.

The notations used to develop equations for the model are as follows. i is the ith state of the vehicle: $i = 0$ (vehicle operating in normal environment), $i = 1$ (vehicle operating in abnormal environment), $i = 2$ (vehicle failed). $P_i(t)$ is the probability that the vehicle is in state i at time t, for $i = 0,1,2$. α is the environment changeover constant rate from state 1 to state 0. β is the environment changeover constant rate from state 0 to state 1. λ_s is the constant failure rate of the vehicle from the abnormal environment state. λ is the constant failure rate of the vehicle from the normal environment state. μ_s is the constant repair rate of the vehicle from state 2 to state 1. μ_n is the constant repair rate of the vehicle from state 2 to state 0. s is the Laplace transform variable.

The following differential equations are associated with Fig. 10.11:

$$\frac{dP_0(t)}{dt} + (\beta + \lambda_n)P_0(t) = P_1(t) \cdot \alpha + P_2(t) \cdot \mu_n \tag{10.101}$$

$$\frac{dP_1(t)}{dt} + (\alpha + \lambda_s)P_1(t) = P_0(t) \cdot \beta + P_2(t) \cdot \mu_s \tag{10.102}$$

$$\frac{dP_2(t)}{dt} + (\mu_n + \mu_s)P_2(t) = P_0(t) \cdot \lambda_n + P_1(t) \cdot \lambda_s \tag{10.103}$$

At time $t = 0$, $P_0(0) = 1$ and $P_1(0) = P_2(0) = 0$.

The following steady-state probability equations were obtained by setting the derivatives of Eqs. (10.101–10.103) equal to zero and utilizing the relationship $P_0 + P_1 + P_2 = 1$:

$$P_0 = A_1/c_1c_2 \tag{10.104}$$

where

$$A_1 \equiv \mu_n \cdot \alpha + \lambda_s \cdot \mu_n + \alpha\mu_s \tag{10.105}$$

$$c_1, c_2 = [-D \pm \sqrt{D^2 - 4(A_1 + A_2 + A_3)}]/2 \tag{10.106}$$

$$A_3 \equiv \beta \cdot \mu_n + \beta \cdot \mu_s + \lambda_n \cdot \mu_s \tag{10.107}$$

$$A_2 \equiv \beta \cdot \lambda_s + \alpha \cdot \mu_n + \lambda_n \cdot \lambda_s \tag{10.108}$$

$$D \equiv \alpha + \mu_n + \mu_s + \beta + \lambda_n + \lambda_s \tag{10.109}$$

$$P_1 = A_3/c_1c_2 \tag{10.110}$$

$$P_2 = A_2/c_1c_2 \tag{10.111}$$

The vehicle steady-state availability A_{ss} is given by

$$A_{ss} = P_0 + P_1 \tag{10.112}$$

Setting $\mu_s = \mu_n = 0$ in Eqs. (10.101–10.103), taking the Laplace transforms, and solving for $P_0(s)$ and $P_1(s)$ yields

$$\begin{aligned} \text{MTTF}_v &= \lim_{s \to 0} R(s) = \lim_{s \to 0} [P_0(s) + P_1(s)] \\ &= \frac{\lambda_s + \beta + \alpha}{(\lambda_n + \beta)(\lambda_s + \alpha) - \beta\alpha} \end{aligned} \tag{10.113}$$

where MTTF_v is the vehicle mean time to failure, $R(s)$ the Laplace transform of the vehicle reliability function, $P_0(s)$ the Laplace transform of the probability that the vehicle is in state 0, and $P_1(s)$ the Laplace transform of the probability that the vehicle is in state 1.

Example 10.4

Assume that we have the following data for λ_s, β, α, and λ_n: $\lambda_s = 2$ failures/month, $\lambda_n = 1$ failure/month, $\beta = 4$ times/month, and $\alpha = 1$ time/month. Calculate the mean time to failure of the vehicle using Eq. (10.113).

Substituting the given data into Eq. (10.113) yields

$$\begin{aligned} \text{MTTF}_v &= \frac{2 + 4 + 1}{(1 + 4)(2 + 1) - (4) \cdot (1)} \\ &= 0.6364 \text{ month} \end{aligned}$$

Thus, the mean time to failure of the vehicle is 0.6364 month.

References

[1]Dhillon, B. S., "Bibliography of Literature on Transit System Reliability," *Microelectronics and Reliability*, Vol. 22, 1982, pp. 641–651.

[2]Edward-Anderson, J., *Transit Systems Theory*, Lexington Books, D. C. Heath, Lexington, MA, 1978, pp. 204–205.

[3]Dhillon, B. S. and Rayapati, S. N., "Reliability and Availability Analysis of Transit Systems," *Microelectronics and Reliability*, Vol. 25, 1985, pp. 1073–1085.

[4]Dhillon, B. S. and Rayapati, S. N., "Transit System Availability Modelling: A Study of the Erlangian Repair-Time Distribution," *International Journal of Systems Science*, Vol. 18, 1987, pp. 9–24.

[5]Dhillon, B. S., *Reliability Engineering in Systems Design and Operation*, Van Nostrand Reinhold, New York, 1983, pp. 50–55.

[6]Dhillon, B. S., "RAM Analysis of the Vehicles in Changing Weather," *Proceedings of Annual Reliability and Maintainability Symposium*, IEEE, New York, 1984, pp. 48–53.

Exercises

10.1 Define the following two terms: service availability and service unavailability.

10.2 Prove that the mean time to failure ($MTTF_{TS}$) of a transit system is given by

$$MTTF_{TS} = \lim_{s \to 0} R(s) \tag{10.114}$$

where s is the Laplace transform variable and $R(s)$ the Laplace transform of the transit system reliability function.

10.3 Assume that the failure times of a transit system are exponentially distributed with a mean value of 500 h. Calculate the failure rate of the transit system.

10.4 Using Eq. (10.29), obtain an expression for the transit system hazard rate.

10.5 Using the given data of Example 10.3 in Eqs. (10.37–10.40) prove that their sum $(P_0 + P_1 + P_2 + P_3)$ is equal to unity.

10.6 Prove, with the aid of Eq. (10.29), that the transit system mean time to failure (MTTF) is given by

$$\text{MTTF} = \int_0^\infty R(t)\,\mathrm{d}t = \frac{\lambda_1 + \lambda_2 + \lambda_4}{(\lambda_1 + \lambda_3)(\lambda_2 + \lambda_4)} \tag{10.115}$$

where t is the time and $R(t)$ the reliability function of the transit system.

11
AUTOMOTIVE VEHICLE RELIABILITY AND ROBOT SAFETY AND RELIABILITY

This chapter presents various different aspects of automotive vehicle reliability and robot safety and reliability. The chapter is divided into two major parts, i.e., automotive vehicle reliability and robot reliability and safety. The introductory aspects of both these areas are discussed. Subsystems of a motor vehicle are listed, along with areas needing attention by design engineers. The Ford and Volvo reliability assurance programs are described. A number of reliability measures concerned with a motor vehicle are presented, including reliability, mean mileage between failures, mean mileage to first failure, mean repair time, and availability. The fault tree concept is discussed along with the brake, rim, and steering system defects. The history of robots is presented and the classifications of industrial robots are outlined. Several different aspects of robot safety are discussed, including the causes for near-accidents, major sources of hazard, safety measures, and guidelines for safety precautions. Robot reliability is another area discussed in this chapter. Robot problems along with methods for robot reliability analysis are discussed. Measures such as mean time to robot failure, mean time to robot problem, and reliability are presented.

11.1 INTRODUCTION

The importance of reliability in the automobile industry has been increasingly realized in recent years. The result of a survey[1] conducted some years ago showed that the reliability of commercial vehicles was the most important factor affecting customer satisfaction—over any other measure of performance, e.g., fuel consumption, maximum speed, or power. Therefore, the reliability factor cannot be ignored simply because automobile producers are operating in competitive environments for a share of the world market. For example, in 1979, 22% of the United States automobile market was captured by the imports from other countries.[2] More specifically, the Japanese giant, Toyota Motor Company Ltd., captured 7.1% of the market in the United States in that year. In fact, it was behind only the General Motors Corporation and Ford Motor Company with respect to automobile market share in the United States. Obviously, these facts indicate even stiffer competition for American automobile manufacturers in the future. In turn, greater emphasis on the reliability of automobile vehicles, as competitive marketplace is one of the environmental parameters influencing product reliability. Other environmental parameters influencing product reliability are operational requirements, legislation, status of the technology, maintainability, and available resources. In subsequent sections, this chapter presents various aspects of reliability assurance in automobile vehicles.

Robots are increasingly being used in industry to perform a variety of tasks. In comparison to other automated machinery, it can possibly be stated that industrial robots are the most flexible in operation and use.[3] Some of the tasks performed by robots in industry are spot welding, routing, arc welding, adhesive bonding, and materials handling. The use of robots is growing rapidly in several parts of the world. According to a report published by the British Robot Association in 1984,[4,5] the distribution of the robot population was 13,000 in the United States, 64,600 in Japan, and 20,500 in Europe.

As with automobile vehicles and other engineering products, the reliability and safety of robots is important. The subsequent sections of this chapter discuss various aspects of robot reliability and safety. The chapter is divided into two main parts, i.e., automotive vehicle reliability and robot safety and reliability.

11.2 AUTOMOTIVE VEHICLE RELIABILITY

Vehicle Subsystems and Guidelines for Design Engineers

A motor vehicle is composed of many parts and subsystems. For the reliability analysis purposes, the vehicle can be broken down into several functional subsystems such as[6] engine and transmission; clutch, brakes, and steering; body; electrical power supply and radio; exhaust; frame and mounting; climate control; suspension; fuel; driveline; and gage and warning devices.

The design engineers have to consider many things in the design of a motor vehicle and its parts. The use of guidelines and standards helps to produce a safe, efficient, and economical product. Thus, during the motor vehicle design the design engineer must pay close attention to the five areas (or guidelines) given in Fig. 11.1.[7] Close attention to these areas will help to result in an effective end product.

Reliability Assurance Programs

The manufacturers of automotive vehicles have developed various types of programs to assure reliability of the products they manufacture. This section briefly describes two of these programs, as practiced by Ford Motor Company and Volvo Car Corporation.

Ford Reliability Assurance Program. This program is largely engineering oriented and considers reliability, just like cost and weight, an important parameter of design.[6,8,9] The program is composed of five major steps as

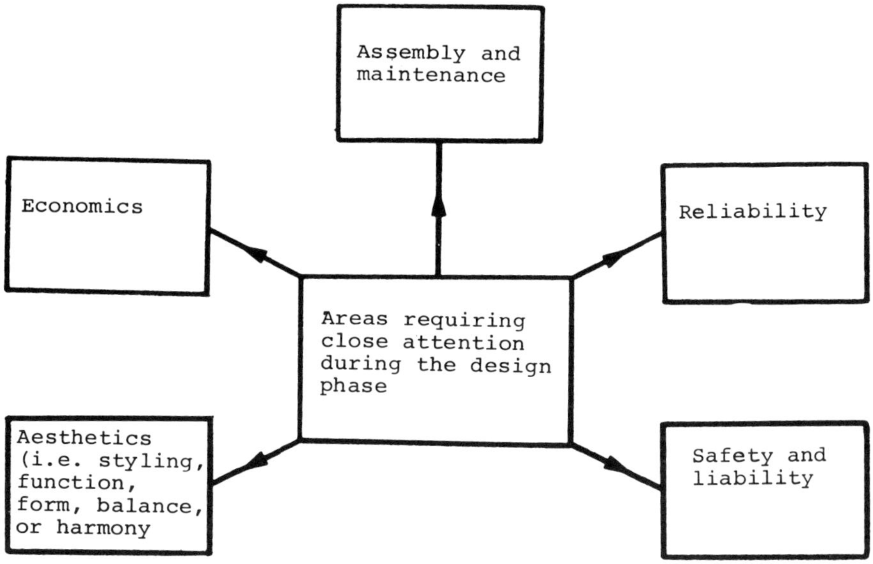

Fig. 11.1 Areas needing close attention during the design of a motor vehicle.

shown in Fig. 11.2. The first step, i.e., define reliability, is concerned with outlining reliability requirements for each vehicle system and subsystem. The objective of step 2 is to design systems, subsystems, and parts so that they meet the specified reliability requirements. In the third step, sufficient testing is conducted to verify the design adequacy. Step 4, i.e., specify parts requirements, deals with clearly explaining parts requirements to inhouse manufacturing and outside suppliers or subcontractors. In the final step, information from field sources is monitored and a quick response to any problem is provided. In addition, feedback of the relevant information is provided to make changes in future designs and processes.

Volvo Reliability Assurance Program. The reliability assurance program of the Volvo Car Division is concerned with three major areas of activity[10]: a reliability field follow-up program, the setting of reliability goals, and analysis and verification of reliability in the preproduction phase. The reliability field follow-up program is basically concerned with evaluations of reliability field data on a continuous basis. This is carried out with the aid of a computerized evaluation and data bank system. More information on the field follow-up program is given in Ref. 10. The second major area of activity is the setting of reliability goals. Of neccessity, this aspect receives much attention in the car division. The reliability goals are set for the system, subsystem, unit, and component levels. The third and the last major area of activity starts after the establishment of the reliability goals. This is concerned

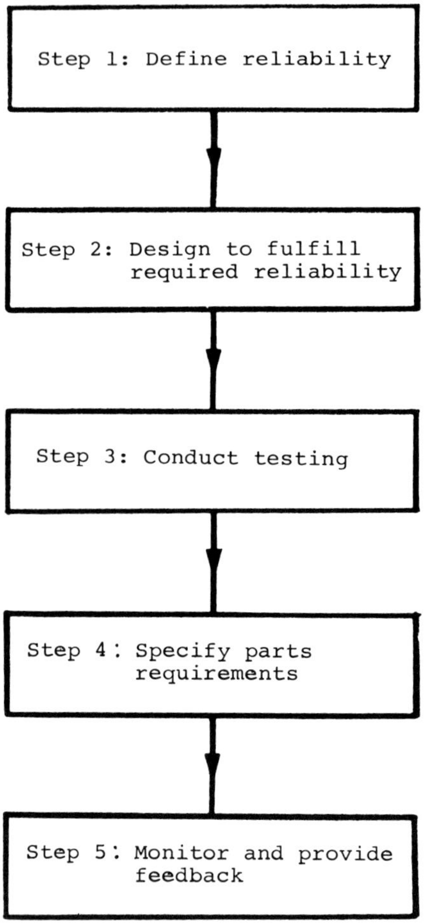

Fig. 11.2 Steps of the Ford reliability assurance program.

analysis and verification of reliability during the preproduction phase. Several activities take place during this phase, as follows[10]:

1) Predicting reliability during the preliminary design.

2) Performing failure mode and effect analysis when new designs are involved.

3) Reliability test planning.

4) Predicting reliability during the final design phase. This prediction is based on such information as the results of component testing, design changes since the last analysis, new modes of failure, previous analysis, and field follow-up.

5) Predicting reliability during the tooling phase. This reliability prediction is based on prototype cars, tryout cars, etc.

6) Predicting reliability before starting production. This is the final reliability prediction and is based on items such as field follow-up, tryout car experience, and earlier analysis.

Reliability Analysis of Motor Vehicles

This section presents various aspects of motor vehicle reliability analysis.

Vehicle Mean Mileage between Failures. This measure is frequently used in reliability analysis of vehicles and is defined as[11]

$$\text{MMBF}_v = \frac{N_v \cdot m}{F_{cn}} \tag{11.1}$$

where MMBF_v is the vehicle mean mileage between failures, F_{cn} the cumulative number of vehicle failures, m the total mileage at the termination of the test, and N_v the total number of vehicles tested.

Vehicle Reliability. For constant failure rate (i.e., $\lambda_v = 1/\text{MMBF}_v$), the vehicle reliability is given by

$$R_v(m) = e^{-\lambda_v m} = e^{-(1/\text{MMBF}_v)m} \tag{11.2}$$

where m is the distance, λ_v the constant failure rate of the vehicle, MMBF_v the vehicle mean mileage between failures, and $R_v(m)$ the vehicle reliability.

Example 11.1

The constant failure rate of a motor vehicle is 0.00006 failures/km. Calculate the vehicle reliability for a 500 km mission distance. In this case, the defined values are $\lambda_v = 0.00006$ failures/km and $m = 500$ km. By substituting these data into Eq. (11.2), we get

$$R_v(500) = e^{-(0.00006)(500)} = 0.9705$$

Thus, the reliability of the motor vehicle is 0.9705.

Vehicle Mean Mileage to First Failure. This is another important parameter that can be used in vehicle reliability analysis. Its value indicates the continuous running distance of the vehicle without repair. It is defined by

$$\text{MMFF}_v = \sum_{j=1}^{k} m_j + (N_v - k)m \tag{11.3}$$

where MMFF_v is the vehicle mean mileage to first failure, N_v the total number of vehicles tested, m_j the mileage of vehicle j to first failure, m the mileage at the termination of the test, and k the accumulated number of vehicles failed by the termination of the test.

Mean Repair Time. For the purpose of vehicle reliability analysis, this is expressed as

$$T_{mr} = \frac{T_r + T_{pr}}{N_v \cdot m \cdot L_p} \tag{11.4}$$

where T_{mr} is the mean repair time (h/ton · m), N_v the number of vehicles tested, L_p the rated payload (in tons), T_{pr} the preventive repair time (in hours), T_r the repair time (in hours) after failure, m the mileage (in miles) at the end of the test.

Vehicle Availability. This is an important parameter that takes into consideration the reliability and repairability of the vehicle. The economic benefits of transport enterprises are directly affected by this parameter. The availability of a vehicle is defined as

$$A_v = \frac{T_{a0}}{T_{a0} + T_{mv}} \tag{11.5}$$

where A_v is the vehicle availability, T_{a0} the actual operating time of the vehicle, and T_{mv} the vehicle maintenance time (vehicle not in operation).

Vehicle Average Economic Loss. The cost factor plays an important role in reliability decisions concerning motor vehicles. The average economic loss expressed in dollars/ton · m is a useful measure. This is defined as[11]

$$L_{ac} = (k_m + P_\ell + C_e)/(N_v \cdot m \cdot L_p) \tag{11.6}$$

where L_{ac} is the average economic loss (in \$/ton · m), C_e the cost of loss in environment damage expressed (in \$), P_ℓ the profit loss (in \$), k_m the cost of maintenance (in \$). The symbols N_v, m, and L_p are already defined in Eq. (11.4).

In Eq. (11.6), it is assumed that the total economic loss due to poor vehicle reliability is comprised of three cost components: profit loss, maintenance cost, and cost of loss in environment damage.

Fault Tree Approach Application to Motor Vehicles. The fault tree concept is being widely used to analyze various kinds of reliability-related problems in industry. This concept was developed in the early 1960's to analyze the Minuteman launch control system at Bell Laboratories. The concept is described in detail in Chapter 3. Some of the fault tree symbols are:

OR gate. This denotes that, for the occurrence of an output event, one or more of the input events must occur. The symbol for this gate is shown in Fig. 11.3.

Fig. 11.3 Symbol for *n* input OR (logic) gate.

Fig. 11.4 Symbol for *m* input AND (logic) gate.

AND gate. This denotes that, for the occurrence of an output event, all of the input events must occur. The symbol for this gate is shown in Fig. 11.4.

Circle. This symbol denotes a basic fault event that requires no further development. More specifically, there is sufficient information known for the event in question.

Diamond. This symbol denotes a fault event known as an incomplete event. In this case, the fault event causes have not been fully developed due to factors such as lack of interest or information. However, it could be developed further.

Rectangle. This symbol represents a fault event resulting from the combination of fault events through the input of a logic gate. Examples of logic gates are AND gate and OR gate.

Four of the above five symbols were used to develop a fault tree[12] for a selected event, i.e., a running motor vehicle suddenly leaving the highway. This "undesired event" is the top event of the fault tree shown in Fig. 11.5. In this figure, each event is labeled by the symbol E_i for $i = 1,2,3,4,...,11$. For example, E_8 denotes the fault event "driver failed to act." The events E_7 (vehicle has a mechanical problem) and E_9 (environmental problem) can be developed further.

Example 11.2

In Fig. 11.5, assume that the fault events E_1, E_2, E_3, E_4, E_5, E_6, E_7, and E_9 occurrence probabilities are $P(E_1) = P(E_2) = P(E_3) = P(E_4) = P(E_5) = P(E_6) = 0.001$, $P(E_7) = 0.02$, and $P(E_9) = 0.01$, respectively. Figure 11.6

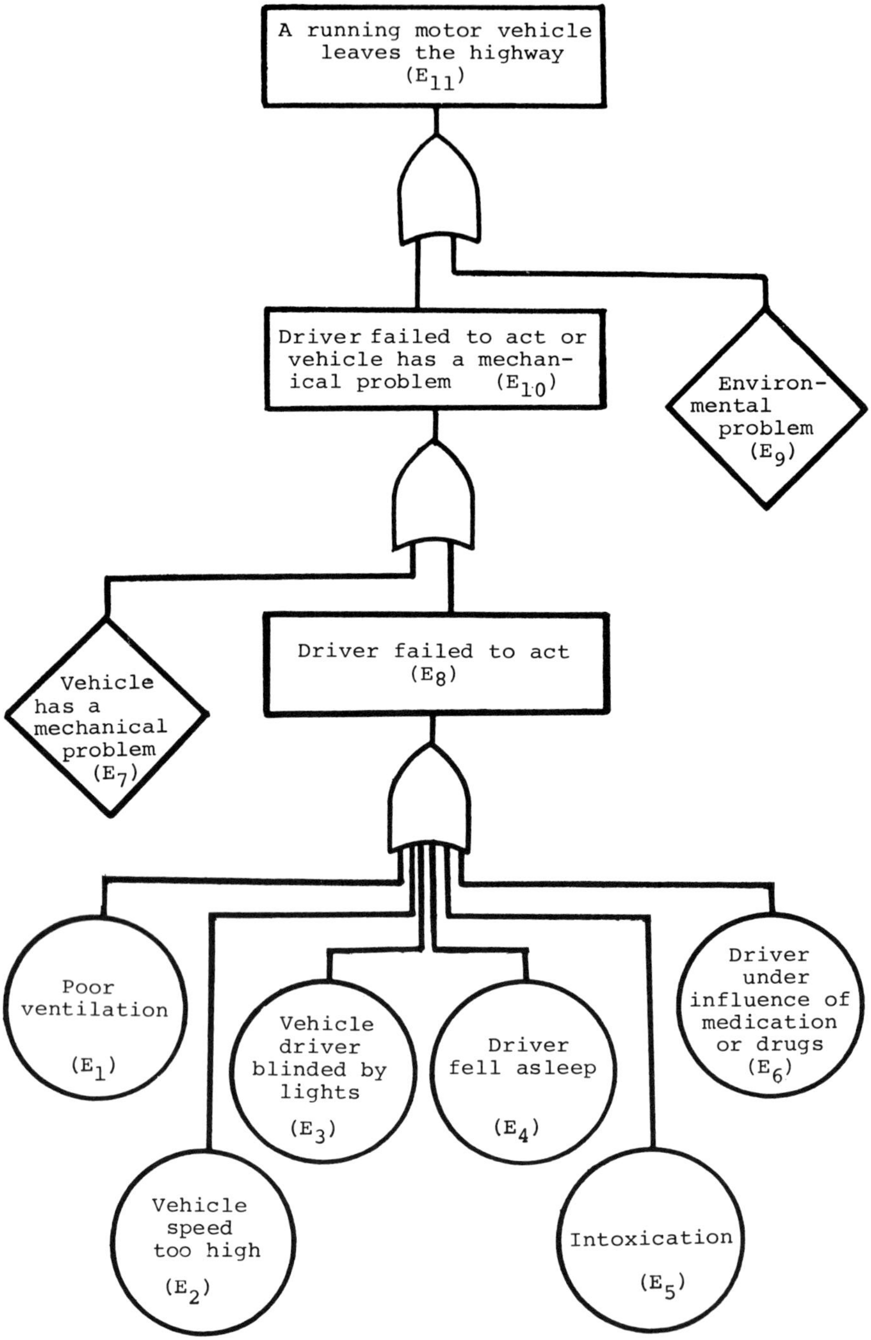

Fig. 11.5 Simplified fault tree for a running vehicle suddenly leaving the highway.

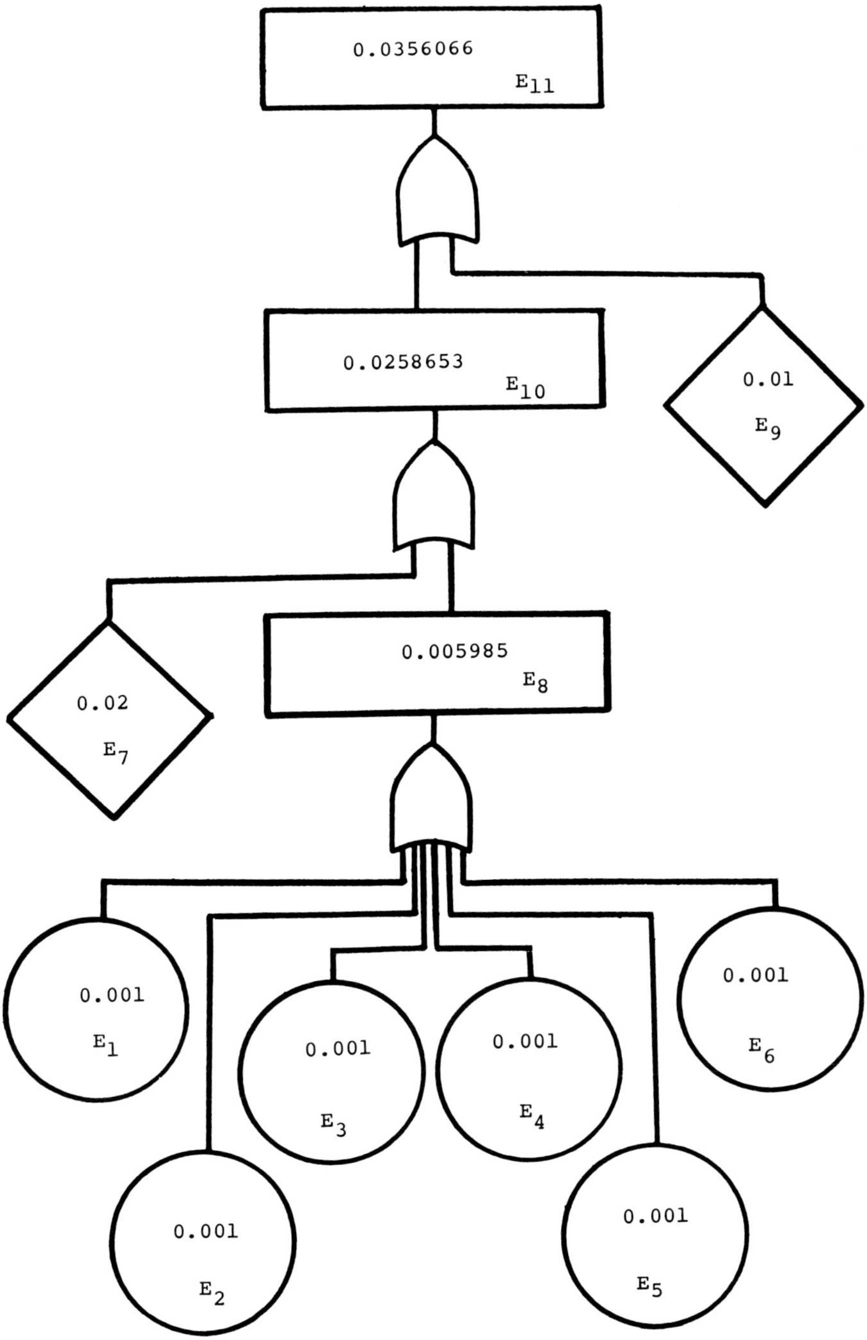

Fig. 11.6 Fault tree with specified probability of occurrence data for basic and incomplete fault events.

shows the same fault tree as that of Fig. 11.5, with assumed probability values for the basic and incomplete fault events. Calculate the occurrence probability of the top fault event (E_{11}) "a running motor vehicle leaves the highway."

The occurrence probability of event E_8 (i.e., the driver failed to act) is given by

$$\begin{aligned} P(E_8) &= 1 - [1 - P(E_1)][1 - P(E_2)][1 - P(E_3)] \\ &\quad \times [1 - P(E_4)][1 - P(E_5)][1 - P(E_6)] \\ &= 1 - (0.001)^6 = 0.005985 \end{aligned} \tag{11.7}$$

Thus, in Fig. 11.6, the occurrence probability of event E_8 is 0.005985. Similarly, the occurrence probability of event E_{10} is

$$\begin{aligned} P(E_{10}) &= 1 - [1 - P(E_7)][1 - P(E_8)] \\ &= 1 - (0.98)(0.994015) \\ &= 0.0258653 \end{aligned}$$

Thus, in Fig. 11.6 the occurrence probability of event E_{10} is 0.0258653.

Finally, in similar fashion, we calculate the occurrence probability of the top event E_{11} as

$$\begin{aligned} P(E_{11}) &= 1 - [1 - P(E_{10})][1 - P(E_9)] \\ &= 1 - (0.9741347)(0.99) \\ &= 0.0356066 \end{aligned}$$

Thus, the probability that a running motor vehicle leaves the highway is 0.0356066. In Fig. 11.6, the calculated probability values are shown in rectangles.

Defects in Vehicle Parts

This section briefly discusses defects in selective automobile parts, such as the brakes, rim, and steering system.[7]

Brake Defects. Under ordinary driving conditions, the parts failure in the vehicle braking system is likely to occur only if the parts become defective, degraded, or severely worn. There are various causes of brake failure, which can be grouped into three distinct categories: excessive part wear, driver inconvenience, and inadequate brake force. For the pupose of discussion, we have grouped the defects into four systems: disk brake, drum brake, common disk and drum brakes, and air brake.

There are three types of defects associated with the disk brake system: low or no brake force, excessive wear of the pad, and excessive brake pedal travel. Typical defects associated with the drum brake system are low braking performance and hard pedal, brake imbalance, brake jams, brake pedal touching floor, increasing heat in the brakes while driving the vehicle, noise generation during braking, brake seizure without action by the vehicle operator, etc. Defects such as brake pedal vibrations, soft pedal, brake fade, and excessive pedal force are associated with brake failures common to disk and drum brake systems. Finally, the defects associated with brake failures of air brake systems are slow brake response or release, no or low brake force, slow pressure buildup in reservoir, etc.

When portions of the primary or service brake system fail, the failure is known as the partial brake failure, some examples of which are brake-line pressure failure in dual-circuit braking systems, ineffectiveness because of brake fade, the power-boost element without a brake force assist, and brake imbalance.

The causes such as broken brake lines, leakage in the wheel cylinder, and leaking master cylinders are associated with hydraulic brake failure. The frequency of leakage in wheel cylinders is much higher in comparison to the occurrence of broken brake lines and leaking master cylinders. Brake fade results from a decrease in the brake factor. The reasons for brake factor decrease are surface contamination, temperature, and rubbing speed. Failures in the vacuum assists result from such causes as defective check valves and master-cylinder attachment bolt welds.

Rim and Steering System Defects. Rim defects are as important as defects in any other important part of a motor vehicle. A study concerned with truck accidents reported that one in approximately 1300–2200 tire failures end up in an accident.[7] Furthermore, the findings of U.S. Bureau of Motor Carrier Safety indicate that 7–13% of all tractors and trailers inspected had a minimum of one defective tire. Rim defects may result due to abusive operation, poor design, or faulty manufacturing operations.

According to Ref. 7, most of the rim defects are due to fatigue fractures at high-stress locations. Normally, these locations are found close to sharp changes in the rim contour.

Steering system defects are also important because they may lead to severe accidents. Such defects can be caused by one or more of the following factors: poor maintenance, faulty changes made to the steering system, faulty manufacturing, faulty design, or inadequate inspection.

11.3 ROBOT SAFETY AND RELIABILITY

Brief History of Robots

The history of robots goes back to the 18th century, when Swiss tradesmen developed human-like automatons. These automatons were able to play musical instruments, write, and draw.[13] In 1818, Mary Shelley, the mistress (and later wife) of the poet Percy Bysshe Shelley published her novel *Frankenstein*. This was one of the earliest works on the "machine monster" and probably the first true science-fiction novel. The word "robot" was coined by Karel Capek, a Czechoslovakian science-fiction writer. He wrote his play entitled *Rossums Universal Robots* (*RUR*) in 1920. In the Czechoslovak language, the word "rozum" (rossum) means "reason" and the word "robot" means "worker."

In 1939, Isaac Asimov (at age 19) wrote a series of stories on robots. In 1950, nine of his stories were published in a book entitled *I, Robot*. Asimov's "three laws of robotics" were devised in 1942 as[14,15] 1) a robot must not injure a person nor, through inaction, allow a person to come to harm; 2) a robot must always obey orders from humans except in situations in which such orders are in conflict with the above (first) law; and 3) a robot must protect its own existence except in situations when it is in conflict with the above two (first and second) laws.

A significant development in the history of robots took place in 1954 when George C. Devol applied for a patent for a programmable transfer device. This was the first patent for an industrial robot.

In 1960, under the leadership of Joseph J. Engelberger the "father of industrial robotics", the Unimation, Inc., robot first saw the light of day. This robot was based on Devol's patent issued in 1961 as U.S. Patent 2,988,237.

Today the leading user of robots, Japan imported its first robot (Versatran) in 1967.[16] Three years later (1970), the first symposium on industrial robots was held in Chicago. The Japan Industrial Robot Association (JIRA) was established in 1971. It was the first robot association in the world. In the United States, the Robot Institute of America (RIA) was formed in 1975. In the United Kingdom another important robot association was formed in 1977, known as the British Robot Association (BRA). Today, there are many other robot engineering associations in existence in various parts of the world.

Two journals on robots came into being in the United Kingdom (UK) and Japan in 1973. In the UK, the journal called *The Industrial Robot* is published by IFS (Publications) Ltd. The Japanese journal *robot* was the result of the effort by the Japan Industrial Robot Association. Today, there are a number of journals on robots.

Since the early 1980's, a large group of new robot manufacturers have appeared on the scene. The domestic market for robots in the United States was estimated to be $\$92 \times 10^6$ in 1980. According to Ref. 16, the projected forecast for the robot market in the U.S. alone is of order of $\$20 \times 10^9$ by the year 2000.

Industrial Robot Categories

Based on the input signal, industrial robots can be classified into five distinct categories as shown in Fig. 11.7.[17] Category I includes intelligent robots, capable of making decisions for their actions. Category II is composed of manual manipulators, operated by human operators. The third category is concerned with playback robots that are capable of carrying out assigned tasks by repeating the sequence, position, and information of the assigned tasks they have memorized from humans. Category IV includes

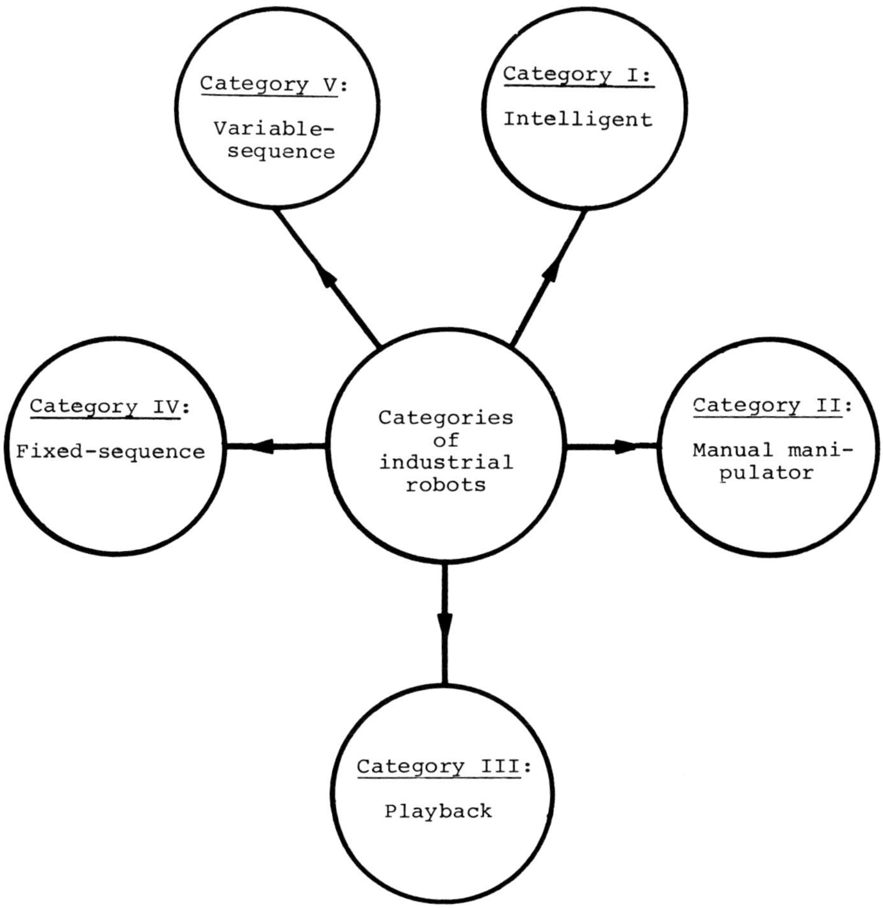

Fig. 11.7 Categorization of industrial robots.

fixed-sequence robots that are capable of conducting successive actions in preset sequences, positions, and/or conditions. However, to make changes in the set information is difficult. Finally, category V is concerned with variable-sequence robots, which are similar to category IV robots, but with one significant difference—the changes in the set information can be made quite easily.

Robot Safety

According to a paper published in 1983,[18,19] there have been five fatal accidents involving robots since 1978. In all of the five cases, workers lost their lives. Four of these workers were Japanese and one American. A Swedish report[20] states that there were 36 accidents in Sweden involving robots that caused human injury. All of these accidents occurred in the years 1979–1983. Furthermore, according to Japanese findings,[21,22] 8% of persons working with robots have been injured and as many as 36% of them have experienced an accident of some sort. Because of such situations, the robot safety issue has become of utmost importance.

Causes for Near-Accidents. There were various causes for near-accidents due to industrial robots,[17] some of which are

1) Sudden accessing to the robot area by human.
2) Robot acted incorrectly in normal operation.
3) Robot acted incorrectly in manual operation.
4) Incorrect move of peripheral equipment while teaching or testing.
5) Incorrect move of peripheral equipment while operating in normal mode.
6) Incorrect action during repairing, adjusting, and checking.
7) Robot acted erroneously during teaching and testing.

Major Sources of Hazard. Basically, there are six major sources of hazard associated with robots,[22] as follows:

1) Faults due to hydraulic, electrical, and pneumatic problems.
2) Control errors due to electrical interference or problems in the control system or software.
3) Environmental factors such as radiation, dust, and fume.
4) Access of personnel without authority to robot enclosures.
5) Mechanical factors, including corrosion, overloading, fatigue, and hazards from the parts and tools accompanied by the robotic unit.
6) Human mistakes.

Robot Safety Measures. To improve robot safety, attention must be focused on such areas as[23]

1) Adequacy of fail-safe design.
2) Adequacy of emergency stop control.
3) Elimination of the need for workers to enter the danger zone while the robot is in an automatic mode of operation.

4) Safe robot moving speed during teaching, testing, setting, and so on.
5) Close cooperation between the robot manufacturer and user.

Guidelines for Safety Precautions. These guidelines concerned with robot safety can be grouped into five distinct classifications[19]: personal, maintenance, electrical, work area, and setup and operation.

The guidelines concerned with personal safety call for effective usage of safety protection equipment, in keeping with safety instructions, and entrance into the work envelope only after switching off the robot power. Maintenance safety is another important area. It calls for thorough understanding of safety instructions and turning off the power before embarking on robot maintenance. The third group of guidelines are concerned with electrical safety and include such admonitions as 1) make use of the emergency stop or switch off the control power when an electrical problem arises for which you are not trained to handle and 2) do not operate and control, make adjustment, or open any cabinet for which you are not fully trained.

The work area safety guidelines include ensuring that the work area is free of hazardous obstructions and has effective warning signs and restrictive devices in place. The set up and operation safety group guidelines are: 1) before switching on the robot make sure that items such as covers, safety guards, and protective devices are in place; 2) before servicing, operating, or setting up the robot in question, make sure that the safety instructions are clearly understood; and 3) pay attention to the movements of the robot in operation.

Robot Reliability

A robot has to be safe and reliable. An unreliable robot may become the cause for unsafe conditions, high maintenance costs, inconvenience, etc. Robots make use of electrical, mechanical, pneumatic, electronic and hydraulic parts. This makes their reliability problem a challenging task because of many different sources of failures. The best mean time between failures (MTBF) achieved for robots is only 2500 h.[17] It means there is definitely room for further improvement in robot reliability. In general, the following factors such as given below affect the robot reliability[24]: human mistakes or errors, errors in software, random failures, and systematic faults in hardware (the cause for these faults is the system elements functional relationships).

Robot Problems. There are various robot problems that may contribute to downtime. Some of these are[25] cabinet faults and collision, fault in power supply, synchronization, stiff robot arm, failure of parts in the arm of the robot, overheating of hydraulics, robot standstill, blown-out fuse, tool problems, and cable faults.

Methods for Robot Reliability Analysis. In reliability theory, there are various methods developed to perform reliability analysis. Each of these methods has its advantages and disadvantages. However, there are two widely used reliability methods to analyze engineering systems. These are known as fault trees and failure modes and effects analysis (FMEA). Both of these concepts can also be used to perform reliability analysis of robots. The application of these two methods to a robot is demonstrated in Ref. 24. Basics of the fault tree concept were discussed earlier in this chapter. The FMEA concept was developed in the early 1950's and is concerned with evaluation of design reliability by taking into consideration the potential failures and their effect on the system. The method is described in detail in Ref. 26.

Mean Time to Robot Failure. This is a good measure to estimate mean productive robot time prior to a failure and is defined[25] by

$$\text{MTRF} = (T_{rp} - T_{drf})/F_r \tag{11.8}$$

where MTRF is the mean time to robot failure, T_{rp} the robot production time, T_{drf} the downtime due to robot failure, and F_r the number of robot failures.

Mean Time to Robot Problem. This is similar to MTRF and is defined[25] as

$$\text{MTRP} = (T_{rp} - T_{drp})/P_r \tag{11.9}$$

where MTRP is the mean time to robot problem, T_{rp} the robot production time, T_{drp} the downtime due to robot related problem, and P_r the number of robot related problems.

Accident Risk. The risk of an accident for an event is given[24] by

$$A_r = P_{oe} \cdot d_e \tag{11.10}$$

where A_r is the risk of an accident for a given hazardous event, d_e the expected damage associated with the outcome of the given event, and P_{oe} the occurrence probability of the given hazardous event over a defined time period.

It is to be noted that in real life the value of d_e is not that easy to estimate because it depends on several factors.

Robot Reliability. For constant failure rate, the time dependent reliability of a robot is given by

$$R_b(t) = e^{-\lambda_r t} = e^{-(1/\text{MTRF})t} \tag{11.11}$$

where $R_b(t)$ is the robot reliability at time t, λ_r the constant failure rate of the robot, and MTRF the mean time to robot failure.

Example 11.3

Two independent robots are performing a specific task at an assembly line. The failure rate of each robot is 0.004 failures/h. Both robots must work normally for accomplishing the task successfully. Compute the probability that the task is going to be performed successfully for an 8 h mission if both robots start operating at time $t = 0$.

By substituting the given data into Eq. (11.11) we get

$$R(8) = e^{-(0.004)8} = 0.9685$$

Thus, with the aid of the above result, the probability $P_s(t)$ of the task success for the given time t is

$$P_s(8) = [e^{-(0.004)8}]^2 = 0.9380$$

Thus, the probability of the task success for the defined mission time is 0.9380.

References

[1]Bruns, H., "Trends in Reliability of Commercial Vehicles and Their Components," *Proceedings of the Second National Reliability Conference*, Birmingham, UK, March 1979, pp. 7/2/1–7/2/11.

[2]Miaoulis, G., O'Brien, V. L., Karns, D., and Horn, P., "The International Automobile: American's Perceptions of Product Reliability and Quality," *Proceedings of the 6th Advances in Reliability Technology Symposium*, National Center of Systems Reliability, Warrington, UK, 1980, pp. 315–327.

[3]Khodabandehloo, K., Duggan, F., and Husband, T.M., "Reliability Assessment of Industrial Robots," *Proceedings of the 14th International Symposium on Industrial Robots*, edited by N. Martensson, Elsevier Scientific, Amsterdam, 1984, pp. 209-220.

[4]*Robot Facts*, British Robot Association, 1984 (available from British Robot Association, 28–30 High Street, Kempston, Bedford MK42 7BT, England).

[5]Bonney, M.C. and Yong, Y.F., *Robot Safety*, Springer-Verlag, Berlin, 1985.

[6]Simpson, B. H., "Reliability and Maintainability, Part 2—Automotive: The Ford Reliability Program," *Mechanical Engineering*, March 1966, pp. 47–51.

[7]Limpert, R., *Vehicle System Components: Design and Safety*, Wiley, New York, 1982.

[8]Simpson, B. H., "The Ford Engineering Reliability Program," *Proceedings of the 4th Annual Reliability and Maintainability Conference*, Spartan Books, Washington, DC, 1965, pp. 171–181.

[9]Knowles, J., "Engineering Reliability into Today's Automotive Vehicles," *Proceedings of the Annual Symposium on Reliability*, 1968, pp. 67–74.

[10]Vikman, S. and Hansson, B., "Reliability Analysis Is a Part of the Volvo Design Engineering Process," *Proceedings of the 6th Advances in Reliability Technology Symposium*, National Centre of Systems Reliability, Warrington, UK, 1980, pp. 279–296.

[11]Binggang, W. and Shuijun, L., "Research on Test and Evaluation Method of Automobile Reliability," *Proceedings of the Third International Pacific Conference on Automotive Engineering*, SAE, Warrendale, PA, 1985, pp. 665–671.

[12]Mateyka, J., Danzeisen, R., and Weiss, D. W., "Fault-Tree Applications to the Automobile Industry," SAE Paper, 1973.

[13]Rooks, B., "The Cocktail Party That Gave Birth to the Robot," *Decade of Robotics*, Springer-Verlag and IFS (Publications) Ltd., Bedford, UK, 1983, pp. 8–11.

[14]Asimov, I. and Frenkel, K. A., *Robots: Machines in Man's Image*, Harmony Books, New York, 1985.

[15]Engelberger, J. F., *Robotics in Practice*, Kogan Page Ltd, London, 1980.

[16]Zeldman, M. I., *What Every Engineer Should Know about Robots*, Marcel Dekker, New York, 1984.

[17]Sugimoto, N., Kawaguchi, K., "Fault Tree Analysis of Hazards Created by Robots," *Proceedings of the 13th International Symposium on Industrial Robots and Robots*, Robotics International of SME, Dearborn, MI, April 1983, pp. 9.13–9.28.

[18]Altamuro, V. M., "Working Safely with the Iron Collar Worker," *National Safety News*, July 1983, pp. 38–40.

[19]Lee, J., "Robotic Safety-Analysis, Consideration, and Answers," *Robots-9 Conference Proceedings*, Vol. 2, Robotics International of SME, Dearborn, MI, June 1985, pp. 19.48–19.53.

[20]Carlsson, J., "Robot Accidents in Sweden," *Robot Safety*, edited by M. C. Bonney and Y. F. Yong, Springer-Verlag, Berlin, 1985, pp. 49–64.

[21]Sugimoto, N., "Safety Engineering on Industrial Robots and Their Draft Standard Safety Requirements," Mechanical Div., Research Institute of Industrial Safety, Tokyo, 1980.

[22]Percival, N., "Is Robot Technology Safe?" *Decade of Robotics*, Springer-Verlag and IFS (Publications) Ltd., Bedford, UK, 1983, pp. 82–85.

[23]Hasegawa, Y. and Sugimoto, N., "Industrial Safety and Robots, *Proceedings of the 12th International Symposium on Industrial Robots and 6th International Conference on Industrial Robot Technology*, Published by IFS (Publications) Ltd., Bedford, UK, 1982, pp. 9–15.

[24]Khodabandehloo, K., Duggan, F., and Husband, T. M., "Reliability of Industrial Robots: A Safety Viewpoint," *Proceedings of the 7th British Robot Association Annual Conference*, 1984, pp. 234–242.

[25]Jones, R. and Dawson, S., "People and Robots: Their Safety and Reliability," *Proceedings of the 7th British Robot Association Annual Conference*, 1984, pp. 243–258.

[26]Dhillon, B. S., *Systems Reliability, Maintainability and Management*, Petrocelli Books, New York, 1983.

Exercises

11.1 Describe the history of the automotive vehicle.

11.2 Discuss the differences and similarities between the Ford and Volvo reliability assurance programs.

11.3 Assume that the constant failure rate of a motor vehicle is 0.0002 failures/km. Compute reliability of the vehicle for a 200 km drive.

11.4 What are the causes for the steering system defects?

11.5 Write an essay on the future uses of robots.

11.6 Define the following: robot, robot reliability, and robot availability.

11.7 Discuss in detail the history of robots.

11.8 What are Asimov's three laws of robotics?

11.9 What are the major sources of hazard associated with robots?

11.10 Assume that two independent robots are jointly performing a certain task. The failure rate of each robot is 0.008 failures/h. At least one robot must work normally for accomplishing the task successfully. Calculate the probability that the task is going to be performed successfully for a 15 h mission if both the robots start operating at time $t = 0$.

12
RELIABILITY OF SPECIFIC MECHANICAL STRUCTURES, EQUIPMENT, PARTS, AND DEVICES

This chapter discusses reliability of specific mechanical structures, equipment, components, and devices. These specific items are offshore structures, pipelines, oil tanks, steam turbine blades, shafts, bearings, blowers, and high-speed motors. The different types of offshore platforms are discussed. A list of factors useful in deciding the suitability of the type of platform to be used for a specific application is presented. The topic of uncertainty treatment in offshore structures is briefly outlined. The basic methods of computing fatigue damage are presented along with two mathematical models. Model I is concerned with estimating the total expected cost, whereas model II deals with approximating the annual failure probability. Benefits of reliability analyses in offshore structure design are listed. Reliability of pipelines (oil and natural gas) is discussed. The reliability performance of pipelines used in Western Europe, the United States, and the Gulf of Mexico is described in detail. A number of causes of oil tank failures are listed, along with causes of steam turbine blade failures. The topic of shaft failures is discussed. Causes for most bearing failures are presented. A method to estimate bearing average failure rate is presented. The method is specifically concerned with grease-lubricated rolling element bearings. In addition, a reliability prediction method for oil-lubricated rolling bearings is given. Equations to estimate failure rate and reliability of blowers and of high-speed motors are presented.

12.1 INTRODUCTION

In recent years, increasing attention has been paid to the reliability of many specific mechanical structures, equipment, parts, and devices. There are many contributing factors for such increased attention, some of which are the benefits derived from the electronic items' reliability studies, stringent requirements and cost.

Because mechanical items take many different forms, shapes, and sizes, performing reliability studies on all of them is a very difficult task. Other reasons are the cost of collecting the data, lack of interest, lack of relevant information, etc. However, there are some mechanical items that have been studied from the reliability viewpoint for a number of years. Examples of such items are offshore oil platforms and structures, pipelines, oil tanks, steam turbine blades, gears, rolling bearings, fluid-flow valves, shafts, tractors, and automobiles. Many of these items are discussed in this chapter from the reliability aspect. The reliability of such items is important and cannot simply be ignored in real-life environments. In some cases, reasonably good field failure data and well-established methods and procedures are available to predict their reliability with some confidence. However, there is still lot of room for further improvements.

12.2 RELIABILITY AND RISK IN OFFSHORE STRUCTURES

In recent years, requirements for risk assessment and reliability criteria are gaining ground in the offshore oil industry because engineers are faced with a multiplicity of risks. There are various reasons for the magnification of uncertainties in offshore structures, some of which are 1) the lack of necessary data about loads and materials, 2) unfamiliarity with the ocean environments, 3) the cost of collecting data, and 4) the severity of ocean environments relative to those of on-land structures.[1]

Offshore Platforms

There are several types of platforms developed for offshore purposes. Generally, they may be broken down into two classes. Class I contains two basic types of conventional platforms: gravity-based platforms and pile-supported, template-type platforms.[2] Gravity-based platforms are constructed with a combination of steel and concrete (or either steel or concrete alone) and rest on the ocean bed without pile support. Pile-supported platforms are generally built using steel and depend upon pile foundations to transmit the environmental and operational loads to the ocean floor. Class I platforms have normally been used in water depths up to approximately 1000 ft. Class II platforms have been developed for use in water depths greater than 1000 ft. In such deep waters, class I platforms become technically less feasible. Three promising types of platforms belonging to class II are the tension leg platform, the guyed tower, and the buoyant tower, as described in Ref. 2. In deciding the suitability of the type of platform to be used for a specific application, such factors as the following play an important role: cost, depth of water, type of soil, functional requirements (i.e., deck area, operating loads, and storage), environmental factors (ice, waves, earthquakes), construction logistics, and strength.

Uncertainty Treatment in Offshore Structures

There are several methods that can be used in risk assessments of offshore structures. These methods vary from a totally subjective and implicit risk determination to the usage of applied probability concepts. The factors[1] on which the appropriate level of uncertainty treatment depends are the quantity of the available data, the risk assessment objective, the amount of funds involved, the quality of the available data, the type of available data, and the importance of the decision situation.

Fatigue Reliability

Because offshore structures are subject to various kinds of environmental loads, metal fatigue is a possible mode of failure. The welded joints at the brace/chord connections are probably the most vulnerable to fatigue. The prediction of fatigue in offshore structures is not a simple process, because it involves a number of factors, some of which are[3]:

1) Uncertainty of any corrosion effects on the joint fatigue strength.
2) Enormous statistical scatter in the basic fatigue data.
3) Poor knowledge of the force of a specific sea wave on a structure.
4) Utilization of approximate equations to describe the evolution of the fatigue crack subject to random stresses.
5) Statistical summaries of the ocean wave processes that cannot be considered to be always accurate.
6) Uncertainty about the cathodic protection effects on the joint fatigue strength.
7) Determination of propagation and initiation of fatigue cracks made complex by defects in the discontinuities in the welded joints.

There are several methods employed in the offshore oil drilling industry to compute fatigue damage. Two methods of computing fatigue damage are the deterministic and probabilistic. These two methods are described in detail in Ref. 3.

The fatigue damage d_f at time t is given by

$$d_f = F^{c_1} \cdot t\theta / c_2 \tag{12.1}$$

where θ is the stress parameter, F the factor that takes into consideration the existence of uncertainties in estimating fatigue stresses from oceanographic data, and c_1,c_2 the empirical constants associated with the S-N curve. (Briefly, when fatigue data displayed on a plot of cyclic stress level versus the logarithm of life, the plot is called S-N curve.) The deterministic and probabilistic methods can be used to evaluate θ. According to the deterministic method (i.e., wave exceedance diagram), θ is given by

$$\theta = F_{as} \cdot \sum_j \beta_j R_j^{c_1} \tag{12.2}$$

where β_j is the fraction of the stress range in which R_j is acting, F_{as} the mean frequency of stresses, and R_j the stress range.

Similarly, according to the probabilistic approach, θ is given by

$$\theta = \mu(c_1)(2\sqrt{2})^{c_1}\Gamma\left(\frac{c_1}{2}+1\right)\cdot X \tag{12.3}$$

$$X \equiv \sum_j \alpha_{tj}\cdot F_{wj}\cdot(\sigma_j)^{c_1} \tag{12.4}$$

where $\Gamma(\cdot)$ is the gamma function, $\sigma_j^{c_1}$ the rms of stress process in sea state j, α_{tj} the percent of time in sea state j, F_{wj} the wave loading frequency in sea state j, and $\mu(c_1)$ the rain flow correction (given by the ratio of the rain flow damage to the narrow-band damage).

Mathematical Models

This section presents two models that can be used in analyses of offshore structures.

Model I. This model estimates the total expected cost. The end objective of this study is to determine the minimum total cost. The model is useful for situations where a failure leads to economic loss. The total expected cost is defined as[1]

$$E(k_t) = k_i + E(k_f) \tag{12.5}$$

where $E(k_t)$ is the total expected cost, k_i the initial cost, and $E(k_f)$ the expected cost of failure.

The expected failure cost is given by

$$E(k_f) = p\cdot E(k/f) \tag{12.6}$$

where $E(k/f)$ is the expected cost given that a failure occurs, k the cost, f the failure, and p the probability of failure during life of the design. This model is described in detail in Ref. 1.

Model II. This model approximates the yearly failure probability when the combination of initial cost and potential losses associated with the platform serviceability loss is at a minimum. Thus, the annual probability of failure can be approximated from[2,4,5]

$$P_{af} = [(2.3)\cdot(PV_f)\cdot\lambda]^{-1} \tag{12.7}$$

$$\lambda \equiv \Delta k_i/k_f \tag{12.8}$$

where P_{af} is the yearly probability of failure. This probability is estimated on the basis of minimizing the triangle costs concerned with placement of the facility and those potential costs resulting from platform loss of serviceability, λ the cost ratio (i.e., $k_f : \Delta k_i$), k_f the cost of failure (expected cost of serviceability loss), Δk_i the initial cost curve slope (when plotted against the common log of the yearly failure probability), and PV_f the present value function. The purpose of this function is to bring the potential losses cost to its present value.

For a detailed description of this model, the reader should consult Ref. 2.

Advantages of Reliability Analyses in Offshore Structure Design

There are several benefits of performing reliability analyses in offshore structure design, some of which are[2]

1) Help in producing safe designs.
2) Realistic treatment of uncertainties with probabilistic methods.
3) Incorporation of experience into the design process through risk analysis methods.
4) Optimal distribution of material to elements such as joints, foundation, and braces within a single structure.
5) Effective studies of the tradeoffs between increased platform reliability and use of safety equipment.

12.3 RELIABILITY OF SELECTED MECHANICAL ITEMS

This section discusses the reliability aspects of several selected mechanical items: pipelines, oil tanks, steam turbine blades, shafts, and bearings.

Pipeline Reliability

Pipelines are increasingly being used around the world to transport crude oil, natural gas, etc. They represent a fairly reliable and safe mode of transportation, as verified[6] by the extremely small quantity of oil spilled per unit transported and the much lower fatalities per ton-kilometer transported relative to other means of transportation. Even though pipelines are a safe and reliable mode of transportation, various kinds of pipeline failure occur each year, causing high repair costs, pollution, and losses of transport capacity, and gas availability.

Generally, a pipeline failure may be described as an incident causing a significant leakage or needing urgent repair. The important causes of pipeline failure are[6] corrosion, mechanical failures (i.e., defects associated with construction and/or materials), and third-party damage/outside force (i.e., sabotage or natural disasters).

As mentioned earlier, the pipelines are being used extensively in many parts of the world. Pipeline failures in several of these areas are briefly discussed here.[6]

Western Europe: Oil Pipelines. According to statistics, the oil pipelines in Western Europe reached the total length of 12,000 miles in 1979. In addition, a total of 4070×10^6 barrels of refined products and crude oil were transported through these pipelines that year. During the period 1971–1980, the gross spillage of oil was 6290–31,450 barrels/year. One study[6] concerned with West European oil pipelines grouped the failures into the following five distinct categories: natural hazards, corrosion, operating error, outside-force/third-party damage, and materials failure and construction defects.

During the period 1966–1978, a total of 177 incidents occurred. The following significant facts were noted with these failures:

1) Out of the total of 177 incidents, 130 were associated with the line pipe and the remaining 47 with the pipeline stations and auxiliary equipment. Examples of the auxiliary equipment are valves and flanges.

2) Out of the 129 line pipe failures, 43 were on crude oil pipelines and 86 on product pipelines.

3) Corrosion was the cause for more than half of the failures associated with product pipelines.

4) The most frequent causes of crude pipeline failures were third-part damage and mechanical failure (materials failure and construction defects).

United States: Transmission Pipelines for Natural Gas. There are over 280,000 miles of natural-gas transmission pipelines in the United States. During the period 1970–1978, a total of 3609 failures occurred on these pipelines.[7] The additional facts and figures are as follows:

1) Over 85% of the reported failures were due to materials failure, corrosion, and outside forces.

2) The overall failure rate was 0.9 failures/620 mile · years (1000 km · years).

3) For pipelines with diameters equal to or greater than 20 in., the approximate average failure rate was 0.3 failures/620 mile · years (1000 km · years).

4) Uncoated or unwrapped pipes accounted for 40–50% of the total corrosion-related failures.

5) External corrosion was the cause for over 80% of the corrosion-related failures.

6) Internal corrosion accounted for less than 20% of the corrosion-related failures.

7) The estimated failure rate of the cathodically protected pipe is 0.08/year/1000 km (620 miles) as opposed to 0.45/year/1000 km (620 miles) for the one not cathodically protected.

United States: Oil Pipelines. During the period 1965–1974, the estimate of the length of pipelines transporting liquid hydrocarbons is 210,000–220,000 miles. There were a total of 2861 failures during the time frame 1968–1976. The causes for such failures were corrosion, third-party activity, mechanical failure, natural hazards, operational error, and miscellaneous.

The number of failures for each of these six causes were 1084, 700, 583, 65, 205, and 224, respectively.

The average failure rate was in order of 0.9 failures/620 miles · year. This figure is quite close to that for oil pipelines in Western Europe and for natural gas transmission pipelines in the United States.

Gulf of Mexico: Pipelines. The Gulf of Mexico is another major area served by many miles of pipelines. During the period from 1967–1977, there were a total of 136 failures. The additional facts and figures are as follows:

1) Pipelines of less than 10 in. diameter accounted for 94 of the total failures.

2) Pipelines of more than 10 in. diameter accounted for 42 of the total failures.

3) Out of the 42 failures associated with pipelines of >10 in. diameter, 31 occurred on oil pipelines and 11 on gas pipelines.

The causes for the failures associated with pipelines with diameters >10 in. were mechanical failure, operational error, corrosion, natural hazards, third-party damage, and "miscellaneous." The distribution of failures in percentages for each of these classifications was 19, 5, 31, 11, 24, and 9, respectively. The two largest causes of failure were corrosion and third-party damage. In addition, internal corrosion was the cause for 10 of the 13 total failures due to corrosion.

Oil Tank Failures

Just like for any other mechanical item, the large oil storage tanks are also prone to failures. Many such failures have been reported,[8–12] for example, the 1925 crude oil tank failure in Oklahoma[11,12] and the 1980 failure of a storage tank containing crude oil in Moose Jaw, Saskatchewan, Canada.[9] Eighteen similar failures are listed in Ref. 8. The most frequently reported causes of oil tank failures are 1) steel with inadequate low-temperature toughness, 2) poor quality welds missed during inspection, 3) low temperatures, and 4) combinations of these three causes.

An official inquiry was conducted into the Moose Jaw oil tank failure by the Province of Saskatchewan, chaired by the Province's Fire Commissioner. The inquiry team made a number of recommendations to prevent any reoccurrence of similar failures in the future, some of which were[13]

1) Take steps to preclude brittle failures.

2) Perform ultrasonic, radiographic, or another concealed fault detection tests and checks on tanks already in operation.

3) Conduct studies to determine the adequacy of the tanks for continued use.

4) Change or amend any appropriate present government regulations to make oil tanks more safe and reliable.

Steam Turbine Blade Failures

The core of any electric power generation system is the steam turbine. Consequently, the reliability of its blades is very important. A half-century ago, turbine blades used to fail by the hundreds, resulting in poor system reliability. The reliability of such blades have improved considerably since then. In current designs, the older types of blades (i.e., small, thin, and 1 cm wide) have been replaced with fewer blades that are thicker, stronger, and wider.[14] However, these blades still fail, resulting in poor reliability. According to the U.S. National Electric Reliability Council's generating availability data system,[15] during the period 1971–1980, blade failures resulted in the loss of 71,303 GW · h. The annual cost of these failures is more than $100 million (it is estimated that an average blade may produce 50 kW of power). The causes of such failures are vibration fatigue, foreign objects, overspeed, water, and erosion.[15]

The vibration of the blade structure itself is probably the most frustrating and greatest cause of failure. Failure occurs at some point on the blade where the vibratory stress exceeds the fatigue strength. From telemetry tests, the following observations have been made:

1) High levels of stress are experienced at resonance.

2) The highest degree of vibratory stresses may be experienced at a single resonance frequency or at combined frequencies.

3) Only low levels of vibration stress occur in most operating conditions.

4) A wide spectrum of stresses is experienced in any row of blades. However, this is not the case when a torsional or disk resonance is the causing factor.

5) During transient conditions, the locking of multiple or single resonances often results in maximum blade vibration stress.

6) At any instance of time, blades in a row have very different stresses and high load flow.

Shaft Failures

The reliability of shafts is important because of the severe consequences of their failure. Shafts operate under various conditions, e.g., extremely high and low temperatures, corrosive environments, and high vibratory stresses. In addition, they are subjected to torsion, tension, bending, and compression loads.[16]

Metal fatigue is the most common cause of shaft failures. Generally, such fatigue can be classified into three distinct categories as torsional, axial, or bending.

The origin of all shaft fractures is the point of stress concentration. Such points result from either design or fabrication. Examples of design features that concentrate stress are oil hole edges, keyway ends, and fillets at shoulders. Similarly, examples of stress concentrators introduced during fabrication are damage resulting from grinding, quench cracks, and machining marks. In addition, the stress concentrators such as pits, laps, cold shuts, and seams are produced during the shaft forging stage.

Bearing Failures

Just like any other mechanical item, bearings are prone to failure. There are various causes for such failures—for example, according to a survey,[16,17] 1400 rolling bearing failures were traced to the vendor (30.1%), to the user (65.9%), and to external causes (4.0%). The vendor problems include incorrect material used for construction, design errors, and workmanship. User-induced failure results from wear, maintenance shortcomings, operational errors, and failure of the monitoring equipment. The external problems comprise such things as oil supply system failure and contaminated lubricants. One or more of the causes presented in Fig. 12.1 explain most bearing failures. These causes are described in detail in Ref. 16.

Bearing Average Failure Rate Estimation Method. This method can be used to estimate the mean failure rates of grease-lubricated rolling element bearings.[18] In this case, the mean failure rate of a bearing is obtained by

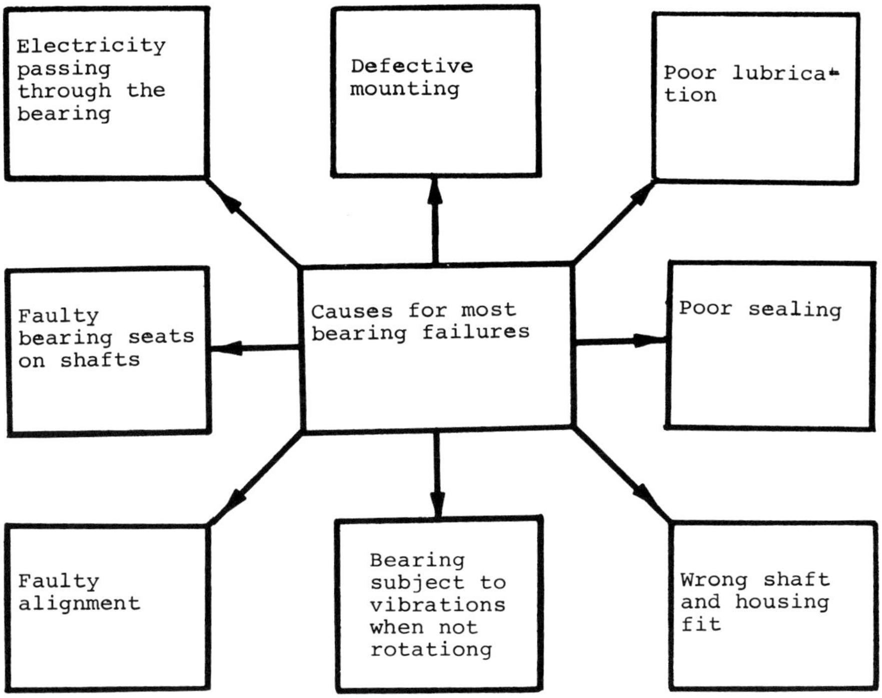

Fig. 12.1 Causes of most bearing failures.

averaging the Weibull cumulative hazard function over the specified time. The average failure rate of a grease-lubricated rolling element bearing is given by

$$\lambda_m = (T)^{\alpha_1}/\theta^{\alpha_2} \tag{12.9}$$

where $\alpha_1 = 1.878$, $\alpha_2 = 2.878$, T is the time over which the failure rate is to be estimated (in hours), θ the Weibull (bearing) characteristic life (in hours), and λ_m the mean failure rate of the grease-lubricated rolling element bearing.

The value of θ can be estimated from

$$\theta = \frac{1.241}{(1/10^{y_1}) + [1/(10^{y_2} + 300)]} \tag{12.10}$$

where

$$y_1 \equiv \left\{\frac{2342}{t_m} + Q - [(4.32)(dS)(10^{-6})] + G_c - 0.001S\left(\frac{L}{C_{dc}}\right)^{1.5}\right\} \tag{12.11}$$

$$y_2 \equiv 19.7 - (4760/t_m) \tag{12.12}$$

The nomenclature in Eqs. (12.11) and (12.12), is as follows:

C_{dc} = specific dynamic capacity, lb (at $33\frac{1}{3}$ rpm—as noted in the bearing manufacturer's catalog)
d = bearing bore diameter, mm
G_c = grease constant (the value differs for different oil and thickener, e.g., $G_c = 1.41$ for mineral oil and sodium thickener, 1.81 for silicon oil and lithium thickener, and 1.55 for diester oil and lithium thickener
L = load, lb
Q = quality factor [two different values are assigned to this factor for military and commercial applications, i.e., $Q = 0.12$ (military), $Q = -0.27$ (commercial)]
S = speed, rpm
t_m = operating temperature of bearing, K

Example 12.1

For an operating time of 5000 h, calculate the mean failure rate of a grease-lubricated rolling element bearing. The bearing operating temperature is 27°C, $G_c = 1.41$, $Q = 0.12$, $L = 12$ lb, $C_{dc} = 500$ lb, $S = 3600$ rpm, $d = 14$ mm, and $t_m = 27 + 273 = 300$ K. Substituting these data into Eqs. (12.11) and (12.12), we get

$$y_1 = \left\{ \frac{2342}{300} + 0.12 - [(4.32)(14)(3600)(10^{-6})] + 1.41 - (0.001)(3600)\left(\frac{12}{500}\right)^{1.5} \right\}$$

$$= (7.8067 + 0.12 - 0.2177 + 1.41 - 0.0134) = 9.1055$$

and

$$y_2 = 19.7 - (4760/300) = 3.8333$$

Using the above results in Eq. (12.10), we get

$$\theta = \frac{1.241}{1/10^{9.1055} + 1/(10^{3.8333} + 300)}$$

$$= 8826.436 \text{ h}$$

Substituting the above result and the given value for T in Eq. (12.9) results in

$$\lambda_m = \frac{(5000)^{1.878}}{(8826.436)^{2.878}}$$

$$= 3.8967 \times 10^{-5} \text{ failures/h}$$

$$= 38.967 \text{ failures}/10^6 \text{ h}$$

Reliability Prediction Method for Rolling Bearings. This method can be used to predict the life of a rolling bearing (oil lubricated). The life of a bearing in question is defined by

$$L_f = (L_b/L_r)^\alpha \times 10^6 \text{ cycles} \tag{12.13}$$

where α is the factor with an assigned value of 3 (i.e., $\alpha = 3$) for ball bearings and of 10/3 (i.e., $\alpha = 10/3$) specifically for roller bearings, L_b the basic load rating expressed (in pounds), L_r the equivalent radial load expressed (in pounds), and L_f the life given in revolutions. It means that, on average, 90% of the bearing population will complete or exceed this value without experiencing failure.

In order to obtain the value of L_f in hours, use the following formula[18]:

$$L_f = \frac{16{,}666.67}{S_r}\left(\frac{L_b}{L_r}\right)^\alpha \tag{12.14}$$

where S_r is the rotational speed given in revolutions/minute (rpm).

For detailed description of this method, consult Refs. 18 and 19.

12.4 RELIABILITY EVALUATION OF ELECTROMECHANICAL DEVICES

This section presents two mathematical models[20] to evaluate reliability and failure rates of blowers and high-speed motors.

Model I

This model is concerned with failure rate and reliability evaluation of blowers. The failure rate of a blower is given by

$$\text{(Total failure rate of a blower)} = \text{(blower mechanical failure rate)} + \text{(blower electrical failure rate)}$$

$$\lambda_{BT} = (\lambda_{BM} + \lambda_{BE}) \text{ failures}/10^6 \text{ h} \tag{12.15}$$

where λ_{BT} is the total failure rate of a blower, λ_{BM} the blower mechanical failure rate, and λ_{BE} the blower electrical failure rate.

The blower mechanical failure rate λ_{BM} is expressed as

$$\lambda_{BM} = F_{op}(10)^4/T_0 \tag{12.16}$$

where T_0 is the blower operational time (in hours) and F_{op} the percentage of mechanical failures during time T_0. The value of F_{op} can be estimated from the plot given in Ref. 20.

Similarly, the blower electrical failure rate λ_{BE} is given by

$$\lambda_{BE} = \lambda_B f_f \cdot f_w \cdot f_e \tag{12.17}$$

where λ_B is the electrical base failure rate and f_f, f_w, and f_e the family, windings, and environmental factors, respectively. The values of these factors are tabulated in Ref. 20.

The electrical base failure rate λ_B can be estimated from

$$\lambda_B = Ce^{((t^h + 273)/k)^\beta} \tag{12.18}$$

where t_h is the hot-spot temperature during operation and C,K,β the constants. Values of these parameters are defined in Ref. 20.

The reliability of the blower is given by

$$R_{BT}(t) = e^{-(\lambda_{BT})t} \tag{12.19}$$

where $R_{BT}(t)$ is the reliability of the blower at time t and t the time.

Model II

This model is concerned with failure rate and reliability evaluation of high-speed motors. The following equation can be used to estimate failure rate of a high-speed motor[20]:

$$\lambda_M = \lambda_T \cdot f_e \text{ failures}/10^6 \text{ h} \tag{12.20}$$

where λ_M is the high-speed motor failure rate, f_e the environmental factor (the value for fixed ground, inhabited airborne, sheltered naval, mobile ground, unsheltered naval, and uninhabited airborne environments are 1.5, 2.5, 4.5, 6, 7.7, and 9.3 respectively), and λ_T the total of electrical and mechanical failure rates.

The total failure rate λ_T is expressed as

$$\begin{aligned}\lambda_T &= \lambda_{\mathrm{E}} + \lambda_{\mathrm{ME}} \\ &= (\lambda_B \cdot f_{Qf}) + [F_{pm}(10{,}000)/T_{m0}]\end{aligned} \tag{12.21}$$

where λ_{E} is the electrical failure rate, λ_{ME} the mechanical failure rate, λ_B the electrical base failure rate, f_{Qf} the quality and family factor (values of this factor are tabulated in Ref. 20), F_{pm} the percentage of mechanical failures[20] during time T_{m0}, and T_{m0} the operational time (in hours) of the motor for which the overall (motor) failure rate is to be computed.

The electrical base failure rate λ_B is defined by Eq. (12.18). The reliability of the high-speed motor is defined by

$$R_M(t) = e^{-\lambda_M t} \tag{12.22}$$

where t is the time and $R_M(t)$ the reliability of the high-speed motor at time t.

References

[1]Vanmarcke, E. and Angelides, S., "Risk Assessment for Offshore Structures: A Review," *ASCE Journal of Structural Engineering*, Vol. 109, 1983, pp. 555–571.

[2]Committee on Reliability of Offshore Structures of the Committee on Structural Safety and Reliability of the Structural Division, "Application of Reliability Methods in Design and Analysis of Offshore Platforms," *ASCE Journal of Structural Engineering*, Vol. 109, 1983, pp. 2265–2291.

[3]Wirsching, P. H., "Fatigue Reliability for Offshore Structures," *ASCE Journal of Structural Engineering*, Vol. 110, 1984, pp. 2340–2356.

[4]Bea, R. G., "Earthquake Criteria for Platforms in the Gulf of Alaska," *Journal of Petroleum Technology*, March 1978, pp. 325–340.

[5]Stahl, B., "Probabilistic Methods for Offshore Platforms," Paper presented at meeting of American Petroleum Institute, Dallas, TX, April 1975.

[6]Andersen, T. and Misund, A., "Pipeline Reliability: An Investigation of Pipeline Failure Characteristics and Analysis of Pipeline Failure Rates for Submarine and Cross-Country Pipelines," *Journal of Petroleum Technology*, April 1983, pp. 709–717.

[7]Gideon, D. N. and Smith, R. B., "Analysis of Reportable Incidents for Natural Gas Transmission and Gathering Lines 1970 Through 1978," American Gas Association, Rept. 121 (NG-18), Sept. 1980.

[8]May, I. L., "Oil Tank Failures and Changing Technical Specifications," *Proceedings of the First International Conference on Structural Failure, Product Liability, and Technical Insurance*, edited by H. P. Rossmanith, North-Holland Publishing, Amsterdam, 1984, pp. 115–120.

[9]May, I. L., "Failure Analysis of a Crude Oil Storage Tank," *Metal Progress*, No. 3, Aug. 1982, pp. 35–37.

[10]Anon, I., "Oil Tank Fails After Sudden Drop in Temperature," *Engineering News-Record*, Vol. 94, No. 6, April 16, 1925.

[11]Parker, G. R., *The Brittle Behavior of Engineering Structures*, Wiley, New York, 1957.

[12]Biggs, W. D., *The Brittle Fracture of Steel*, Macdonald and Evans, London, 1960.

[13]Fisher, M. G. G., "Moose Jaw Inquiry Report," Final Report of the Fire Commissioner, Fire Safety Unit, Saskatchewan Labor Dept., Regina, Saskatchewan, Canada, 1981.

[14]Trumpler, W. E., "The Statistics of Steam Turbine Blade Failures," *Proceedings of the Structural Failure, Product Liability and Technical Insurance*, edited by H. P. Rossmanith, North-Holland Publishing, Amsterdam, 1984, pp. 121–129.

[15]Olds, F. C., "Major Rotating Equipment Design Developments," *Power Engineering*, Feb. 1983, pp. 42–50.

[16]Bloch, H. P. and Geitner, F. K., *Practical Machinery Management for Process Plants: Machinery Failure Analysis and Troubleshooting*, Gulf Publishing, Houston, TX, 1983.

[17]*Handbook of Loss Prevention*, edited by Allianz Versicherungs-AG, translated from German by P. Cahn-Speyer, Springer-Verlag, New York, 1978.

[18]Schafer, R. E., Angus, J. E., Finkelstein, J. M., Yerasi, M., and Fulton, D. W., "RADC Non-Electronic Reliability Notebook," Rome Air Development Center, Griffiss Air Force Base, NY, Rept. RADC-TR-85-194, 1985.

[19]*Marks' Standard Handbook for Mechanical Engineers*, McGraw-Hill, New York, 1977.

[20]"Reliability Prediction of Electronic Equipment," Rome Air Development Center, Griffiss Air Force Base, NY, U.S. MIL-HBK-217, 1978 (available from the National Technical Information Service, Springfield, VA).

Exercises

12.1 Discuss the types of offshore platforms and their reliability performance.

12.2 What are the basic methods of computing fatigue damage employed in offshore industry?

12.3 Compare the reliability performance of oil pipelines used in Western Europe with the one used (for oil) in the United States.

12.4 Compare the reliability performance of oil pipelines used in Western Europe with the one used for natural gas in the United States.

12.5 What are the causes of oil tank failures?

12.6 What are the reasons for steam turbine blade failures?

12.7 What are the causes of most bearing failures?

12.8 For an operating time of 8000 h, compute the mean failure rate of grease lubricated rolling element bearing. The bearing operating temperature is 28°C. In addition, the following data are known: $Q = 0.12$, $L = 11$ lb, $C_{dc} = 503$ lb, $G_c = 1.55$, $d = 12$ mm, $t_m = 28 + 273 = 301$ K, and $S = 3600$ rpm. Q, L, C_{dc}, G_c, d, t_m, and S are defined in the chapter.

APPENDIX BIBLIOGRAPHY OF LITERATURE ON MECHANICAL RELIABILITY AND RELATED AREAS

A1 INTRODUCTION

Over the past number of years, a large number of publications on mechanical reliability and related areas have appeared in the published literature. This Appendix presents selected references grouped into three separate categories, i.e., mechanical reliability, aerospace systems reliability, and structural reliability. There can be some overlap in the listings and, therefore, the same reference may appear in more than one of these three categories. The purpose of such a listing is that the same reference fits quite well in each of the categories. However, every effort was made to keep this duplication to a minimum.

These references will provide a useful service to readers in the event they desire to probe deeper into the area of their interest. Sections A2, A3, and A4 present references on mechanical reliability, aerospace system reliability, and structural reliability, respectively.

Following are the addresses of the organizations frequently cited in the subsequent sections.

AIAA: American Institute of Aeronautics and Astronautics, 370 L'Enfant Promenade, SW, Washington, DC 20024-2518
ASCE: American Society of Civil Engineers, 345 East 47th St., New York, NY 10017
ASM: American Society of Metals, Metals Park, OH 44073
ASME: American Society of Mechanical Engineers, 345 East 47th St., New York, NY 10017
ASTM: American Society for Testing and Materials, 1916 Race St., Philadelphia, PA 19103
Battelle Memorial Institute, 505 King St., Columbus, OH 43201
Human Factors Society, P.O. Box 1369, Santa Monica, CA 90406
IEEE: Institute of Electrical and Electronics Engineers, 345 East 47th St., NY 10017
IIT: Illinois Institute of Technology, 10 W. 35th St., Chicago, IL 60616
NTIS: See below
Robotics International of SME, One SME Dr., P.O. Box 930, Dearborn, MI 48121

SAE: Society of Automotive Engineers, 400 Commonwealth Dr., Warrendale, PA 15096
SIAM: Society of Industrial and Applied Mathematics, 117 S. 17th Street, Philadelphia, PA 19103
Southwest Research Institute, Bldg. 139, 62222 Culebra Rd., San Antonio, TX 78284

Note: The Institute of Aeronautical Sciences (IAS) merged with AIAA and the Institute of Radio Engineers (IRE) with IEEE.

In addition to being available from the sponsoring organizations, many of the publications cited are available from one or more of the following information services:

AIAA Technical Information Service, 555 West 57th St., New York, NY 10018
Engineering Societies Library, 345 East 47th St., New York, NY 10017
National Technical Information Service (NTIS), U.S. Department of Commerce, 5285 Port Royal Rd., Springfield, VA 22151

When requesting documents from these organizations, give as much information as possible: i.e., author(s), article title, conference name, journal or publication title, volume number, data, page numbers, sponsoring organization (if known), etc.

A2 MECHANICAL RELIABILITY

Absher, G., "Reliability Approach to Ship Systems," *Proceedings of the Annual Symposium on Reliability*, IEEE, New York, Jan. 1971, pp. 194–212.

American Society for Civil Engineers, "Fatigue Reliability: Introduction," *ASCE Journal of the Structural Division*, Vol. 108, No. ST1, June 1982, pp. 3–23.

American Society for Mechanical Engineers, *Mechanical Reliability Concepts*, ASME, New York, 1965.

American Society for Metals, *Metals Handbook*, 8th ed., Vol. 1, ASM, Metals Park, OH, 1967.

American Society for Metals, *Case Histories in Failure Analysis*, ASM, Metals Park, OH, 1979.

American Society for Testing and Materials, *A Guide for Fatigue Testing and Statistical Analysis of Fatigue Data*, STP, 91-A, 2nd ed., ASTM, Philadelphia, 1963.

American Society for Testing and Materials, *Symposium on Statistical Aspects of Fatigue*, STP 121, ASTM, Philadelphia, 1952.

Amirikian, A., "Reliability in Navy Welded Steel Shore Structures," *Proceedings of the Annual Symposium on Reliability*, IEEE, New York, Jan. 1969, pp. 426–428.

Anderson, J. H. Jr. and Cox, W. P., "Failure Modes in Gold-Aluminum Thermocompression Bonds," *IEEE Transactions on Reliability*, Vol. R-18, No. 4, Nov. 1969, pp. 206–207.

Anderson, J. H. Jr., Maple, T. G., and Cox, W. P., "Aging Effects in Gold Thermocompression Bonds to Complex Metallizations," *IEEE Transactions on Reliability*, Vol. R-19, No. 1, Feb. 1970, pp. 32–34.

Apple, R. E., Simpson, H. M., and Hogstad, J. E., "Applying Reliability in Marine Engineering Design," *Proceedings of Annual Symposium on Reliability*, IEEE, New York, Jan. 1969, pp. 108–117.

Arno, R. G., "Nonelectronic Parts Reliability Data," IIT Research Institute, Chicago, 1981 (available from NTIS).

Arsham, H., "A Generalized Confidence Region for Stress-Strength Reliability," *IEEE Transactions on Reliability*, Vol. R-35, No. 5, Dec. 1986, pp. 586–589.

Asoklis, R. A., "Failure Scoring Criteria for Tank-Wheeled Systems," *Proceedings of Annual Reliability and Maintainability Symposium*, IEEE, New York, 1975, pp. 423–426.

Babecki, A. J., "Metallurgical Aspects of Reliability for Small Nitrided Parts," *Proceedings of Sixth Reliability and Maintainability Conference* (*Annals of Reliability and Maintainability*, Vol. 6), SAE, Warrendale, PA, July 1967, pp. 392–398.

Bajaria, H. J., "Communicating Reliability to Design Engineers," SAE Paper 790392, Feb. 1979.

Bamberger, E. N. et al., *Life Adjustment Factors for Ball and Roller Bearings, An Engineering Design Guide*, Rolling Element Committee, ASME, New York, 1971.

Battelle Memorial Institute, Defense Metals Information Center, Columbus, OH, number of reports through 1971.

Bartholomew, C. S., "Some Economic Implications of Reliability, Maintainability and Safety in Transportation Systems—Introduction," *Proceedings of Seventh Reliability and Maintainability Conference* (*1968 Annals of Assurance Sciences*), ASME, New York, July 1968, pp. 142–145.

Baum, J. V., "Reliability Audit for Improving Mechanical Systems," *Proceedings of ASME Design Engineering Conference*, ASME, New York, May 1965, pp. 110–130.

Bazovsky, I., "Reliability of Rifles, Machine Guns and Other Small Arms," *Proceedings of Annual Symposium on Reliability*, IEEE, New York, 1971, pp. 71–73.

Bazovsky, I. and Benz, G. E., "Effectiveness of Ship Systems and Machinery," *Proceedings of Annual Symposium on Reliability*, IEEE, New York, Jan. 1968, pp. 131–136.

Bazovsky, I. Sr. and Benz, G. E., "Mechanical R&M Modeling and Simulation Methods," *Proceedings of Annual Reliability and Maintainability Symposium*, IEEE, New York, 1984, pp. 27–32.

Bell, R. and Mioduski, R., "Extension of Life of U.S. Army Trucks," *Proceedings of Annual Reliability and Maintainability Symposium*, IEEE, New York, 1976, pp. 200–205.

Bennet, S. B., Ross, A. L., and Zemanick, P. Z. (eds.), *Failure Prevention and Reliability*, ASME, New York, 1977.

Berk, J. H. and Stewart, L. M., "Failure Analysis of Ordnance Systems," *Proceedings of Annual Reliability and Maintainability Symposium*, IEEE, New York, 1987, pp. 141–145.

Bhattacharyya, G. K. and Fries, A., "Fatigue Failure Models-Birnbaum-Saunders vs. Inverse Gaussian," *IEEE Transactions on Reliability*, Vol. R-31, No. 5, Dec. 1982, pp. 439–441.

Bhattacharyya, G. K. and Johnson, R. A., "Stress-Strength Models for System Reliability," *Reliability and Fault Tree Analysis*, SIAM, Philadelphia, 1975, pp. 509–532.

Bild, C. F., "Failure From a Materials Point of View," *Proceedings of Tenth National Symposium on Reliability and Quality Control*, IEEE, New York, Jan. 1964, pp. 330–337.

Bilikam, J. E., "Some Stochastic Stress-Strength Processes," *IEEE Transactions on Reliability*, Vol. R-34, No. 3, Aug. 1985, pp. 269–274.

Bills, K. W. Jr. and Wiegand, J. H., "The Application of an Integrated Structural Analysis to the Prediction of Reliability," *Proceedings of Ninth Reliability and Maintainability Conference* (*Annals of Reliability and Maintainability*, Vol. 9), SAE, Warrendale, PA, July 1970, pp. 514–526.

Birnbaum, Z. W., "A Survey of Some Recent Results on Reliability of Structures," *Proceedings of Fifth Reliability and Maintainability Conference* (*Annals of Reliability and Maintainability*, Vol. 5), AIAA, Washington, DC, July 1966, pp. 845–850.

Birnbaum, Z. W. and Saunders, S. C., "A Probabilistic Interpretation of Miner's Rule," *SIAM Journal on Applied Mathematics*, Vol. 16, No. 3, May 1968, pp. 637–652.

Birnie, C. Jr., "The Status of the Incorporation of Reliability Techniques into Mechanical Engineering Education," *Proceedings of Eighth Reliability and Maintainability Conference* (*1969 Annals of Assurance Sciences*), Gordon and Breach Science Publishers, New York, July 1969, pp. 392–402.

Blake, K. R., "Some Effects of Machining Processes on the Reliability of Engine Components," *Proceedings of Fourth Reliability and Maintainability Conference*, (*Annals of Reliability and Maintainability*, Vol. 4), Spartan Books, Washington, DC, July 1965, pp. 639–660.

Bocchi, W. J. "Predicting Mechanical Reliability," *Proceedings of Annual Reliability and Maintainability Symposium*, IEEE, New York, 1981, pp. 33–37.

Boe, C. and Tveit, O. J., "Reliability Engineering and Safety at Sea," *IEEE Transactions on Reliability*, Vol. R-23, No. 3, Aug. 1974, pp. 174–178.

Bogdanoff J. L. and Kozin, F., "On a New Cumulative Damage Model for Fatigue," *Proceedings of Annual Reliability and Maintainability Symposium*, IEEE, New York, 1981, pp. 9–18.

Bompass-Smith, J. H., "The Determination of Distributions That Describe the Failures of Mechanical Components," *Proceedings of Eighth Reliability and Maintainability Conference* (*1969 Annals of Assurance Sciences*), Gordon and Breach Science Publishers, New York, 1969, pp. 343–356.

Bompass-Smith, J. H., *Mechanical Survival: The Use of Reliability Data*, McGraw-Hill, London, 1973.

Bonnett, G. M., "Reliability in Shipbuilding Through Nondestructive Testing," *Proceedings of Annual Symposium on Reliability*, IEEE, New York, Jan. 1969, pp. 333–339.

Bourne, A. J., "System Reliability Analysis Techniques and Their Relationship to Mechanical/Structural Reliability Problems," *Nuclear Engineering and Design*, Vol. 60, 1980, pp. 133–136.

Braland, R. E. and Horne, R. C., "Reliability Prediction for Ship Machinery," *Proceedings of Annual Symposium on Reliability*, IEEE, New York, Jan. 1971, pp. 219–226.

Brammer, J. E., "Activities of the Mechanical-System Design Subgroup of the Joint Technical Coordinating Group for RAM," *Proceedings of Annual Reliability and Maintainability Symposium*, IEEE, New York, 1981, pp. 275–277.

Bratt, M. J., Reethof, G., and Weber, G. W., "A Model of Time Varying and Interfering Stress-Strength Probability Density Distributions with Consideration for Failure Incidence and Property Degradation," *Proceedings of Third Aerospace Reliability and Maintainability Conference*, SAE, Warrendale, PA, June 1964, pp. 566–575, and 592.

Bratt, M. J., Truscott, H. A., and Weber, G. W., "Probabilistic Strength Mapping Reliability Versus Life Prediction Tool," *Proceedings of Seventh Reliability and Maintainability Conference* (*1968 Annals of Assurance Sciences*), ASME, New York, July 1968, pp. 501–510.

Brewer, J. W., "Interim Scale Reliability Statements Consistent With Conventional Materials Strength Criteria," *Proceedings of Tenth Reliability and Maintainability Conference* (*Annals of Reliability and Maintainability*, Vol. 10), ASME, New York, 1971, pp. 63–71.

Broding, W. C., Diederich, F. W., and Parker, P. S., "Structural Optimization and Design Based on a Reliability Design Criterion," *Journal of Spacecraft and Rockets*, Vol. 1, Jan.–Feb. 1964, pp. 56–61.

Brook, R. H. W., *Reliability Concepts in Engineering Manufacture*, Halsted Press, division of Wiley, New York, 1972.

Browning, G. V. and Bester, M. H., "Experimental Evaluation of Reliable Soldering Processes," *Proceedings of Ninth National Symposium on Reliability and Quality Control*, IEEE, New York, 1963, pp. 211–221.

Buckland, W. R., *Statistical Assessment of the Life Characteristics*, Griffin, London, 1964.

Burger, M., "Reliability Program for a Shipboard Mechanical System," *Proceedings of Annual Symposium on Reliability*, IEEE, New York, Jan. 1967, pp. 789–802.

Burns, J. J. Jr., "Reliability of Nuclear Mechanical Systems," *Proceedings of Annual Reliability and Maintainability Symposium*, IEEE, New York, 1975, pp. 163–169.

Bury, K. V., "On Product Reliability under Random Field Loads," *IEEE Transactions on Reliability*, Vol. R-27, No. 4, Oct. 1978, pp. 258–260.

Bussiere, R., "A Method for Critiquing Designs and Predicting Reliability in Advance of Hardware Availability," SAE Paper 343A, April 1961.

Bussiere, R., "Predict Reliability First . . . Build It Later," *SAE Journal*, Vol. 69, July 1961, pp. 74–76.

Cable, C. W. and Virene, E. P., "Structural Reliability with Normally Distributed Static and Dynamic Loads and Strength," *Proceedings of Annual Symposium on Reliability*, IEEE, New York, Jan. 1967, pp. 329–336.

Canterbury, J. and Lowther, J. D., "Application of Dimensional Analysis to the Prediction of Mechanical Reliability," Sept. 1976 (available from NTIS as AD-A035295).

Carter, A. D. S., *Mechanical Reliability*, Wiley, London, 1972.

Cavalleri, R., "Reliability Improvement of a Positive Displacement Compressor Expander," *Proceedings of Annual Reliability and Maintainability Symposium*, IEEE, New York, 1983, pp. 30–36.

Cheng, K. F. and Chao, A., "Confidence Intervals for Reliability From Stress-Strength Relationships," *IEEE Transactions on Reliability*, Vol. R-33, No. 3, Aug. 1984, pp. 246–249.

Chenoweth, H. B., "Vibration-Thermal Screening Reliability-Prediction," *Proceedings of Annual Reliability and Maintainability Symposium*, IEEE, New York, 1982, pp. 91–96.

Chernowitz, G., "Electromechanical Component Reliability," Rome Air Development Center, Griffiss Air Force Base, NY, Rept. RADC-TR-63-295, May 1963 (available from NTIS).

Chester, L. B., Nolf, C. F., and Kececioglu, D. "Combined Bending-Torsion Fatigue Reliability—III," *Proceedings of Annual Reliability and Maintainability Symposium*, IEEE, New York, 1975, pp. 511–518.

Chiu, Y. P. et al., "Development of a Mathematical Model for Predicting Life of Rolling Bearings," Rome Air Development Center, Griffiss Air Force Base, NY, Rept. RADC-TR-68-54, April 1968 (available from NTIS).

Chopra, P. S. and Wolosewicz, R. M., "Application of Reliability, Maintainability and Availability Engineering to Solar Heating and Cooling Systems," *Proceedings of Annual Reliability and Maintainability Symposium*, IEEE, New York, 1980, pp. 248–253.

Cohen, A. C., "Maximum Likelihood Estimation in the Weibull Distribution," *Technometrics*, Vol. 17, 1965, pp. 347–351.

Colangelo, V. J. and Heiser, F. A., *Analysis of Metallurgical Failures*, Wiley, New York, 1974.

Conover, J. C., Kaeckel, H. R., and Keppola, W. T., "Simulation of Field Loadings in Fatigue Testing," SAE Paper 660102, Jan. 1966.

Cornog, D. Y. and Ruder, A. H., "A Postal Service Field Evaluation of Letter Sorting Machine Keyboards," *Proceedings of Annual Reliability and Maintainability Symposium*, IEEE, New York, 1977, pp. 43–48.

Cottrell, D. F., "Dormancy Effects on Nonelectronics," *Proceedings of Annual Reliability and Maintainability Symposium*, IEEE, New York, 1977, pp. 7–11.

Crow, L. H., "Reliability Analysis for Complex, Repairable Systems," *Reliability and Biometry: Statistical Analysis of Lifelength*, edited by F. Proschan and R. J. Serfling, SIAM, Philadelphia, 1974, pp. 379–410.

Czechowski, A. and Lenk, A., "Miner's Rule in Mechanical Test of Electronic Parts," *IEEE Transactions on Reliability*, Vol. R-27, No. 3, Aug. 1978, pp. 183–190.

D'Agostino, R. B. and Stephens M. (eds.), *Goodness-of-Fit Techniques*, Marcel Dekker, New York, 1984.

Davidson, J. R., "Reliability After Inspection," *Proceedings of ASTM Symposium on the Fatigue of Composite Materials*, ASTM, Philadelphia, 1973, pp. 110–111.

Davis, D. J., "An Analysis of Some Failure Data," *Journal of American Statistical Association*, 1952, pp. 113–150.

Dehardt, J. H. and McLaughlin, H. D., "Using Bayesian Methods to Select a Design with Knoron Reliability Without a Confidence Coefficient," *Proceedings of Fifth Reliability and Maintainability Conference* (*Annals of Reliability and Maintainability*, Vol. 5), AIAA, Washington, DC, 1966, pp. 611–617.

Demsky, S., Fedele, M., and Frassmann, F., "Portable Respirator Reliability Demonstration Test—Case History," *Proceedings of Annual Reliability and Maintainability Symposium*, IEEE, New York, 1976, pp. 234–238.

Dennis, N. G., "PMR, NDE, Design Practices: Present and Future," *Proceedings of Annual Reliability and Maintainability Symposium*, IEEE, New York, 1977, pp. 164–170.

Desmond, A. F., "On the Relationship between Two Fatigue-Life Models," *IEEE Transactions on Reliability*, Vol. R-35, No. 2, June 1986, pp. 167–169.

DesPlas, E. P., "Reliability in the Manufacturing Cycle," *Proceedings of Annual Reliability and Maintainability Symposium*, IEEE, New York, 1986, pp. 139–144.

Deutschman, A. P., Michels, H. J., and Wilson, C. E., *Machine Design—Theory and Practice*, Macmillan, New York, 1975.

Dey, K. A., "Practical Statistical Analysis for the Reliability Engineer," IIT Research Institute, Chicago, 1983 (available from NTIS).

Dhillon, B. S., *Reliability Engineering in Systems Design and Operation*, Van Nostrand Reinhold, New York, 1982.

Dhillon, B. S., *Quality Control, Reliability and Engineering Design*, Marcel Dekker, New York, 1985.

Dhillon, B. S., *Human Reliability: With Human Factors*, Pergamon Press, New York, 1986.

Dhillon, B. S., "Mechanical Reliability: Interference Theory Models," *Proceedings of Annual Reliability and Maintainability Symposium*, IEEE, New York, 1980, pp. 462–467.

Dhillon, B. S., "Interference Theory Application to Mechanical Systems," *Proceedings of Annual Reliability and Maintainability Symposium*, IEEE, New York, 1981, pp. 24–27.

Dhillon, B. S., "RAM Analysis of Vehicles in Changing Weather," *Proceedings of Annual Reliability and Maintainability Symposium*, IEEE, New York, 1984, pp. 48–53.

Dhillon, B. S., "Stress-Strength Reliability Models," *Microelectronics and Reliability*, Vol. 20, 1980, pp. 513–516.

Dhillon, B. S., "Unified Availability Modeling: A Redundant System with Mechanical, Electrical, Software, Human and Common-Cause Failures," *Microelectronics and Reliability*, Vol. 21, No. 5, 1981, pp. 653–659.

Dhillon, B. S., "Transit Vehicle Reliability Models," *Microelectronics and Reliability*, Vol. 22, No. 3, 1982, pp. 619–624.

Dhillon, B. S., "Bibliography of Literature on Transit System Reliability," *Microelectronics and Reliability*, Vol. 22, No. 3, 1982, pp. 641–651.

Dhillon, B. S., "Mechanical Component Reliability under Environmental Stress," *Microelectronics and Reliability*, Vol. 20, 1980, pp. 153–160.

Dhillon, B. S., and Natesan, J., "Probabilistic Analysis of a Pulverizer System with Common-Cause Failures," *Microelectronics and Reliability*, Vol. 22, 1982, pp. 1121–1134.

Dhillon, B. S., Proctor, C. L., and Elsayed, E. A. R., "Stress-Strength Reliability Analysis of a Redundant System," *Reliability, Stress Analysis and Failure Prevention Methods in Mechanical Design*, edited by W. D. Milestone, ASME, New York, 1980, pp. 9–12.

Dhillon, B. S. and Rayapati, S. N., "Reliability and Availability Analysis of On-Surface Transit Systems," *Microelectronics and Reliability*, Vol. 24, No. 6, 1984, pp. 1029–1033.

Dhillon, B. S. and Rayapati, S. N., "Availability Modeling of On-Surface Transit Systems with Partial Failures," *Maintenance Management International*, Vol. 6, 1986, pp. 205–214.

Dhillon, B. S. and Rayapati, S. N., "Transit System Availability Modelling: A Study of the Erlangian Repair Time Distribution," *International Journal of Systems Science*, Vol. 18, No. 1, 1987, pp. 9–24.

Dhillon, B. S. and Rayapati, S. N., "Stochastic Analysis of Transit Systems," *Stochastic Analysis and Applications*, Vol. 4, No. 1, 1986, pp. 49–75.

Dhillon, B. S. and Rayapati, S. N., "Reliability and Availability Analysis of Transit Systems," *Microelectronics and Reliability*, Vol. 25, No. 6, 1985, pp. 1073–1085.

Dhillon, B. S. and Rayapati, S. N., "Reliability Modeling of On-Surface Transit Systems with Human Errors," *Microelectronics and Reliability*, Vol. 25, No. 6, 1985, pp. 1087–1098.

Dhillon, B. S. and Singh, C., *Engineering Reliability: New Techniques and Applications*, Wiley, New York, 1981.

Dillin, A. L., "Reliability Assessment of Army Weapons and Weapons Systems," *Proceedings of Annual Symposium on Reliability*, IEEE, New York, 1971, pp. 74–76.

Dillin, A. L., "Military Vehicle Reliability Assessment," *Proceedings of Annual Symposium on Reliability*, IEEE, New York, Jan. 1969, pp. 276–279.

Disney, R. and Lipson, C., "The Determination of the Probability of Failure by Stress/Strength Interference Theory," *Proceedings of Annual Symposium on Reliability*, IEEE, New York, Jan. 1968, pp. 417–422.

Disney, R. L., Lipson, C., and Sheth, N. J., "The Determination of the Probability of Failure by Stress/Strength Interference Theory," *Proceedings of Annual Symposium on Reliability*, IEEE, New York, Jan. 1968, pp. 417–422.

Distefano, R., "Development of Ship Equipment Maintenance History," *Proceedings of Annual Reliability and Maintainability Symposium*, IEEE, New York, 1987, pp. 268–273.

DiTrapani, A. R. and Dumbleton, J. J., "Reliability and Availability of the DD963 Destroyer Program," *Proceedings of Ninth Reliability and Maintainability Conference* (*Annals of Reliability and Maintainability*, Vol. 9), SAE, Warrendale, PA, July 1970, pp. 344–352.

Dunkel, W. E., "Long Term Component Analysis Utilizing Short Term Materials and Design Studies," *Proceedings of Eighth National Symposium on Reliability and Quality Control*, IEEE, New York, 1962, pp. 520–523.

Dunn, W. W., "An Insulation Materials Design to Achieve a High Reliability Medium-Weight Wire," *Proceedings of Seventh Reliability and Maintainability Conference* (*1968 Annals of Assurance Sciences*), ASME, New York, July 1968, pp. 537–546.

Dwyer, H. I. Jr., "Review of the Mechanical Aspects of Electronic Equipment Design as a Tool for Missile Reliability," *Proceedings of Fifth National Symposium on Reliability and Quality Control*, IEEE, New York, 1959, pp. 235–242.

Earles, D. R. and Eddins, M. F., "Failure Stresses," *Proceedings of Third Aerospace Reliability and Maintainability Conference*, AIAA, Washington, DC, May 1963, pp. 318–325.

Ebrahimi, N., "Estimation of Reliability For a Series Stress-Strength System," *IEEE Transactions on Reliability*, Vol. R-31, No. 2, June 1982, pp. 202–205.

Ebrahimi, N., "On Estimating the Reliability for a Stress-Strength System," *IEEE Transactions on Reliability*, Vol. R-34, No. 3, Aug. 1985, p. 280.

Einbinder, S. K., "Reliability Models and Estimation in Terms of Stress-Strength," Doctoral Dissertation, Polytechnic Institute of New York, New York, 1968.

Etherton, J. R., "Automated Maintainability Records and Robot Safety," *Proceedings of Annual Reliability and Maintainability Symposium*, IEEE, New York, 1987, pp. 135–140.

Evans, R. A., "Stress vs. Damage," *Proceedings of Annual Symposium on Reliability*, IEEE, New York, Jan. 1967, pp. 633–635.

Fiderer, L., "Design For Reliability In Hostile Environment," *Microelectronics and Reliability*, Vol. 15 Supp., 1976, pp. 75–85.

Flage, D. and Peterson, M., "Mechanical Failure Technology: A Coordinated Government Program," *Proceedings of Annual Symposium on Reliability*, IEEE, New York, Jan. 1969, pp. 127–132.

Forrestor, E. R. and Thevenow, V. H., "Designing For Expected Fatigue Life," *Proceedings of Seventh Reliability and Maintainability Conference* (1968 *Annals of Assurance Sciences*), ASME, New York, July 1968, pp. 511–519.

Frankel, H. E., Babecki, A. J., and Predmore, R. E., "Fatigue Phenomena in Small Metallurgically-Bonded Joints," *Proceedings of Annual Symposium on Reliability*, IEEE, New York, Jan. 1966, pp. 380–394.

Freudenthal, A. M., "Safety, Reliability and Structural Design," *ASCE Journal of the Structural Division*, Vol. 87, No. ST3, 1961, pp. 304–323.

Freudenthal, A. M., "Fatigue Sensitivity Measures Reliability," *SAE Journal*, Vol. 70, May 1962, pp. 37–41.

Frey, H. H., "Safety Evaluation of Mass Transit Systems by Reliability Analysis," *IEEE Transactions on Reliability*, Vol. R-23, No. 3, Aug. 1974, pp. 161–168.

Frost, E. C. and Stacey, A. B., "Reliability and Maintainability Considerations for the Next Generation Subsonic Transport," *Proceedings of Seventh Reliability and Maintainability Conference* (*1968 Annals of Assurance Sciences*), ASME, New York, July 1968, pp. 177–182.

Fulton, D., Holtz, J. N., and Schafer, R. E., "Nonelectronic Reliability Part Data Collection and Analysis," *Proceedings of Annual Symposium on Reliability*, IEEE, New York, Jan. 1968, pp. 114–121.

Furman, T. T., *Approximate Methods in Engineering Design*, Academic Press, London, 1981.

Galetto, F., "Mechanics Reliability," *Proceedings of International Symposium on Automotive Technology and Automation*, Automotive Automation Ltd., Croyden, England, Sept.–Oct. 1976, Vol. 1, pp. 1–11.

Gallagher, J. P. and Ratchford, G. J., "Statistical Testing of Manufacturing Equipment," *Proceedings of Annual Symposium on Reliability*, IEEE, New York, Jan. 1967, pp. 230–239.

Garrick, B. J., "Reliability Analysis of Bulk Mail Handling Facilities," *Proceedings of Tenth Reliability and Maintainability Conference* (*Annals of Reliability and Maintainability*, Vol. 10), ASME, New York, June 1971, pp. 394–399.

Gauthey, J. R. and Ventriglio, F. J., "Reliability of Navy Pollution Abatement Equipment," *Proceedings of Annual Reliability and Maintainability Symposium*, IEEE, New York, 1976, pp. 182–187.

Gibson, W. R., "Reliability Achievements in Sonar Systems," *Proceedings of Annual Symposium on Reliability*, IEEE, New York, Jan. 1968, pp. 149–154.

Giordano, P. J. and Sacks, J., "Reliability and Maintainability Considerations for Total Ship Systems," *Proceedings of Annual Symposium on Reliability*, IEEE, New York, Jan. 1968, pp. 155–163.

Goode, H. P. and Kao, J. H. K., "Weibull Tables for Bio-assaying and Fatigue Testing," *Proceedings of Ninth National Symposium on Reliability and Quality Control*, IEEE, New York, 1963, pp. 270–286.

Green, A. E. and Bourne, A. J., "Reliability Technology," Wiley, New York, 1972.

Gregory, L. D. and Spruill, C. E., "Structural Reliability on Re-Entry Vehicles Using Brittle Materials in Primary Structure," *Proceedings of IAS Aerospace Systems Reliability*, April 1962, pp. 33–45 (available from AIAA, Washington, DC).

Hahn, P. G., "VLS Hatch Actuation System Dynamic Analysis," *Proceedings of Annual Reliability and Maintainability Symposium*, IEEE, New York, 1984, pp. 39–43.

Hamilton, D. O. et al., "Investigation of Reliability of Mechanical Systems," Lockheed-Georgia Co., Marietta, GA, Oct. 1965 (available from NTIS as AD-475977).

Hamilton, R. L., "Ship Propulsion Machinery Reliability and Availability Analysis," *Proceedings of Annual Symposium on Reliability*, IEEE, New York, Jan. 1968, pp. 137–148.

Harrington, R. L., Coats, J. W., and Farley, F. E., "Assessment of Marine Systems' Reliability and Maintainability," *Proceedings of Ninth Reliability and Maintainability Conference* (*Annals of Reliability and Maintainability*, Vol. 9), SAE, Warrendale, PA, July 1970, pp. 298–318.

Hartman, P., Luetjen, P., and Buckberg, M., "Lower Limits for Total Ship Reliability," *Proceedings of Annual Reliability and Maintainability Symposium*, IEEE, New York, 1984, pp. 210–215.

Haugen, E. B., *Probabilistic Approaches to Design*, Wiley, New York, 1968.

Haugen, E. B., "Statistical Methods for Structural Reliability Analysis," *Proceedings of Tenth National Symposium on Reliability and Quality Control*, IEEE, New York, Jan. 1964, pp. 97–121.

Haugen, E. B., "Implementing a Structural Reliability Program," *Proceedings of Eleventh National Symposium on Reliability and Quality Control*, IEEE, New York, Jan. 1965, pp. 158–168.

Haugen, E. B., *Probabilistic Mechanical Design*, Wiley, New York, 1980.

Haugen, E. B., "Electromechanical Analysis by Statistics," *Proceedings of Seventh Annual West Coast Reliability Symposium*, Western Periodicals, North Hollywood, CA, May 1966, pp. 31–50.

Haugen, E. B. and Wirsching, P., "Probabilistic Fatigue Design Alternative to Miner's Cumulative Damage Rule," *Proceedings of Annual Reliability and Maintainability Symposium*, IEEE, New York, 1973, pp. 428–436.

Haviland, R. P., *Engineering Reliability and Long Life Design*, Van Nostrand, Princeton, NJ, 1964.

Heimann, D. I., "The Determination of Transit-System Dependability," *Proceedings of Annual Reliability and Maintainability Symposium*, IEEE, New York, 1979, pp. 314–322.

Heller, R. A. and Heller, A. S., "Analysis of Early Failures in Unequal Size Samples," *Proceedings of Annual Reliability and Maintainability Symposium*, IEEE, New York, 1973, pp. 198–202.

Heller, R. A. and Heller, A. S., "Fatigue Life and Reliability of a Redundant Structure," *Proceedings of Fourth Reliability and Maintainability Conference* (*Annals of Reliability and Maintainability*, Vol. 4), Spartan Books, Washington, DC, July, 1965, pp. 881–896.

Heller, A. S. and Oelkers, E., "Reliability Predictions for a Pressurized Water Reactor during the Design Process," *Proceedings of Annual Reliability and Maintainability Symposium*, IEEE, New York, 1978, pp. 348–354.

Heller, R. A., and Shinozuka, M., "State-of-the-Art: Reliability Techniques in Materials and Structures," *Proceedings of Ninth Reliability and Maintainability Conference* (*Annals of Reliability and Maintainability*, Vol. 9), SAE, Warrendale, PA, July 1970, pp. 635–636.

Higgins, P. E. and Constantinides, A., "MK 30 Mod 1, Reliability Enhancement Program," *Proceedings of Annual Reliability and Maintainability Symposium*, IEEE, New York, 1987, pp. 146–151.

Hjorstvang, H. and Overeen, J. V., "Reliability and Maintainability Provisions for a Rapid Transit Interurban System," *Proceedings of Seventh Reliability and Maintainability Conference* (*1968 Annals of Assurance Sciences*), ASME, New York, July 1968, pp. 153–166.

Hopkinson, K., "Reliable Valves and Their Performance in Service Equipments," *Proceedings of Fifth National Symposium on Reliability and Quality Control*, IEEE, New York, 1959, pp. 251–264.

Horger, O. J. (ed.), *Metals Engineering Design*, 2nd ed. (publication of ASME), McGraw-Hill, New York, 1965.

Hurd, W. L. Jr., "Full Design Definition is Essential to Missile Reliability," *Proceedings of Eighth National Symposium on Reliability and Quality Control*, IEEE, New York, 1962, pp. 177–180.

Ichikawa, M., "Confidence Limit of Probability of Failure Based on Stress-Strength Model," *Reliability Engineering*, Vol. 8, 1984, pp. 75–83.

Inabinette, R. A., "Reliability Requirements for Truck Systems and Parts," *Proceedings of Annual Reliability and Maintainability Symposium*, IEEE, New York, 1974, pp. 450–455.

Innis, C. L. and Hammand, T., "Predicting Mechanical Design Reliability Using Weighted Fault Trees," *Failure Prevention and Reliability*, ASME, New York, 1977, pp. 213–228.

Isaacs, R. E., "Implementation of Reliability and Maintainability Requirements for Postal Hardware Systems," *Proceedings of Tenth Reliability and Maintainability Conference* (*Annals of Reliability and Maintainability*, Vol. 10), ASME, New York, June 1971, pp. 392–393.

Iuculano, G. and Zanini, A., "Evaluation of Failure Models Through Step-stress Tests," *IEEE Transactions on Reliability*, Vol. R-35, No. 4, Oct. 1986, pp. 409–413.

Johnson, W. F. Jr., "Evaluation of Mechanical Systems Service Experience and Its Application to Reliability in Preliminary Design," *Proceedings of Third Aerospace Reliability and Maintainability Conference*, AIAA, Washington, DC, May 1963, pp. 419–428.

Juvinall, R. C., *Engineering Considerations of Stress, Strain, and Strength*, McGraw-Hill, New York, 1967.

Kalivoda, F. E. and Yun, K. W., "Modeling Mechanical System Accelerated Life Tests," *Proceedings of Annual Reliability and Maintainability Symposium*, IEEE, New York, 1976, pp. 206–212.

Kao, J. H. K., "Statistical Models in Mechanical Reliability," *Proceedings of 11th National Symposium on Reliability and Quality Control*, IEEE, New York, 1965, pp. 240–247.

Kapur, K. C., "Reliability Bounds in Probabilistic Design," *IEEE Transactions on Reliability*, Vol. R-24, No. 3, Aug. 1975, pp. 193–195.

Kapur, K. C. and Lamberson, L. R., *Reliability in Engineering Design*, Wiley, New York, 1977.

Karr, C. H. and Wagner, G. L., "R&M—Today's Heating and Cooling vs. Solar Energy," *Proceedings of Annual Reliability and Maintainability Symposium*, IEEE, New York, 1976, pp. 491–499.

Katz, M. D., Jaffe, H., and Rosenthal, S. A., "Designing Reliability into the B-58 Bombing-Navigation System," *Proceedings of Seventh National Symposium on Reliability and Quality Control*, IEEE, New York, 1961, pp. 273–282.

Kececioglu, D., "Why Design by Reliability?," *Proceedings of Seventh Reliability and Maintainability Conference* (*Annals of Assurance Sciences*, Vol. 7), ASME, New York, July 1968, p. 491.

Kececioglu, D., "Reliability Analysis of Mechanical Components and Systems," *Nuclear Engineering and Design*, North-Holland, Amsterdam, 1972, pp. 259–290.

Kececioglu, D., "Fundamentals of Mechanical Reliability Theory and Applications to Vibroacoustic Failures," *Proceedings of Reliability Design for Vibroacoustic Environments*, ASME, New York, 1974, pp. 1–38.

Kececioglu, D., "Probabilistic Design Methods for Reliability and Their Data and Research Requirements," *Failure Prevention and Reliability Conference*, ASME, New York, 1977, pp. 285–309.

Kececioglu, D. and Chester, L. B., "Fatigue Reliability under Combined Mean and Alternating Axial Stresses for AISI 1018 and 1038 Steels," ASME Paper 75-DET-128, Sept. 1975.

Kececioglu, D. and Chester, L. B., "Combined Axial Stress Fatigue Reliability of AISI 4130 and 4340 Steels," *Transactions of ASME, Journal of Engineering for Industry*, Vol. 98, Feb. 1976, pp. 153–160.

Kececioglu, D., Chester, L. B., and Dodge, T. M., "Alternating Bending-Steady Torque Fatigue Reliability," *Proceedings of Annual Reliability and Maintainability Symposium*, IEEE, New York, 1974, pp. 163–173.

Kececioglu, D., Chester, L. B., and Dodge, T. M., "Alternating Bending-Steady Torque Fatigue Reliability," *Proceedings of Annual Reliability and Maintainability Symposium*, IEEE, New York, 1974, pp. 153–173.

Kececioglu, D., Chester, L. B., and Dodge, T. M., "Combined Bending-Torsion Fatigue Reliability of AISI 4340 Steel Shafting with $K_t = 2.34$-II," *Transactions of ASME, Journal of Engineering for Industry*, Vol. 97, May 1975, pp. 748–761.

Kececioglu, D., Chester, L. B., and Gardner, E. O., "Sequential Cumulative Fatigue Reliability, *Proceedings of Annual Reliability and Maintainability Symposium*, IEEE, New York, Jan. 1974, pp. 533–539.

Kececioglu, D. and Cormier, D., "Designing a Specified Reliability Directly Into a Component," *Proceedings of Third Aerospace Reliability and Maintainability Conference*, SAE, Warrendale, PA, June 1964, pp. 546–565.

Kececioglu, D. and Haugen, E. B., "A Unified Look at Design Safety Factors, Safety Margins, and Measures of Reliability," *Proceedings of Seventh Reliability and Maintainability Conference* (*Annals of Assurance Sciences*, Vol. 7), ASME, New York, July 1968, pp. 520–528.

Kececioglu, D. and Haugen, E. B., "Fatigue Reliability Design Data for Dynamic Rotary Machinery," ASME Paper 70-AV/SpT-36, June 1970.

Kececioglu, D. and Hughes, R. C., "Reliability and Relative Cost of Steam Turbine Governing and Stop Valve Arrangements," *Proceedings of Third Aerospace Reliability and Maintainability Conference*, AIAA, Washington, DC, May 1963, pp. 98–105.

Kececioglu, D. and Lamarre, G., "Designing Mechanical Components to a Specified Reliability and Confidence Level," *Nuclear Engineering and Design*, Vol. 50, 1978, pp. 149–162.

Kececioglu, D., Smith, R. E., and Felsted, E. A., "Distributions of Cycles-to-Failure in Simple Fatigue and the Associated Reliabilities," *Proceedings of Eighth Reliability and Maintainability Conference* (*1969 Annals of Assurance Sciences*, Vol. 8), Gordon and Breach Science Publishers, New York, July 1969, pp. 357–374.

Kececioglu, D., Smith, R. E., and Felsted, E. A., "Distributions of Strength in Simple Fatigue and the Associated Reliabilities," *Proceedings of Ninth Reliability and Maintainability Conference* (*Annals of Reliability and Maintainability*, Vol. 9), SAE, Warrendale, PA, July 1970, pp. 659–672.

Kececioglu, D., Stultz, J. D., Chester, L. B., and Nolf, C. F., "Fatigue Reliability with Notch Effects for AISI 4130 and 1038 Steels," *Transactions of ASME, Journal of Engineering for Industry*, Vol. 97, Feb. 1975, pp. 359–370.

Kiciman, M. O., "Prediction of Failure for a Multiple Load-Path System under Random Loading," *Operations Research and Reliability*, edited by D. Grouchko, Gordon and Breach Science Publishers, New York, 1971, pp. 567–580.

Kilgore, T. D., "Reliability Models in Mechanical Design," Rept. USAMC-ITC-02-08-73-002, March 1974 (available from NTIS as AD-787297).

Kimball, D. E., "Reliability of Merchant-Marine Turbine-Gear Propulsion Sets," *Proceedings of Sixth Annual Reliability and Maintainability Conference* (*Annals of Reliability and Maintainability*, Vol. 6), SAE, Warrendale, PA, July 1967, pp. 272–281.

King, J. H. Jr., "Reliability Comes to the Transit Industry," *Proceedings of Annual Reliability and Maintainability Symposium*, IEEE, New York, 1976, pp. 314–318.

Kirkpatrick, I., "Predicting Reliability of Electromechanical Devices," *Proceedings of Sixth National Symposium on Reliability and Quality Control*, IEEE, New York, 1960, pp. 272–281.

Kitchel, G. E., "Maintainability Considerations During Design and Development of the XM561, 1-1/4 Ton Cargo Truck," *Proceedings of Fifth Reliability and Maintainability Conference* (*Annals of Reliability and Maintainability*, Vol. 5), AIAA, Washington, DC, July 1966, pp. 811–817.

Kobayashi, A. S. (ed.), *Manual of Engineering Stress Analysis*, Prentice-Hall, Englewood Cliffs, NJ, 1982.

Konno, K., Nakano, K., and Yashimura, Y., "A New Accelerated Fatigue Test: The Effective Random Peak Method," *Proceedings of Annual Reliability and Maintainability Symposium*, IEEE, New York, 1975, pp. 263–268.

Kozoriz, E. P., "Planetary Quarantine and Its Challenge to Materials Reliability," *Proceedings of Sixth Reliability and Maintainability Conference* (*Annals of Reliability and Maintainability*, Vol. 6), SAE, Warrendale, PA, July 1967, pp. 399–404.

Kullman, L. W. and Phillips, G. W., "Reliability Engineering Disciplines Applied to Commercial Weapon Systems—Guns and Ammunition," *Proceedings of Tenth National Symposium on Reliability and Quality Control*, IEEE, New York, 1964, pp. 122–137.

Kurtz, P. H., "A Strength-Stress Interference Model for Time-Varying Probability of Failure and Its Application to Electromechanical System," Naval Sea Systems Command, Rept. TM-75-122, 1975 (available from NTIS as AD-A016299).

Kurtz, P. H., "Reliability Study of a Hydraulic Control System Using the Hybrid Computer," *Proceedings of Tenth Reliability and Maintainability Conference* (*Annals of Reliability and Maintainability*, Vol. 10), ASME, New York, June 1971, pp. 81–85.

Kurtz, P. H., "A Probabilistic Model of Fatigue Failure," *Proceedings of Annual Reliability and Maintainability Symposium*, IEEE, New York, 1974, pp. 528–532.

Lalli, V. R., "Ion Engine Subsystem Reliability Procedure," *Proceedings of Eleventh National Symposium on Reliability and Quality Control*, IEEE, New York, Jan. 1965, pp. 361–379.

Lalli, V. R. and Kececioglu, D., "An Approach to Reliability Determination of a Rotating Component Subjected to Complex Fatigue," *Proceedings of Ninth Reliability and Maintainability Conference* (*Annals of Reliability and Maintainability*, Vol. 9), SAE, Warrendale, PA, July 1970, pp. 534–548.

Lambert, R. G., "Mechanical Reliability for Low Cycle Fatigue," *Proceedings of Annual Reliability and Maintainability Symposium*, IEEE, New York, 1978, pp. 179–183.

Lambert, R. G., "Low-Cycle Fatigue Reliability Expressions," *Proceedings of Annual Reliability and Maintainability Symposium*, IEEE, New York, 1982, pp. 47–50.

Landgraf, R. W. and Conle, A., "Trends in Assuring the Mechanical Durability of Automotive Structures," *Proceedings of XXth FISITA Congress*, Pub. P-143, SAE, Warrendale, PA, May 1984, pp. 4.124–4.130 (also designated SAE Paper 845124).

Lazar, L. S., "Accelerated Fatigue of Plastics," Bull. 220, ASTM, Philadelphia, Feb. 1957, p. 67.

Leach, K. E., "NAD/MCAIR F-18 APU Components—A Case History of Mechanical Reliability," *Proceedings of Annual Reliability and Maintainability Symposium*, IEEE, New York, 1983, pp. 37–42.

Liebowitz, H., "Navy Reliability Research," *Proceedings of Sixth Reliability and Maintainability Conference* (*Annals of Reliability and Maintainability*, Vol. 6), SAE, Warrendale, PA, July 1967, pp. 34–53.

Lemaitre, J. and Plumtree, A., "Application of Damage Concepts to Predict Creep-Fatigue Failures," *Journal of Engineering Materials and Technology*, Vol. 101, July 1979, pp. 284–292.

Lennon, J. R., "Reliability Analysis of Ship Systems During Contract Definition," *Proceedings of Annual Symposium on Reliability*, IEEE, New York, Jan. 1968, pp. 545–557.

Levine, S., "Reliability Data for Improving Ship Systems," *Proceedings of Annual Symposium on Reliability*, IEEE, New York, Jan. 1971, pp. 238–244.

Lewis, G. A., "Engineered Marine System Maintenance Extends Life Cycle," *Proceedings of Annual Reliability and Maintainability Symposium*, IEEE, New York, 1982, pp. 283–290.

Lieblein, J. and Zelen, M., "Statistical Investigation of the Fatigue Life of Deep-Groove Ball Bearings," *Journal of Research of the National Bureau of Standards*, Vol. 5, 1956, pp. 273–316.

Ling, A. H. K., "Reliability Experience in the Development of an Intermediate Capacity Transit System," *Proceedings of Annual Reliability and Maintainability Symposium*, IEEE, New York, 1981, pp. 38–43.

Lipson, C., "Reliability Prediction-Mechanical Stress/Strength Interference Models," Rome Air Development Center, Griffiss Air Force Base, NY, Rept. RADC-TR-66-710, March 1967.

Lipson, C., Kerawalla, C., and Mitchell, L., *Engineering Applications of Reliability*, Engineering Summer Conferences, College of Engineering, University of Michigan, Ann Arbor, 1963.

Lipson, C. and Sheth, N. J., *Statistical Design and Analysis of Engineering Experiments*, McGraw-Hill, New York, 1973, pp. 311–343.

Lipson, C., Sheth, N. J., and Disney, R. L., "Reliability Prediction-Mechanical Stress/Strength Interference," Rome Air Development Center, Griffiss Air Force Base, NY, Rept. RADC-TR-66-710, March 1967.

Lipson, C., Sheth, N. J., Disney, R. L., and Altun, M., "Reliability Prediction-Mechanical Stress/Strength Interference (Nonferrous)," Rome Air Development Center, Griffiss Air Force Base, NY, Rept. RADC-TR-68-403, Feb. 1969.

Little, R. E., *Tables for Estimating Median Fatigue Limits*, STP 731, ASTM, Philadelphia, 1981.

Little, R. E. and Ekvall, Y. (eds.), *Statistical Analysis of Fatigue Data*, STP 744, ASTM, Philadelphia, 1981.

Lloyd, D. K., "Reliability Demonstration of Clustered Engines," *Proceedings of Sixth Reliability and Maintainability Conference* (*Annals of Reliability and Maintainability*, Vol. 6), SAE, Warrendale, PA, July 1967, pp. 192–199.

Lloyd, F. H. and Mokkapati, C., "Reliability Demonstration in a Transit Environment," *Proceedings of Annual Reliability and Maintainability Symposium*, IEEE, New York, 1987, pp. 35–37.

Lloyd, G. J. and Wareing, J., "Life-Prediction Methods for Combined Creep-Fatigue Endurance," *Metals Technology*, Vol. 8, Aug. 1981, pp. 297–305.

MacGregor, D. S. and Paul, F. A., "Reliability Inputs to Mariner 9 Data Explosion," *Proceedings of Annual Reliability and Maintainability Symposium*, IEEE, New York, 1974, pp. 293–301.

Marble, Q. G., "Improving Mechanical Reliability of Digital Computers," *Proceedings of Eleventh National Symposium on Reliability and Quality Control*, IEEE, New York, 1965, pp. 136–143.

Marks, C. H., "Commercial and Government R&M Requirements for Shipboard Incinerators," *Proceedings of Annual Reliability and Maintainability Symposium*, IEEE, New York, 1976, pp. 196–199.

Martin, P., "Reliability in Mechanical Design and Production," *Proceedings of Generic Techniques in Systems Reliability Assessment*, Noordhoff-Leyden, Amsterdam, 1976, pp. 267–271.

Maruvada, S. N., Weise, K. E., and Chamow, M. F., "Failure Rate as a Design Parameter: Possibilities and Limitations," *Proceedings of Annual Reliability and Maintainability Symposium*, IEEE, New York, 1978, pp. 342–347.

Matney, V. D., "Reliability, Pollutants and Aluminum Raw Material," *Proceedings of Annual Reliability and Maintainability Symposium*, IEEE, New York, 1974, pp. 174–178.

Mayer, J. E. Jr., "Prediction of Ball Bearing Assembly Life from the Two-Parameter Weibull Life Distributions of the Components," *Proceedings of Ninth Reliability and Maintainability Conference* (*Annals of Reliability and Maintainability*, Vol. 9), SAE, Warrendale, PA, July 1970, pp. 549–561.

McAdams, P., "Reliability Continuing Challenge in Construction Machinery," *SAE Journal*, Vol. 74, Aug. 1966, pp. 87–89.

McCall, J. L. and French, P. M. (eds.), *Metallography in Failure Analysis*, Plenum Press, New York, 1978.

McCall, J. L. and French, P. M. (eds.), *Metallography as a Quality Control Tool*, Plenum Press, New York, 1980.

McCarthy, G. E., McCarthy, R. L., and Swanger, L. A., "Reliability and Service Life of Steel Truck Wheels," *Proceedings of Annual Reliability and Maintainability Symposium*, IEEE, New York, 1986, pp. 327–331.

McCrory, R.J., "The Design Method—A Scientific Approach to Valid Design," ASME Paper 63-MD-4, May 1963.

McGinnis, N. F. and Vadeboncoeur, J. R., "An Approach to the Assessment of Transportation Service Reliability," *Proceedings of Tenth Reliability and Maintainability Conference* (*Annals of Reliability and Maintainability*, Vol. 10), ASME, New York, June 1971, pp. 360–371.

Mechanical Properties Data Center, Belfour Stulen, Inc., 13919 W. Bay Shore Dr., Traverse City, MI 49684.

Mercado, J. D., "Reliability Prediction Studies of Degradation-Prone Systems," *Proceedings of Annual Reliability and Maintainability Symposium*, IEEE, New York, 1972, pp. 200–204.

Mesloh, R., "Reliability Design Criteria for Mechanical Creep," *Proceedings of Fifth Reliability and Maintainability Conference* (*Annals of Reliability and Maintainability*, Vol. 5), AIAA, Washington, DC, July 1966, pp. 590–597.

Mesloh, R. E. and Mark, D. G., "Variability Analysis of Precision Mechanical Devices," *Proceedings of Third Aerospace Reliability and Maintainability Conference*, SAE, Warrendale, PA, June 1964, pp. 576–588.

Metals and Ceramics Information Center, Battelle Columbus Laboratories, 505 King Ave., Columbus, OH, 43201.

Mihalasky, J., "The Impact of Robots on Product Reliability," *Proceedings of Annual Reliability and Maintainability Symposium*, IEEE, New York, 1985, pp. 464–466.

Miller, C. R., "A Reliability Program for Commercial 'Heating, Ventilating, Air-Conditioning' (HVAC) Controls," *Proceedings of Annual Reliability and Maintainability Symposium*, IEEE, New York, 1983, pp. 485–490.

Mischke, C. R., "Some Tentative Weibullian Descriptions of the Properties of Steels, Aluminums, and Titaniums," ASME Paper 71-Vibr-64, Sept. 1971.

Mittenbergs, A. A., "Fundamental Aspects of Mechanical Reliability," *Mechanical Reliability Concepts*, ASME, New York, 1965, pp. 17–34.

Morris, B. V. and Welch, R. C., "Integrated Reliability Program for Scout Research Vehicle," *Proceedings of Sixth Reliability and Maintainability Conference* (*Annals of Reliability and Maintainability*, Vol. 6), SAE, Warrendale, PA, July 1967, pp. 642–647.

Mukkayee, B., "Residual Fatigue Life Determination of Anchor Bolts," *Journal of Engineering Materials and Technology*, Vol. 101, Jan. 1979, pp. 47–52.

Neal, D. et al., "A Simulation Process for Determining Reliability of Cyclic Random Loaded Structures," U.S. Army Materials and Mechanic Research Center (available from NTIS as AD-P003 849).

Necsulescu, D. S. and Krausz, A. S., "A Multi-Step Stress-Strength Model of a Parallel System," *IEEE Transactions on Reliability*, Vol. R-35, No. 1, April 1986, pp. 119–122.

Neff, J. M. and Kapur, K. C., "Methodology for R&M Comparison of Foreign and US Military Trucks," *Proceedings of Annual Reliability and Maintainability Symposium*, IEEE, New York, 1982, pp. 398–403.

Nenoff, L., "A New Reliability Design Approach for Electro-Mechanical Systems," *Proceedings of Annual Symposium on Reliability*, IEEE, New York, Jan. 1967, pp. 542–552.

Nenoff, L., "Critical Mechanical Design Evaluation of Electronic Communication Systems," *Proceedings of Annual Symposium on Reliability*, IEEE, New York, Jan. 1971, pp. 362–367.

Neogy, R., "Reliability for N/C Machines—A Must," *Proceedings of Annual Symposium on Reliability*, IEEE, New York, Feb. 1970, pp. 101–105.

Nilsson, S. O., "Reliability Data on Automotive Components," *Proceedings of Annual Reliability and Maintainability Symposium*, IEEE, New York, 1975, pp. 276–279.

North, R. E., "Solderability—The Key to Solder Joint Reliability," *Proceedings of Fifth Reliability and Maintainability Conference* (*Annals of Reliability and Maintainability*, Vol. 5), AIAA, Washington, DC, July 1966, pp. 648–652.

Ogawa, K., Suzuki J., and Sano, K., "Automatically Controlled 2-Vessel Pressure-Cooker Test-Equipment," *IEEE Transactions on Reliability*, Vol. R-32, No. 2, June 1983, pp. 164–167.

Okubo, T., Tanaka, S., and Goromaru, K., "Reliability Evaluation of Bolts Tightened in Plastic Region and Their Application to Diesel Engines," SAE Paper 840505, Feb. 1984.

Olsson, D. A., "The Assurance of Reliability in Fabricated Steel Structures through Nondestructive Testing," *Proceedings of Annual Symposium on Reliability*, IEEE, New York, Jan. 1969, pp. 323–328.

Onodera, K., Miki, M., and Nukada, K., "Reliability Assessment for Heavy Machinery by HI-FMECA Method," *Proceedings of Annual Reliability and Maintainability Symposium*, IEEE, New York, 1977, pp. 346–352.

Osgood, C. C., *Fatigue Design*, Wiley-Interscience, New York, 1970.

Padgett, W. J., "On Bayes Estimation of Reliability for the Birnbaum-Saunders Fatigue Life Model," *IEEE Transactions on Reliability*, Vol. R-31, No. 5, Dec. 1982, pp. 436–438.

Paez, T. L., "A Probabilistic Mechanical Design and Analysis Technique," AFWL-TR-81-111, Sept. 1982 (available from NTIS as AD-A 120978).

Pandey, M. and Upadhyay, S. K., "Reliability Estimation in Stress Strength Models: A Bayes Approach," *IEEE Transactions on Reliability*, Vol. R-35, No. 1, April 1986, p. 98.

Parker, E. R., *Materials Data Book for Engineers and Scientists*, McGraw-Hill, New York, 1967.

Payne, A. O. and Graham, A. D., *Reliability Analysis for Optimum Design*, Pergamon Press, Oxford, England, 1979, pp. 329–345.

Peterson, R. E., *Stress Concentration Design Factors*, Wiley, New York, 1953.

Peterson, R. E., *Stress Concentration Factors*, Wiley, New York, 1974.

Pierce, D. L., "Design for Reliability of NC Machine Tool Control," *Proceedings of Annual Symposium on Reliability*, IEEE, New York, Jan. 1966, pp. 245–259.

Plait, A. O., "The Navy's New Maintenance Data Collection Subsystem," *Proceedings of Ninth Reliability and Maintainability Conference* (*Annals of Reliability and Maintainability*, Vol. 9), SAE, Warrendale, PA, July 1970, pp. 319–335.

Popp, H. G., "Reliability Through Statistical Material Property Definition," SAE Paper 580B, Oct. 1962.

Prot, M. E., "Fatigue Testing under Progressive Loading: A New Technique for Testing Materials," *Metallurgical Reviews*, Vol. XLV, No. 12, 1948, p. 481.

Quandt, E. R., "Developments in Machinery Reliability at the Marine Engineering Laboratory," *Proceedings of Annual Symposium on Reliability*, IEEE, New York, Jan. 1967, pp. 428–441.

Quinn, M. F., "Reliability Improvement in the Commercial Truck Industry," *Proceedings of Annual Reliability and Maintainability Symposium*, IEEE, New York, 1987, pp. 378–380.

Rabon, L. M. Jr., "Mechanical-System Reliability Testing," *Proceedings of Annual Reliability and Maintainability Symposium*, IEEE, New York, 1981, pp. 278–282.

Rabon, L. M. Jr., "Application of Reliability Growth to Mechanical Systems," *Proceedings of Annual Reliability and Maintainability Symposium*, IEEE, New York, 1982, pp. 439–443.

Ratnaparkhi, M. V. and Park, W. J., "Lognormal Distribution-Model for Fatigue Life and Residual Strength of Composite Materials," *IEEE Transactions on Reliability*, Vol. R-35, Aug. 1986, pp. 312–315.

Raze, J. D., Nelson, J. J., and Simard, D. J., "Reliability Models for Mechanical Equipment," *Proceedings of Annual Reliability and Maintainability Symposium*, IEEE, New York, 1986, pp. 322–326.

Raze, J. D., Nelson, J. J., Simard, D. J., and Bradley, M., "Reliability Models for Mechanical Equipment," *Proceedings of Annual Reliability and Maintainability Symposium*, IEEE, New York, 1987, pp. 130–134.

Redler, W. M., "Mechanical Reliability Research in the NASA," *Proceedings of Fifth Reliability and Maintainability Conference* (*Annals of Reliability and Maintainability*, Vol. 5), AIAA, New York, 1966, pp. 763–768.

Redler, W. M., "Reliability and Maintainability Requirements for Motor Vehicles," *Proceedings of Tenth Reliability and Maintainability Conference* (*Annals of Reliability and Maintainability*, Vol. 10), ASME, New York, June 1971, pp. 372–381.

Reethof, G., "Relating Probabilistic Methods to Reliability Considerations in Product Design," ASME Paper 70-DE-70, May 1970.

Reethof, G., "State-of-the-Art: Mechanical and Structural Reliability," *Proceedings of Reliability and Maintainability Conference* (*Annals of Reliability and Maintainability*, Vol. 10), ASME, New York, June 1971, p. 62.

Reliability Analysis Center (RAC), "Non-Electronic Parts Reliability Data," RAC, Rome Air Development Center, Griffiss Air Force Base, NY, Rept. NPRD-2, 1981.

Reliability Analysis Center (RAC), "Analysis Techniques for Mechanical Reliability," Fall 1985 (available from IITRI/Reliability Analysis Center, RADC/RAC, Griffiss Air Force Base, NY 13441-5700).

Retterer, B. L., "State of the Art Assessment of Reliability and Maintainability as Applied to Ship Systems," *Proceedings of Annual Symposium on Reliability*, IEEE, New York, Jan. 1969, pp. 133–145.

Riddick, R. P. Jr., "The Effect of Scheduled Repair Cycle on Marine Equipment Reliability," *Proceedings of Annual Symposium on Reliability*, IEEE, New York, Jan. 1967, pp. 457–465.

Rigney, D. A. (ed.), *Fundamentals of Friction and Wear of Materials*, ASM, Metals Park, OH, 1980.

Roark, R. J., *Formulas for Stress and Strain*, McGraw-Hill, New York, 1965.

Robinson, C., "The Influence of Gear Bore Deflections on Bearing Reliability," SAE Paper 821046, Sept. 1982.

Rolfe, S. T. and Barson, J. M., *Fracture and Fatigue Control in Structures—Applications of Fracture Mechanics*, Prentice-Hall, Englewood Cliffs, NJ, 1977.

Rome Air Development Center & Hughes Aircraft Co., "Thermal Guide for Reliability Engineers," Rome Air Development Center, Griffiss Air Force Base, NY, Rept. RADC-TR-82-172, 1982 (available from NTIS).

Rothstein, A. A., "New Concepts in the Prediction of Mechanical and Structural Reliability," *Proceedings of Third Aerospace Reliability and Maintainability Conference*, AIAA, Washington, DC, May 1963, pp. 91–97.

Sacks, J., "Why Reliability and Maintainability in the Marine Industry," *Proceedings of Ninth Reliability and Maintainability Conference* (*Annals of Reliability and Maintainability*, Vol. 9), SAE, Warrendale, PA, July 1970, pp. 294–297.

Salt, T. L., "A Designer's Approach to the Fatigue Failure Mechanism," *Proceedings of Annual Reliability and Maintainability Symposium*, IEEE, New York, 1973, pp. 437–446.

Sarver, S. J., "Reliability Evaluation of a Containment Fan Cooler System," *Proceedings of Annual Reliability and Maintainability Symposium*, IEEE, New York, 1975, pp. 154–162.

Savage, M., Paridon, C. A., and Coy, J. J., "Reliability Model for Planetary Gear Trains," U.S. Army Aviation Research and Development Command, Washington, DC, Rept. AV RADCOM TR-82-C-6, 1982.

Schade, G. R. and Mischke, C. R., "Deterministic vs Probabilistic Models in Slider-Crank Function Generator Design," *Proceedings of Ninth Reliability and Maintainability Conference* (*Annuals of Reliability and Maintainability*, Vol. 9), SAE, Warrendale PA, July 1970, pp. 527–533.

Schatz, R., Shooman, M. L., and Shaw, L., "Application of Time Dependent Stress-Strength Models of Non-electrical and Electrical Systems," *Proceedings of Annual Reliability and Maintainability Conference*, IEEE, New York, 1966, pp. 598–604.

Schertel, H. V., "Safety, Reliability and Maintainability of Supramar Commercial Hydrofoil Boats," *Proceedings of Seventh Reliability and Maintainability Conference* (*1968 Annals of Assurance Sciences*), ASME, New York, July 1968, pp. 146–152.

Schramp, J. M., "Metallization Failures in Semiconductor Devices," *Proceedings of Fifth Reliability and Maintainability Conference* (*Annals of Reliability and Maintainability*, Vol. 5), AIAA, Washington, DC, July 1966, pp. 177–181.

Schriever, F. M., "Ford Instrument Company's Reliability Program," *Proceedings of Fifth National Symposium on Reliability and Quality Control*, IEEE, New York, 1959, pp. 425–430,

Schwartz, H. C., "Reliability Engineering in Transportation—A Pragmatic Approach," *Proceedings of Annual Reliability and Maintainability Symposium*, IEEE, New York, 1976, pp. 319–321.

Sears, R. W. and Bailey, H. J., "Reliability Management for Marine Instrumentation," *Proceedings of Annual Symposium on Reliability*, IEEE, New York, Jan. 1971, pp. 383–388.

Shah, H. C. and Chow, T.-Y., "Use of Maximum Entropy in Estimating the Damage Distribution of a Single Degree of Freedom System Subjected to Random Loading," *Proceedings of Fifth Reliability and Maintainability Conference* (*Annals of Reliability and Maintainability*, Vol. 5), AIAA, Washington, DC, July 1966, pp. 598–604.

Shaw, L., Shooman, M. L., and Schatz, R., "Time-Dependent Stress-Strength Models for Non-Electrical and Electrical Systems," *Proceedings of Annual Reliability and Maintainability Symposium*, IEEE, New York, 1973, pp. 186–197.

Sherman, A. M., Davies, R. G., and Krause, A. R., "Fatigue Life Predictions for High Strength Steels in Automotive Applications," SAE Paper 810435, Feb. 1981.

Shigley, J. E., *Mechanical Engineering Design*, McGraw-Hill, New York, 1972.

Shinozuka, M., "On the Reliability Analysis of Mechanical Systems," *Proceedings of Fifth International Symposium on Space Technology and Science*, Sept. 1963, AGNE Corp, Tokyo, 1964, pp. 943–954.

Shooman, M. L., *Probabilistic Reliability: An Engineering Approach*, McGraw-Hill, New York, 1968.

Simon, R., "Dimensionless Parameter Reliability Analysis and Application to Mechanical Creep," *Transactions of ASME, Journal of Basic Engineering*, Vol. 88, No. 1, March 1966, pp. 87–92.

Sines, G. and Waisman, J. L. (eds.), *Metal Fatigue*, McGraw-Hill, New York, 1959.

Sineath, R. M., "Case History of a Mechanical System Reliability Development Test," *Proceedings of Annual Reliability and Maintainability Symposium*, IEEE, New York, 1981, pp. 19–23.

Singh, C., "Reliability Models for Track-Bound Transit Systems," *Proceedings of Annual Reliability and Maintainability Symposium*, IEEE, New York, 1977, pp. 242–247.

Singh, C. and Kankam, M. D., "Failure-Data Analysis for Transit Vehicles," *Proceedings of Annual Reliability and Maintainability Symposium*, IEEE, New York, 1979, pp. 308–313.

Singh, N., "MVUE of $Pr\{X < Y\}$ for Multivariate Normal Populations: An Application to Stress-Strength Models," *IEEE Transactions on Reliability*, Vol. R-30, No. 2, June 1981, pp. 192–193.

Singh, N., "A Distribution-Free Upper Confidence Bound for $Pr\{Y < X\}$: An Application to Stress-Strength Models," *IEEE Transactions on Reliability*, Vol. R-32, No. 2, June 1983, pp. 214–216.

Singpurwalla, N. D., "Statistical Fatigue Models: A Survey," *IEEE Transactions on Reliability*, Vol. R-20, 1971, pp. 185–189.

Sirull, R. M., "A Technique for Estimating Heat Shield Reliability," *Proceedings of Tenth National Symposium on Reliability and Quality Control*, IEEE, New York, 1964, pp. 567–572.

Skewis, W. H., "Predicting the Reliability of Mechanical Equipment," *Proceedings of Annual Reliability and Maintainability Symposium*, IEEE, New York, 1985, pp. 413–417.

Skewis, W. H. and Mangum, W. C., "Reliability Programs for Nonelectronic Designs," Vols. I and II, Rome Air Development Center, Griffiss Air Force Base, NY, Rept. RADC-TR-83-85, April 1983 (available from NTIS as AD-A 133624 and AD-A 133625).

Smith, C. O., "Probabilistic Design Criteria for Cylinder and Spheres Under Thermal Stresses," *Transactions of ASME, Journal of Vibration, Acoustics, Stress and Reliability in Design*, Vol. 106, Oct. 1984, pp. 523–528.

Smith, C. O., "Failure of a Drawbar," *Proceedings of Annual Reliability and Maintainability Symposium*, IEEE, New York, 1974, pp. 119–122.

Smith, C. O., *Introduction to Reliability in Design*, McGraw-Hill, New York, 1976.

Smith, F. C., "Transit System Acceptance by Sequential Sampling," *Proceedings of Annual Reliability and Maintainability Symposium*, IEEE, New York, 1980, pp. 129–135.

Snyder, R. E., "Application of Reliability to a Naval Weapon System," *Proceedings of Ninth Reliability and Maintainability Conference* (*Annals of Reliability and Maintainability*, Vol. 9), SAE, Warrendale, PA, July 1970, pp. 336–343.

Soboyejo, A. B. O., "Application of Maximum Entropy in Estimating the Reliability Functions for Creep Failure Modes of Engineering Materials at High Temperatures," *Proceedings of Sixth Reliability and Maintainability Conference* (*Annals of Reliability and Maintainability*, Vol. 6), SAE, Warrendale, PA, July 1967, pp. 382–388.

Society of Automotive Engineers, *Fatigue Design Handbook*' McGraw-Hill, New York, 1959.

Solomon, K. A. and Bazovsky, I. Jr., "A Time Dependent Model of a SCRAM System," *Proceedings of Annual Reliability and Maintainability Symposium*, IEEE, New York, 1975, pp. 149–153.

Sorensen, A. Jr., "A Statistical Analysis of Product Reliability Due to Random Vibration," *Proceedings of Annual Reliability and Maintainability Symposium*, IEEE, New York, 1973, pp. 379–387.

Sorensen, A. Jr., "Product Reliability and Random Fatigue," *IEEE Transactions on Reliability*, Vol. R-20, No. 4, Nov. 1971, pp. 244–248.

Sorensen, A. Jr., "Product Reliability Due to Random Load Variations at Elevated Temperatures," *IEEE Transactions on Reliability*, Vol. R-28, No. 1, April 1979, pp. 19–22.

Spoormaker, J. L., "Reliability Prediction of Hairpin Type Springs," *Proceedings of Annual Reliability and Maintainability Symposium*, IEEE, New York, 1977, pp. 142–147.

Spoormaker, J. L., "Design of Reliable Plastic Assemblies," *Proceedings of Annual Reliability and Maintainability Symposium*, IEEE, New York, 1975, pp. 498–503.

Stancampiano, P. A., "Monte Carlo Approaches to Stress Strength Interference," *Failure Prevention and Reliability*, ASME, New York, 1977, pp. 285–309.

Stiehl, G. L., Thomas C. S., and Facey, J. H., "Shipboard Reliability—A Design Program," *Proceedings of Annual Symposium on Reliability*, IEEE, New York, Jan. 1967, pp. 448–456.

Stiles, E. M., "Reliability in Mass-Produced Consumer Products," *Proceedings of National Symposium on Reliability*, IEEE, New York, 1964, pp. 85–96.

Subramanian, R. and Ramesh, K. S., "A Note on Stress-Strength Reliability Models," *IEEE Transactions on Reliability*, Vol. R-18, No. 4, Nov. 1969, pp. 204–205.

Swale, D. L., "Integrated Electro-Mechanical Design as Applied to Electronic Equipment," *British Communication Electronics*, Vol. 7, 1960, pp. 174–180.

Tallian, T. E., "A Unified Model for Rolling Contact Life Prediction," *Journal of Lubrication Technology*, Vol. 104, July 1982, pp. 336–346.

Taraman, S. I. and Kapur, K. C., "Optimization Considerations in Design Reliability by Stress-Strength Interference Theory," *IEEE Transactions on Reliability*, Vol. R-24, No. 2, June 1975, pp. 136–138.

Tenney, J. B. Jr. and Withey, D. J., "Test and Checkout Procedures for the Tektite Ocean Habitat System," *Proceedings of Ninth Reliability and Maintainability Conference* (*Annals of Reliability and Maintainability*, Vol. 9), SAE, Warrendale, PA, July 1970, pp. 96–106.

Thomas, T. M., Derbalian, G., and Bischel, K., "Parametric Sensitivity of Rocket Motor Reliability," *Proceedings of Annual Reliability and Maintainability Symposium*, IEEE, New York, 1980, pp. 121–128.

Tiger, B. and Donald, E., "The Stress-Survival Matrix Test as a Tool for Evaluating Integrated Circuit Effectiveness," *Proceedings of Fourth Reliability and Maintainability Conference* (*Annals of Reliability and Maintainability*, Vol. 4), Spartan Books, Washington, DC, July 1965, pp. 393–404.

Tiger, B. and Weir, K., "Stress-Strength Theory and Its Transformation into Reliability Functions," *Physics of Failure in Electronics*, Vol. 2, Sept. 1964, pp. 94–101.

Todorovic, J., "An Approach to Braking System Reliability," *Proceedings of 14th International Automobile Technical Congress of FISITA*, Institution of Mechanical Engineers, London, June 1972, pp. 3/12–3/18.

Townsend, A. R., "Structural and Mechanical Reliability Technology—Introduction," *Proceedings of Eighth Reliability and Maintainability Conference* (*1969 Annals of Assurance Sciences*), Gordon and Breach Science Publishers, New York, July 1969, pp. 341–342.

Trainer, T. M., "Autoselection—A Process for Improving Predictability and Reliability," *Mechanical Reliability Concepts*, ASME, New York, May 1965, pp. 49–59.

Tullier, P., "Determining Reliability and Degradation of Shipboard Machinery," *Proceedings of Annual Reliability and Maintainability Symposium*, IEEE, New York, 1976, pp. 239–244.

Tumolillo, T. A., "Methods for Calculating the Reliability Function for Systems Subjected to Random Stresses," *IEEE Transactions on Reliability*, Vol. R-23, No. 4, Oct. 1974, pp. 256–262.

Tustin, W., "Quantized Goals for the Design of Vibration and Shock Test Fixtures," *Proceedings of Annual Reliability and Maintainability Symposium*, IEEE, New York, 1973, pp. 239–244.

Tustin, W., "Vibration, Shock and Intense Noise Testing for Reliability," *IEEE Transactions on Reliability*, Vol. R-28, No. 2, June 1979, pp. 129–132.

Tustin, W. and Mercado, R., *Random Vibration in Perspective*, Tustin Institute of Technology, Santa Barbara, CA, 1984.

Venton, A. O. F., Davidson, A. T. G., and Fraser, G. G., "The Effects of Limited Spares and Scheduled Downtime upon the Dependability of a Simple System," *IEEE Transactions on Reliability*, Vol. R-25, No. 5, Dec. 1976, pp. 319–323.

Wagner, G. R. and Mischke, C., "Cycles-to-Failure and Stress-to-Failure Weibull Distributions in Steel Wire Fatigue," *Proceedings of Annual Reliability and Maintainability Symposium*, IEEE, New York, 1973, pp. 447–451.

Wallace, W. D., "Deep Submergence Rescue Vehicle (DSRV) Reliability," *Proceedings of Annual Symposium on Reliability*, IEEE, New York, Jan. 1967, pp. 442–447.

Ward, J. M., "Reproducibility and Reliability of Silica Fabrics and Fibers Made From Leached Glass," *Proceedings of Fourth Reliability and Maintainability Conference* (*Annals of Reliability and Maintainability*, Vol. 4), Spartan Books, Washington, DC, July 1965, pp. 457–459.

Weaver, L. and Scarlett, T., "Reliability and Failure Distributions of Inertial Sensors," *Proceedings of Eleventh National Symposium on Reliability and Quality Control*, IEEE, New York, 1965, pp. 144–153.

Weaver, L. A. and Smith, M. P., "The Life Distribution and Reliability of Electro-Mechanical Parts of an Inertial Guidance System," *Proceedings of Eighth National Symposium on Reliability and Quality Control*, IEEE, New York, 1962, pp. 148–155.

Weber, G. E., "Probabilistic Life Consumed Analysis on High Temperature, Highly Stressed Structural Parts," *Proceedings of Fourth Annual Reliability and Maintainbility Conference* (*Annals of Reliability and Maintainability*, Vol. 4), Spartan Books, Washington, DC, July 1965, pp. 385–392.

Weibull, W., "A Statistical Distribution Function of Wide Applicability," *Journal of Applied Mechanics*, Vol. 18, Sept. 1951, pp. 293–297.

Weichbrodt, B., "Mechanical Signature Analysis, A New Tool for Product Assurance and Early Fault Detection," *Proceedings of Fifth Reliability and Maintainability Conference* (*Annals of Reliability and Maintainability*, Vol. 5), AIAA, Washington, DC, July 1966, pp. 569–581.

Weiss, D. W., "Prediction of Mechanical Reliability—A Review," *Proceedings of Annual Symposium on Reliability*, IEEE, New York, Jan. 1967, pp. 30–39.

Weiss, D. W., Mateyka, J., and Danzeisan, R., "A Reliability and Safety Analysis of Automotive Vehicles," *Proceedings of Annual Symposium on Reliability*, IEEE, New York, Jan. 1971, pp. 150–157.

Welker, E. L., "Uses and Abuses of Transit System Failure Rates," *Proceedings of Annual Reliability and Maintainability Symposium*, IEEE, New York, 1978, pp. 294–299.

Welker, E. L., "Reliability and Availability Assessment Criteria, Data Inputs and Analytic Methods for Mass Transit Systems," *Proceedings of Annual Reliability and Maintainability Symposium*, IEEE, New York, 1976, pp. 309–313.

Welker, E. L. and Buchanan, H. N., "Safety-Availability Study Methods Applied to BART," *Proceedings of Annual Reliability and Maintainability Symposium*, IEEE, New York, 1975, pp. 269–275.

Wellauer, E. J., "Gears: New Rating System Simplifies Design," *Mechanical Engineering*, Vol. 87, Feb. 1965, pp. 21–27.

Whittier, R. J., "Mechanical Considerations Affecting Vacuum Tube Reliability," *Tele-Tech*, Vol. 11, No. 2, 1952, pp. 42–44 and No. 3, pp. 54–56, 102, 104, 106.

Wilson, D. S. and Smith, R., "Electric Motor Reliability Model," Rome Air Development Center, Griffiss Air Force Base, NY, Rept. RADC TR-77-408, 1977, (available from NTIS as AD 050179).

Wilson, D. S. and Wilkinson, J., "Reliability Model for Miniature Blowers," Rome Air Development Center, Griffiss Air Force Base, NY, Rept. RADC-TR-75-178, July 1975.

Wilson, S. A., "Distribution of Time-to-Damage Method in Reliability of Pipes," *Proceedings of Generic Techniques in Systems Reliability Assessment*, Noordhoff-Leyden, Amsterdam, 1976, pp. 463–474.

Woo, T. G., "Reliability Analysis of a Fluidized-Bed Boiler for a Coal-Fueled Power Plant," *IEEE Transactions on Reliability*, Vol. R-29, No. 5, Dec. 1980, pp. 422–424.

Woytowitz, D., "R&M Requirements in Shipboard Machinery Specifications," *Proceedings of Annual Symposium on Reliability*, IEEE, New York, Jan. 1971, pp. 227–237.

Wyatt, S. V. and Kemezis, R. J., "R&M Requirement for Shipboard Incinerators," *Proceedings of Annual Reliability and Maintainability Symposium*, IEEE, New York, 1976, pp. 188–195.

Yadav, R. P. S., "A Reliability Model for Stress Versus Strength Problem," *Microelectronics and Reliability*, Vol. 12, 1973, pp. 119–123.

Yadav, R. P. S., "Component Reliability Under Environmental Stress," *Microelectronics and Reliability*, Vol. 13, 1974, pp. 473–475.

Yao, J. T. P. and Yeh, H. Y., "Safety Analysis of Statistically Indeterminate Trusses," *Proceedings of Sixth Reliability and Maintainability Conference* (*Annals of Reliability and Maintainability*, Vol. 6), SAE, Warrendale, PA, 1967, pp. 54–62.

Young, T. N. and Manning, S. D., "Distribution of Equivalent Initial Flaw Size," *Proceedings of Annual Reliability and Maintainability Symposium*, IEEE, New York, 1980, pp. 112–120.

Yurkowsky, W., "Data Collection for Non-electronic Reliability Handbook (NEDCO I and II)," Rome Air Development Center, Griffiss Air Force Base, Rept. RADC/TR-68-114, Vols. I–V, June 1968 (available from NTIS as AD 841 106–AD 841 110).

Yurkowsky, W., "Nonelectronic Reliability Notebook," Rome Air Development Center, Griffiss Air Force Base, NY, Rept. RADC-TR-69-458, 1969 (available from NTIS).

Yurkowsky, W. et al., "Accelerated Testing Technology-Handbook of Accelerated Test Methods," Rome Air Development Center, Griffiss Air Force Base, NY, Rept. RADC-TR-67-420, Nov. 1967 [available from NTIS as AD 824 449 (Vol. I) and 824 451 (Vol. II)].

Yurkowsky, W. and Schafer, R. E., "Accelerated Reliability Test Methods for Mechanical and Electromechanical Parts," Rome Air Development Center, Griffiss Air Force Base, NY, Rept. RADC-TR-65-46, July 1965 (available from NTIS as AD 621 074).

Yurkowsky, W. and Schafer, R. E., "Accelerated Reliability Testing for Nonelectronic Parts," Rome Air Development Center, Griffiss Air Force Base, NY, Rept. RADC-TR-66-425, Sept. 1966 (available from NTIS as AD 803 484).

Zimmerman, R. H., "Manufacturing Controls for Reliable Welded Electronic Connections," *Proceedings of Third Aerospace Reliability and Maintainability Conference*, AIAA, Washington, DC, May 1963, pp. 252–262.

Zirinsky, S., Switzky, H., and Kenger, A., "Utilization of an Ultrasonic Fatigue Damage Indicator for Inspection and Design," *Proceedings of Annual Symposium on Reliability*, IEEE, New York, Jan. 1966, pp. 419–430.

A3 AEROSPACE SYSTEMS RELIABILITY

Abbot, W. R. and Jenkins, L., "Flight Failure Analysis," *Proceedings of Annual Symposium on Reliability*, IEEE, New York, Jan. 1969, pp. 244–249.

Abernethy, R. B., Medlin, C. H., and Ringhiser, B. G., "Analysis of Turbo Pump Failures Using Advanced Weibull Techniques," *Proceedings of Annual Reliability and Maintainability Symposium*, IEEE, New York, 1983, pp. 367–370.

Abraham, L. H., *Structural Design of Missiles and Spacecraft*, McGraw-Hill, New York, 1962.

Abrams, F. L., "The F-15E R&M Challenge," *Proceedings of Annual Reliability and Maintainability Symposium*, IEEE, New York, 1987, pp. 253–257.

Adkins, L. A. Jr., "C-141 Reliability Flight Test Program," *Proceedings of Annual Symposium on Reliability*, IEEE, New York, Jan. 1967, pp. 668–674.

Adkins, L. A. and Thatro, M. C., "Reliability Requirements for Safe All Weather Landings," *Proceedings of Seventh Reliability and Maintainability Conference* (*1968 Annals of Assurance Sciences*), ASME, New York, July 1968, pp. 328–335.

Allen, R. J., "Reliability Criteria for Experimental Programs," *Proceedings of Annual Reliability and Maintainability Symposium*, IEEE, New York, 1975, pp. 98–101.

Althaus, E. J., Morrison, S. C., and Tate, W. R., "A Method of Testing and Evaluation of Complex Missile Systems," *IRE Convention Record*, Pt. II, March 1954, pp. 23–28 (available from IEEE, New York).

Anderson, R. T., *Reliability Design Handbook*, IIT Research Institute, Chicago, 1976.

Arnold, G. M., "F-16 Centralized Data System (CDS)," *Proceedings of Annual Reliability and Maintainability Symposium*, IEEE, New York, 1982, pp. 326–328.

Aroian, L. A. and Myers, R. H., "Redundancy Considerations in Space and Satellite Systems," *Proceedings of Seventh National Symposium on Reliability and Quality Control*, IEEE, New York, 1961, pp. 292–301.

Ascher, H. and Feingold, H., "The Aircraft Air-conditioner Data Revisited," *Proceedings of Annual Reliability and Maintainability Symposium*, IEEE, New York, 1979, pp. 153–159.

Atkinson, J. E., "Further Studies into Connector Field Failure Data as Related to the Commercial Aviation Industry," *Proceedings of Annual Symposium on Reliability*, IEEE, New York, Jan. 1968, pp. 103–113.

Austin, R. N., "Optimized Operational Payloads for Manned Mars Missions," *Proceedings of Fifth Reliability and Maintainability Conference* (*Annals of Reliability and Maintainability*, Vol. 5), AIAA, Washington, DC, July 1966, pp. 202–214.

Austin, W. H. Jr., "Development of Improved Gust Load Criteria for USAF Aircraft," *Proceedings of Sixth Reliability and Maintainability Conference* (*Annals of Reliability and Maintainability*, Vol. 6), SAE, Warrendale, PA, 1967, pp. 68–74.

Baal, J. B., "Hardware/Software FMEA Applied to Airplane Safety," *Proceedings of Annual Reliability and Maintainability Symposium*, IEEE, New York, 1985, pp. 250–255.

Babcock, P. and Calvo, A. B., "RAM Analysis Requirements for Avionic Systems," *Proceedings of Annual Reliability and Maintainability Symposium*, IEEE, New York, 1983, pp. 205–211.

Babcock, D., DeKanel, A., and Metzner, H., "Failure Distributions in F111D Aircraft Components," *Proceedings of Annual Reliability and Maintainability Symposium*, IEEE, New York, 1983, pp. 398–402.

Babin, R. S., "DC-10 Avionics Parts Reliability in Review," *Proceedings of Annual Reliability and Maintainability Symposium*, IEEE, New York, 1974, pp. 403–408.

Baeker, J. B., Collins, J. D., and Haber, J. M., "Launch Risk Analysis," *Proceedings of Annual Reliability and Maintainability Symposium*, IEEE, New York, 1977, pp. 124–129.

Balaban, H., Cuppett, D., and Harrison, G., "The F-16, RIW Program," *Proceedings of Annual Reliability and Maintainability Symposium*, IEEE, New York, 1979, pp. 79–82.

Balaban, H., Plotkin, R., and Harrison, G., "Reliability Prediction by Function for Avionic Equipment," *Proceedings of Fifth Reliability and Maintainability Conference* (*Annals of Reliability and Maintainability*, Vol. 5), AIAA, Washington, DC, July 1966, pp. 694–703.

Banks, C. R., "Reliability in Airborne Electronic Equipment," *Proceedings of Symposium on Improved Quality Electronic Components*, Trielectro Co., Washington, DC, May 1950, pp. 19–22.

Barnard, R. A. and Matteson, T. D., "Military Aircraft Maintenance—A New Concept," *Proceedings of Annual Reliability and Maintainability Symposium*, IEEE, New York, 1975, pp. 596–600.

Barnett, E. H. and Hiramoto, T., "Similarities Between Space and Commercial Problems," *Proceedings of Annual Symposium on Reliability*, IEEE, New York, Jan. 1971, pp. 91–97.

Bean, E. E. and Bloomquist, C. E., "Reliability Data from In-Flight Spacecraft," *Proceedings of the Annual Symposium on Reliability*, IEEE, New York, Jan. 1968, pp. 271–279.

Bean, E. E. and Bloomquist, C. E., "Field Failure Rates for Selected KSC Ground Support Equipment," *Proceedings of Eighth Reliability and Maintainability Conference* (*1969 Annals of Assurance Sciences*), Gordon and Breach Science Publishers, New York, July 1969, pp. 244–250.

Bean, E. E., Bloomquist, C. E., and Finkelstein, J., "More Reliability Data from In-Flight Spacecraft," *Proceedings of Annual Reliability and Maintainability Symposium*, IEEE, New York, 1973, pp. 224–237.

Bear, J. C., "Approach to Reliability for the SM-2 Missile," *Proceedings of Annual Reliability and Maintainability Symposium*, IEEE, New York, 1973, pp. 79–84.

Beard, K. W. and Blundell, J. K., "Maintenance Strategies for Aero Engines," *Proceedings of Annual Reliability and Maintainability Symposium*, IEEE, New York, 1985, pp. 92–98.

Behmann, F. F., "Improved Reliability for New Satellite Systems," *Proceedings of Annual Reliability and Maintainability Symposium*, IEEE, New York, 1981, pp. 408–413.

Behmann, F. F., "Satellite Travelling Wave Tubes-Reliability Controls," *Proceedings of Annual Reliability and Maintainability Symposium*, IEEE, New York, 1982, pp. 193–198.

Behmann, F. F. and Nawar, G. Y., "Satellite System Availability and Effectiveness," *Proceedings of Annual Reliability and Maintainability Symposium*, IEEE, New York, 1984, pp. 434–439.

Behmann, F. F. and Nawar, G. Y., "Effective Space Systems With Optimized Protection," *Proceedings of Annual Reliability and Maintainability Symposium*, IEEE, New York, 1985, pp. 191–196.

Bell, R. and Bruno, O. P., "Analysis of Testing Risks for an Air Defense System," *Proceedings of Annual Reliability and Maintainability Symposium*, IEEE, New York, 1975, pp. 427–431.

Bell, C. F., "Reliability of Aircraft as Determined by Operational Field Tests: The Need for Proper Test Design and Data Requirements," *Operations Research and Reliability*, edited by D. Grouchko, Gordon and Breach, Science Publishers, New York, 1971, pp. 233–250.

Belless, R. M. Jr. and Anderson, L. R., "Needed: Space Maintenance Tools," *Proceedings of Ninth Reliability and Maintainability Conference* (*Annals of Reliability and Maintainability*, Vol. 9), SAE, Warrendale, PA, July 1970, pp. 362–374.

Benet, H. J. and Shipman, C. H., "Logistics Planning Simulation Model for USAF Spare Engine Management," *Proceedings of Annual Reliability and Maintainability Symposium*, IEEE, New York, 1977, p. 500.

Berry, C. J., "Industry-Government Reliability Program for a High-Density Wooden Round Missile," *Proceedings of Fifth Reliability and Maintainability Conference* (*Annals of Reliability and Maintainability*, Vol. 5), AIAA, Washington, DC, July 1966, pp. 505–514.

Billet, A. B., "Reliability of Hydraulic Controls in Space Vehicles," *Proceedings of Tenth National Symposium on Reliability and Quality Control*, IEEE, New York, 1964, pp. 69–84.

Billet, A. B., "Testing of Space Vehicle Fluid Power Controls in Nuclear Radiation," *Proceedings of Fourth Reliability and Maintainability Conference* (*Annals of Reliability and Maintainability*, Vol. 4), Spartan Books, Washington, DC, July 1965, pp. 747–758.

Billet, A. B., "Space Environments and the Reliability of Hydraulic Controls," *Proceedings of IAS Aerospace Systems Reliability Symposium*, April 1962, pp. 111–119 (available from AIAA, Washington, DC).

Bingham, W. W., "A Theoretical Method of Predicting Airplane Crash Loads," *Proceedings of Fourth Annual Reliability and Maintainability Conference* (*Annals of Reliability and Maintainability*, Vol. 4), Spartan Books, Washington, DC, July 1965, pp. 301–316.

Birdsall, C. R. and Frank, H. J., "Space Station Reliability," *Proceedings of Annual Reliability and Maintainability Symposium*, IEEE, New York, 1985, pp. 197–202.

Bishop, L. L., "Utilization of Army Aviation Experiences via RAMMIT (Reliability and Maintainability Management Improvement Techniques)," *Proceedings of Ninth Reliability and Maintainability Conference* (*Annals of Reliability and Maintainability*, Vol. 9), SAE, Warrendale, PA, July 1970, pp. 149–159.

Boiles, J. H. and Hadel, J. J., "Reliability Growth Measured by AFM 66-1 Data," *Proceedings of Annual Reliability and Maintainability Symposium*, IEEE, New York, 1981, p. 465.

Black, R. O., "Measurement of Army Missile Effectiveness," *Proceedings of Third Aerospace Reliability and Maintainability Conference*, AIAA, Washington, DC, May 1963, pp. 10–13.

Blackburn, F. W. and Letow, A. M., "Aspects of an Effective Reliability Program Implemented for an Advanced Bombing-Navigation System," *Proceedings of Eighth National Symposium on Reliability and Quality Control*, IEEE, New York, 1962, pp. 272–280.

Blackmore, D. M. and Georgiades, M. S., "MD-80 Airplane Service-Maturity Program," *Proceedings of Annual Reliability and Maintainability Symposium*, IEEE, New York, 1986, pp. 224–227.

Blackney, A. B. and Watts, M. W., "UK RAF Approach to Reliability and Maintainability," *Proceedings of Annual Reliability and Maintainability Symposium*, IEEE, New York, 1982, pp. 329–335.

Bloomquist, C. E., "Spacecraft Anomalies and Lifetimes," *Proceedings of the Annual Reliability and Maintainability Symposium*, IEEE, New York, 1984, pp. 186–191.

Bossange, E. R. Jr., "Logistic Support Requirements for Commercial Airline Operations," *Proceedings of Third Aerospace Reliability and Maintainability Conference*, AIAA, Washington, DC, May 1963, pp. 367–374.

Bonney, E. A. et al., *Reliability Concepts: Reliability of Guided Missile Elements*, Van Nostrand, New York, 1956.

Bonis, A. J., "Reliability Growth Curves for One-Shot Devices," *Proceedings of Annual Reliability and Maintainability Symposium*, IEEE, New York, 1977, pp. 181–185.

Bond, A. C. and McLane, J. C. Jr., "Thermal Vaccum Testing on the Apollo Program," *Proceedings of Annual Symposium on Reliability*, IEEE, New York, Jan. 1971, pp. 310–315.

Bouton, I., "Fundamental Aspects of Structural Reliability," *Aerospace Engineering*, Vol. 21, June 1962, pp. 66–67, 82–93.

Bouton, I., "The Implementation of Reliability Concepts in Structural Design Criteria," Paper presented at Fourth Congress of International Council of the Aeronautical Sciences, Aug. 1964 (AIAA Paper 64-571).

Bower, J. N., "Reliability Growth During Flight Test," *Proceedings of Annual Reliability and Maintainability Symposium*, IEEE, New York, 1986, pp. 101–106.

Boylan, R. P., "Foreign Particles and Space Equipment Reliability," *Proceedings of Annual Symposium on Reliability*, IEEE, New York, Jan. 1967, pp. 836–846.

Bozovich, J. F., Bullard, S. A., Cary, R. C., and Morgenthaler, G. E., "A Cost-Effectivness Model for Evaluation of Large Launch Programs," *Proceedings of Fifth*

Reliability and Maintainability Conference (*Annals of Reliability and Maintainability*, Vol. 5), AIAA, Washington, DC, July 1966, pp. 184–192.

Bracha, V. J., "The Air Force Reliability Program: MIL-R-27542," *Proceedings of Eighth National Symposium on Reliability and Quality Control*, IEEE, New York, 1962, pp. 17–23.

Brady, T. V. and Hogg, G. W., "Reliability and Maintainability of a Helicopter Optical Flight-Control System," *Proceedings of Annual Reliability and Maintainability Symposium*, IEEE, New York, 1983, pp. 280–286.

Brode, H. A., "Cost of Ownership—An Overview of Life Cycle Costs: Evaluation of Avionic System Reliability Improvements," *Proceedings of Annual Reliability and Maintainability Symposium*, IEEE, New York, 1975, pp. 212–216.

Bromberg, B. G. and Hill, R. D., "Reliability of Airborne Electronic Components," *Proceedings of the IRE*, Vol. 41, No. 4, 1953, pp. 513–516 (available from IEEE).

Brueschke, E. E., "Reliability Research on Components for Space," *Proceedings of Eighth National Symposium on Reliability and Quality Control*, IEEE, New York, 1962, pp. 281–295.

Brush, G. G., Healy, J. D., and Leibesman, B. S., "A Bayes Procedure for Combining Black Box Estimates and Laboratory Tests," *Proceedings of Annual Reliability and Maintainability Symposium*, IEEE, New York, 1984, pp. 242–246.

Calderon, S. S. and Senator, F. E., "Improvement in Effectiveness of an Earth Orbital Mission Through Utilization of Standby Systems," *Proceedings of Eighth Reliability and Maintainability Conference* (*1969 Annals of Assurance Sciences*), Gordon and Breach Science Publishers, New York, July 1969, pp. 105–114.

Caldwell, B. H., "Voyager's Most Challenging System Requirement," *Proceedings of Sixth Reliability and Maintainability Conference* (*Annals of Reliability and Maintainability*, Vol. 6), SAE, Warrendale, PA, July 1967, pp. 101–103.

Caldwell, B. H., "An Engineering Approach to Long Life Complex Space Systems," *Proceedings of Seventh Reliability and Maintainability Conference* (*1968 Annals of Assurance Sciences*), ASME, New York, July 1968, pp. 2–10.

Campbell, C. C., "High Reliability for Space Launch Vehicles," *Proceedings of Eleventh National Symposium on Reliability and Quality Control*, IEEE, New York, Jan. 1965, pp. 439–448.

Campbell, C. C., "Safety Margins Established by Combined Environmental Test Increase Atlas Missile Component Reliability," *IRE Transactions on Reliability and Quality Control*, Vol. RQC-10, No. 1, March 1961, pp. 1–6 (available from IEEE).

Carhart, R. R., "Reliability in Guided Missile Systems," *Aeronautical Engineering Review*, Vol. 12, No. 2, 1953, p. 15.

Carhart, R. R., "Reliability in Guided Missiles," *Aeronautical Engineering Review*, Vol. 2, Feb. 1943, pp. 23–26.

Carpenter, R. B. Jr., "Apollo Reliability by Demonstration or Assessment," *Proceedings of Tenth National Symposium on Reliability and Quality Control*, IEEE, New York, 1964, pp. 517–524.

Carpenter, R. B. Jr., "Demonstrating Reliability for Long Space Missions," *Proceedings of Eleventh National Symposium on Reliability and Quality Control*, IEEE, New York, Jan. 1965, pp. 223–232.

Carpenter, R. B., "Vehicle Maintenance Time Constraints for Space Missions," *Proceedings of Annual Symposium on Reliability*, IEEE, New York, Jan. 1968, pp. 338–349.

Carpenter, R. B., "Success Assurance for Manned Planetary Exploration," *Proceedings of Annual Symposium on Reliability*, IEEE, New York, Jan. 1968, pp. 622–630.

Carpenter, R. B. Jr., "Reliability for Manned Interplanetary Travel," *Proceedings of Fourth Reliability and Maintainability Conference* (*Annals of Reliability and Maintainability*, Vol. 4), Spartan Books, Washington, DC, July 1965, pp. 415–420.

Carrubba, E. R., "Establishment of Spacecraft Lifetime Requirements," *Proceedings of Annual Symposium on Reliability*, IEEE, New York, Jan. 1968, pp. 323–330.

Carrubba, E. R., "Establishment of a Redundancy Priority for Spacecraft Elements," *Proceedings of Annual Symposium on Reliability*, IEEE, New York, Feb. 1970, pp. 186–191.

Carter, J. W. and Schlechter, R. L., "Some Testing Aspects of Reliability for a Man-rated Launch Vehicle," *Proceedings of Fourth Reliability and Maintainability Conference* (*Annals of Reliability and Maintainability*, Vol. 4), Spartan Books, Washington, DC, July 1965, pp. 679–688.

Caruso, H. J., Silver, W., Cichetti, D. J., and Kubilus, D. M., "CERT Technology Applied to an Airborne RADAR," *Proceedings of Annual Reliability and Maintainability Symposium*, IEEE, New York, 1979, pp. 131–135.

Caruso, P. S. Jr., "Reliability and the Thermal Testing of the OAO2," *Proceedings of Annual Symposium on Reliability*, IEEE, New York, Jan. 1971, pp. 304–309.

Cato, R. E. Jr. and Wheadon, W. C., "The Impact of Failure Data on Management of a Launch Operations Reliability Program," *Proceedings of Eighth Reliability and Maintainability Conference* (*1969 Annals of Assurance Sciences*), Gordon and Breach Science Publishers, New York, July 1969, pp. 238–243.

Charp, S., "Long Life Design Features of the Nimbus II Meteorological Observatory," *Proceedings of Seventh Reliability and Maintainability Conference* (*1968 Annals of Assurance Sciences*), ASME, New York, July 1968, pp. 29–49.

Chasen, S. H., "Optimum Developmental Launch Programs," *Proceedings of IAS Aerospace Systems Reliability Symposium*, April 1962, pp. 98–102 (available from AIAA, Washington, DC).

Chiara, G. and Porru, M., "Cost Minimization Studies for 'One-Shot' Electro-Explosive Device System Reliability Demonstration Tests," *Operations Research and Reliability*, edited by D. Grouchko, Gordon and Breach Science Publishers, New York, 1971, pp. 441–462.

Clark, A. R., "SPRINT Missile Subsystem Reliability Achievements," *Proceedings of Annual Reliability and Maintainability Symposium*, IEEE, New York, 1977, pp. 367–371.

Clifford, D. R., "All-Weather Automatic Landing System Development and Testing," *Proceedings of Ninth Reliability and Maintainability Conference* (*Annals of Reliability and Maintainability*, Vol. 9), SAE, Warrendale, PA, July 1970, pp. 75–81.

Cohen, H., "Long Duration Space Missions Through Use of In-Flight Maintenance," *Proceedings of Sixth Reliability and Maintainability Conference* (*Annals of Reliability and Maintainability*, Vol. 6), SAE, Warrendale, PA, July 1967, pp. 104–112.

Cohen, H., "Space Reliability Technology: A Historical Perspective," *IEEE Transactions on Reliability*, Vol. R-33, No. 1, April 1984, pp. 36–40.

Collins, J. A., Hagan, G. T., and Bratt, H. M., "Helicopter Failure Modes and Corrective Actions," *Proceedings of Annual Reliability and Maintainability Symposium*, IEEE, New York, 1975, pp. 504–510.

Connor, J. A., "Electronic Part Failure Rates in Space Environments," *IEEE Transactions on Reliability*, Vol. R-12, No. 3, Sept. 1963, pp. 5–10.

Cook, T. N. and Condon, T. E., "Assessing the R&M Attributes of Advanced Structures," *Proceedings of Annual Reliability and Maintainability Symposium*, IEEE, New York, 1986, pp. 312–317.

Coy, R. I., "Reliability Program Planning and Avionics Systems," *Proceedings of Annual Reliability and Maintainability Symposium*, IEEE, New York, 1980, pp. 136–140.

Craine, R. L., "Safety Management in Air Force Systems Development—A Progress Report," *Proceedings of Fourth Annual Reliability and Maintainability Conference* (*Annals of Reliability and Maintainability*, Vol. 4), Spartan Books, Washington, DC, July 1965, pp. 329–332.

Crowley, J. J., "Organization for Reliability in the Manufacture of Airborne Military Electronics," *Proceedings of Fourth National Symposium on Reliability and Quality Control*, IEEE, New York, 1958, pp. 233–238.

Culbertson, A. T., "Air Force Turns Heat on Reliability," *SAE Journal*, Vol. 73, March 1965, pp. 79–80.

Cummings, D. G., "Benefits of the Titan II Parts Improvement Program," *IEEE Transactions on Reliability*, Vol. R-12, No. 2, June 1963, pp. 26–29.

Dahlberg, L. and Westlund, S., "Viggen Aircraft Maintainability—Maintainability Evaluation of the Engine Installation for the Viggen Aircraft," *Proceedings of Annual Reliability and Maintainability Symposium*, IEEE, New York, 1975, pp. 30–39.

Dalrymple, W. B., "Data Acquisition and Analysis for Flight Readiness Review," *Proceedings of Fifth Reliability and Maintainability Conference* (*Annals of Reliability and Maintainability*, Vol. 5), AIAA, Washington, DC, July 1966, pp. 965–971.

Darrah, J. T. Jr. and Haley, J. L. Jr., "The Use of U.S. Army Rotary Wing Mishap Experience to Assure Improved Product Effectiveness," *Proceedings of Ninth Reliability and Maintainability Conference* (*Annals of Reliability and Maintainability*, Vol. 9), SAE, Warrendale, PA, July 1970, pp. 136–148.

Davis, D. J. "Effective Space Systems—Introduction," *Proceedings of Seventh Reliability and Maintainability Conference* (*1968 Annals of Assurance Sciences*), ASME, New York, July 1968, p. 1.

Davis, L. I., "Reliability in Missile and Space Operations," *IRE Transactions on Reliability and Quality Control*, Vol. RQC-10, No. 2, Aug. 1961, pp. 1–3 (available from IEEE).

Davison J. W., Fileccia, G. L., Fudge, K. A., Gephart, L. S., and Tinsley, H. G., "Flight Control Safety: A Total System Approach," *Proceedings of Annual Reliability and Maintainability Symposium*, IEEE, New York, 1978, pp. 143–150.

DeCallies, R. N., "Human Reliability in the Operation of V/STOL Aircraft," *Proceedings of Fifth Reliability and Maintainability Conference* (*Annals of Reliability and Maintainability*, Vol. 5), AIAA, Washington, DC, July 1966, pp. 133–145.

Diamond, M. and Essenwanger, O. M., "Statistical Data on Atmospheric Design Properties to 30 km," *Astronautics and Aerospace Engineering*, Vol. 1, Dec. 1963, p. 69.

Dighton, R. D. and McClellan, H. S., "The F/A-18 Hornet-Low Life-Cycle Cost by Design," *Proceedings of Annual Reliability and Maintainability Symposium*, IEEE, New York, 1983, pp. 403–410.

Dodge, H. S., "Physics of Failure in Spacecraft Application," *Proceedings of Eighth Reliability and Maintainability Conference* (*1969 Annals of Assurance Sciences*), Gordon and Breach Science Publishers, New York, July 1969, pp. 100–104.

Donnelly, J. A. and Macedo, A. J., "Spacecraft Maintenance Simulation Model," *Proceedings of Annual Symposium on Reliability*, IEEE, New York, Jan. 1971, pp. 368–375.

Dorsey, R. S., "Evaluation of Turbojet Engine Development Test Plans With Computer Simulation and Monte Carlo Techniques," *Proceedings of Fifth Reliability and Maintainability Conference* (*Annals of Reliability and Maintainability*, Vol. 5), Spartan Books, Washington, DC, July 1966, pp. 528–534.

Dorsey, R. S. and Truscott, H. A., "Simulation Method Applied to Reliability Analysis of Turbojet Engine Components Operating In Airline Service," *Proceedings of Eighth Reliability and Maintainability Conference* (*1969 Annals of Assurance Sciences*), Gordon and Breach Science Publishers, New York, July 1969, pp. 524–529.

Doshay, I. and Shube, D. B., "Advanced Missile Models and Methods for Availability Prediction," *Proceedings of Seventh Reliability and Maintainability Conference* (*1968 Annals of Assurance Sciences*), ASME, New York, July 1968, pp. 203–213.

Dougherty, J. J. and Hogg, G. W., "Army Helicopter Mission—Reliability and Cost Analysis," *Proceedings of Annual Reliability and Maintainability Symposium*, IEEE, New York, 1983, pp. 411–416.

Douglas, W. A. S., "Reliability Requirements for Extended Space Flights," *Proceedings of Seventh Annual West Coast Reliability Symposium*, Western Periodicals, North Hollywood, CA, May 1966, pp. 121–126.

Drose, J. S., "Reliability Requirements for a Ballistic System Process," *Proceedings of Aerospace Reliability and Maintainability Conference*, AIAA, Washington, DC, May 1963, pp. 88–90.

Duhig, J. J. Jr. and South, J. L., "Cost-Effective Improvement of the Timeless C-130 Hercules Airlifter," *Proceedings of Annual Reliability and Maintainability Symposium*, IEEE, New York, 1978, pp. 422–427.

Duhig, J. J. Jr. and Weaver, T. E., "Effects of Temperature on Avionics Reliability," *Proceedings of Annual Reliability and Maintainability Symposium*, IEEE, New York, 1977, pp. 409–413.

Dummer, G. W. A., "The Reliability of Components in Satellites," *Journal of the British Institution of Radio Engineers*, Vol. 21, Dec. 1961, pp. 457–463.

Dunlap, R. A. and O'Keefe, J. K., "Systems Analysis for Space Programs," *Proceedings of Fifth Reliability and Maintainability Conference* (*Annals of Reliability and Maintainability*, Vol. 5), AIAA, Washington, DC, July 1966, pp. 286–294.

Dwyer, D. J., "Reliability Test Planning for 1-Shot Systems," *Proceedings of Annual Reliability and Maintainability Symposium*, IEEE, New York, 1987, pp. 412–415.

Eagle, K. H. and Reh, J. J., "Assessing Performance Readiness with the Saturn V Reliability Analysis Model," *Proceedings of Sixth Reliability and Maintainability Conference* (*Annals of Reliability and Maintainability*, Vol. 6), SAE, Warrendale, PA, July 1967, pp. 487–497.

Ebenfelt, H. and Holmquist, H. R., "Effectiveness Model for an Anti-Aircraft Fire Control System," *Proceedings of Annual Symposium on Reliability*, IEEE, New York, Feb. 1970, pp. 68–74.

Edmonds, D. S. and Samson, D. G., "Cost-Reliability Relationships of Launch Vehicles," *Proceedings of IAS Aerospace Systems Reliability Symposium*, April 1962, pp. 125–129 (available from AIAA, Washington, DC).

Edwards, G. W., "Fault Tolerance Analysis for Space Transportation System Payloads," *Proceedings of Annual Reliability and Maintainability Symposium*, IEEE, New York, 1982, pp. 476–478.

Ehrenpreis, D., "Analysis and Theoretical Investigations of New Military Electronic Missile and Aircraft-Borne Equipment," *IRE Convention Record*, Vol. 6, Pt. 8, 1958, pp. 66–73 (available from IEEE).

Eichna, O. L. Jr., "Concept and System of the Versatile Avionic Shop Test (VAST) System," *Proceedings of Annual Reliability and Maintainability Symposium*, IEEE, New York, 1973, pp. 313–317.

Elliott, T. W., "Accelerated Testing of Air-to-Air Guided Missiles," *Proceedings of Annual Reliability and Maintainability Symposium*, IEEE, New York, 1973, pp. 105–108.

Ellis, R. R. and Pollock, J. J., "Reliability Overstress Testing of the Gemini Inertial Measuring Unit System Electronics Package," *Proceedings of Annual Symposium on Reliability*, IEEE, New York, Jan. 1966, pp. 431–436.

Erdle, F. E. et al., "Reliability Programs for Commercial Communication Satellites," *IEEE Transactions on Reliability*, Vol. R-32, No. 3, Aug. 1983, pp. 236–239.

Evans, A. C. and Stolar, G., "Reliability Testing of Aircraft Instruments," *Proceedings of Seventh National Symposium on Reliability and Quality Control*, IEEE, New York, 1961, pp. 466–473.

Evans, T. M. and Merkord, D. L., "Is There Life After 10,000 Flight Hours," *Proceedings of Annual Reliability and Maintainability Symposium*, IEEE, New York, 1985, pp. 396–401.

Fagan, T. L. and Wilson, M. A., "Maintainability in Space—A Survey," *IEEE Transactions on Reliability*, Vol. R-16, No. 1, May 1967, pp. 52–60.

Fallon, E. R. Jr., "Requirements for Air Force Weapon System Maintainability," *Proceedings of Aerospace Reliability and Maintainability Conference*, AIAA, Washington, DC, May 1963, pp. 326–331.

Fedor, O. H., "Aerospace Mechanical-Reliability Practice," *Proceedings of Annual Reliability and Maintainability Symposium*, IEEE, New York, 1982, pp. 56–61.

Ferguson, D., Kolson, J., Meek, S., and Stracener, J. T., "Development of Conceptual Navy-Aircraft Reliability-Prediction Models," *Proceedings of Annual Reliability and Maintainability Symposium*, IEEE, New York, 1981, pp. 395–401.

Finkelstein, J. M., "A Logarithmic Reliability-Growth Model for Single-Mission Systems," *IEEE Transactions on Reliability*, Vol. R-32, No. 5, Dec. 1983, pp. 508–511.

Fleig, N. G., "The US Air Force Product Performance Agreement Center (PPAC)," *Proceedings of Annual Reliability and Maintainability Symposium*, IEEE, New York, 1986, pp. 449–453.

Fleig, O. T., "Optimizing R&M in the Airline Support Cycle," *Proceedings of Tenth Reliability and Maintainability Conference* (*Annals of Reliability and Maintainability*, Vol. 10), ASME, New York, June 1971, pp. 11–14.

Franciscovich, P. J., "Technique for Reliability Circuit Design Review in Space Electronics," *Proceedings of Sixth Reliability and Maintainability Conference* (*Annals of Reliability and Maintainability*, Vol. 6), SAE, Warrendale, PA, July 1967, pp. 261–271.

Freitag, M. and Cleveland, W. A., "One-Shot System Reliability: Prediction and Demonstration," *Proceedings of IAS Aerospace Systems Reliability Symposium*, April 1962, pp. 26–32 (available from AIAA).

Frumkin, B., "Maintainability on the Lunar Excursion Module," *Proceedings of Fifth Reliability and Maintainability Conference* (*Annals of Reliability and Maintainability*, Vol. 5), AIAA, Washington, DC, July 1966, pp. 834–840.

Fuqua, N. B., "Reliability of Automated Flight Service Stations," *Proceedings of Annual Reliability and Maintainability Symposium*, IEEE, New York, 1976, pp. 113–119.

Furia, T., "Reliability Improvement of the Tiros Momentum Wheel Assembly through Testing and Simulation," *Proceedings of Annual Symposium on Reliability*, IEEE, New York, Jan. 1971, pp. 292–297.

Furlaw, R. N., "In-Flight Testing—A Parameter-Dependent Approach," *Proceedings of Fourth Annual Reliability and Maintainability Conference* (*Annals of Reliability and Maintainability*, Vol. 4), Spartan Books, Washington, DC, July 1965, pp. 535–544.

Gai, E., Harrison, J. V., and Luppold, R. H., "Reliability Analysis of a Dual-Redundant Engine Controller," *IEEE Transactions on Reliability*, Vol. R-32, No. 1, April 1983, pp. 14–20.

Gatzek, L. E., "Advanced Aerospace Technology for Joining Dissimilar Materials," *Proceedings of Seventh Annual West Coast Reliability Symposium*, Western Periodicals, North Hollywood, CA, May 1966, pp. 51–76.

Gean, J. E. and Knudsen, G. E., "The Huey/Cobra M&R Field Program Results and Significance," *Proceedings of Ninth Reliability and Maintainability Conference* (*Annals of Reliability and Maintainability*, Vol. 9), SAE, Warrendale, PA, July 1970, pp. 160–173.

Gephart, L. S., Fuchs, W. P., Fileccia, G. L., Johani, T., and Tinsley, H. G., "A Systems Approach to All-Weather Landings," *Proceedings of Annual Reliability and Maintainability Symposium*, IEEE, New York, 1976, pp. 25–30.

Gephart, L. S. and Keegan, R. H., "Flight Test Safety Analysis of the All Weather Landing System (AWLS) Program," *Proceedings of Annual Reliability and Maintainability Symposium*, IEEE, New York, 1974, pp. 436–442.

Gersch, W. C., "A-7A Aircraft Maintenance Guarantee and the Role of Training," *Proceedings of Sixth Reliability and Maintainability Conference* (*Annals of Reliability and Maintainability*, Vol. 6), SAE, Warrendale, PA, July 1967, pp. 458–463.

Gevarter, W. B., "Analytical Evaluation of Launch Strategy Using Past Countdown Hold Data," *Proceedings of Eighth Reliability and Maintainability Conference* (*1969 Annals of Assurance Sciences*), Gordon and Breach Science Publishers, New York, July 1969, pp. 426–434.

Gilles, D. L., "Development of Douglas Commercial Aircraft Reliability Programs," *Proceedings of Annual Reliability and Maintainability Symposium*, IEEE, New York, 1972, pp. 578–585.

Gilles, D. L., Buscher, R. G., Christie, R. J., Fickeisen, F., and Waterman, H. E., "Civil Aircraft System Safety Panel—The Regulations and Design Assessment," *Proceedings of Annual Reliability and Maintainability Symposium*, IEEE, New York, 1985, pp. 296–299.

Gindorf, T. E. and DeVries, T. R., "The Role of Space Simulation Testing in the Development of Mariner Spacecraft," *Proceedings of Annual Symposium on Reliability*, IEEE, New York, Jan. 1971, pp. 298–303.

Goldfine, M. T., "Mariner' 69 Testing and Flight Results," *Proceedings of Ninth Reliability and Maintainability Conference* (*Annals of Reliability and Maintainability*, Vol. 9), SAE, Warrendale, PA, July 1970, pp. 48–62.

Golovin, N. E., "Reliability Planning for Space Systems," *IRE Transactions on Reliability and Quality Control*, Vol. RQC-10, No. 2, Aug. 1961, pp. 20–23 (available from IEEE).

Golovin, N. E., "Reliability Programs and the Problem of Attaining High Probabilities of Mission Success in Space Exploration," *IEEE Transactions on Reliability*, Vol. R-13, No. 2, June 1964, pp. 1–4.

Gomberg, L. and Thierfelder, H., "Assuring Reliability in Aerospace Batteries," *Proceedings of Eleventh National Symposium on Reliability and Quality Control*, IEEE, New York, Jan. 1965, pp. 459–469.

Gordon, G. S., "Failure Reporting on Satellite Programs," *Proceedings of Annual Symposium on Reliability*, IEEE, New York, Jan. 1967, pp. 128–136.

Gordon, T. W., "F/A-18 Hornet Reliability Program—Status Report," *Proceedings of Annual Reliability and Maintainability Symposium*, IEEE, New York, 1986, pp. 228–231.

Grainger, G. R., Faragher, W. E., and Philipson, L. L., "A Realistic Measure of Spacecraft Reliability," *Proceedings of Ninth National Symposium on Reliability and Quality Control*, IEEE, New York, 1963, pp. 125–131.

Gray, L. B., "Minuteman Flight Control System—The Approach to and the Results of a Formal Reliability Program," *Proceedings of Aerospace Reliability and Maintainability Conference*, AIAA, Washington, DC, May 1963, pp. 398–405.

Griffin, P., "Analysis of the F/A-18 Hornet Flight-Control-Computer Field-Mean-Time-Between-Failure," *Proceedings of Annual Reliability and Maintainability Symposium*, IEEE, New York, 1985, pp. 108–114.

Griswold, G. H. and Topmiller, D., "Maintainability Prediction for Avionic Equipment," *Proceedings of the Eleventh National Symposium on Reliability and Quality Control*, IEEE, New York, Jan. 1965, pp. 49–58.

Grose, V. L., "Prelaunch Calculation of Space System Reliability," *Proceedings of Ninth National Symposium on Reliability and Quality Control*, IEEE, New York, 1963, pp. 539–550.

Grose, V. L., "Man as a Control Element in Space Systems," *Proceedings of Sixth Reliability and Maintainability Conference* (*Annals of Reliability and Maintainability*, Vol. 6), SAE, Warrendale, PA, July 1967, pp. 510–515.

Gross, H., "Reliability Efforts in Large European Programs for Military and Commercial Aircraft Development," *IEEE Transactions on Reliability*, Vol. R-23, No. 3, Aug. 1974, pp. 169–173.

Gross, T. W. and Werner, H. C., "Some Aspects of Satellite and Space Probe Reliability," *IRE Transactions on Reliability and Quality Control*, Vol. RQC-10, No. 3, Nov. 1961, pp. 7–14 (available from IEEE).

Grubin, C., Irvine R. B., and Trunnell, D. W., "An Approach to High Reliability for a Spacecraft Inertial-Reference Unit," *Proceedings of Annual Reliability and Maintainability Symposium*, IEEE, New York, 1982, pp. 479–487.

Hadel, J. J., "The Eagle—A Classic Study in Reliability Growth," *Proceedings of Annual Reliability and Maintainability Symposium*, IEEE, New York, 1984, pp. 476–483.

Hall, C. F., "The Pioneer 10/11 Program: From 1969 to 1994," *IEEE Transactions on Reliability*, Vol. R-32, No. 5, Dec. 1983, pp. 414–416.

Hall, H. D., "Computer-Monitored Inspection Program (CMIP)—A Key to Increased Aircraft and Personnel Productivity," *Proceedings of Annual Reliability and Maintainability Symposium*, IEEE, New York, 1982, pp. 408–416.

Hall, H. D. and Ness, W. G., "Maintenance and Reliability Simulation Modeling for Airfreighters," *Proceedings of Ninth Reliability and Maintainability Conference* (*Annals of Reliability and Maintainability*, Vol. 9), SAE, Warrendale, PA, July 1970, pp. 354–361.

Hamilton, C. W., "Bayesian Procedures and Reliability Information," *Proceedings of Aerospace Reliability and Maintainability Conference*, AIAA, New York, May 1963, pp. 278–283.

Hamilton, C. W. and Drennan, J. E., "Research Toward a Bayesian Procedure for Calculating System Reliability," *Proceedings of Third Aerospace Reliability and Maintainability Conference*, SAE, Warrendale, PA, June 1964, pp. 614–620.

Hamilton, D. O. and Ness, W. G., "Aircraft Reliability Growth Characteristics," *Proceedings of Annual Symposium on Reliability*, IEEE, New York, Jan. 1969, pp. 465–471.

Hamilton, K. R. and Chopin, M. H., "Diagnostic Engine Monitoring for Military Aircraft," *Proceedings of Annual Reliability and Maintainability Symposium*, IEEE, New York, 1975, pp. 88–91.

Haniuk, E. S. and Baird, B. L., "Reliability Through Propellant Research and Development," *Proceedings of IAS Aerospace Systems Reliability Symposium*, April 1962, p. 241 (available from AIAA, Washington, DC).

Harding, D. G. and Mack, J. C., "Design of Helicopter Transmission for 'On-Condition' Maintenance," *Proceedings of Ninth Reliability and Maintainability Conference* (*Annals of Reliability and Maintainability*, Vol. 9), SAE, Warrendale, PA, July 1970, pp. 123–135.

Hardy, A. W. and Holmes W. W., "Missile Propulsion Subsystem Effectiveness," *Proceedings of Fourth Annual Reliability and Maintainability Conference* (*Annals of Reliability and Maintainability*, Vol. 4), Spartan Books, Washington, DC, 1965, pp. 689–704.

Hardy, C. A., Allen, R. J., and Fondon, M. C., "F-111A Reliability," *Proceedings of Annual Symposium on Reliability*, IEEE, New York, Feb. 1970, pp. 128–134.

Harter, W. W., "Reliability Application of AFM 66-1 Maintenance Data," *Proceedings of Fifth Reliability and Maintainability Conference* (*Annals of Reliability and Maintainability*, Vol. 5), AIAA, Washington, DC, July 1966, pp. 626–631.

Hassett, R. P. and Jonson, E. C., "AFSATCOM Terminal Segment Reliability Test Program," *Proceedings of Annual Reliability and Maintainability Symposium*, IEEE, New York, 1977, pp. 372–377.

Haus, M., Fuchs, J. L., Denning, E, and Schierman, R. H., "Airforce System Effectiveness and Reliability Data Systems," *Proceedings of Sixth Reliability and Maintainability Conference* (*Annals of Reliability and Maintainability*, Vol. 6), SAE, Warrendale, PA, July 1967, pp. 116–139.

Hecht, H., "Economics of Reliability for Spacecraft Computers, *Proceedings of Annual Reliability and Maintainability Symposium*, IEEE, New York, 1972, pp. 554–564.

Hecht, H. and Fiorentino, E., "Reliability Assessment of Spacecraft Electronics," *Proceedings of Annual Reliability and Maintainability Symposium*, IEEE, New York, 1987, pp. 341–346.

Heggerston, H. E. and Brazier, T. E., "A Decion Table for Determining a Base-Depot Maintenance Policy for Aircraft Engines," *Proceedings of Fifth Reliability and Maintainability Conference* (*Annals of Reliability and Maintainability*, Vol. 5), AIAA, Washington, DC, 1966, pp. 84–91.

Herron, D. P., "Optimizing Trade-Offs of Reliability Vs. Weight," *IEEE Transactions on Reliability*, Vol. R-12, No. 4, Dec. 1963, pp. 50–54.

Hevesh, A. H., "Mission Reliability for Planetary Surveys," *Proceedings of Annual Symposium on Reliability*, IEEE, New York, Jan. 1968, pp. 516–521.

Hicks, J. E., "Analysis of the Economic Benefits of Utility Helicopter Safety Design Features," *Proceedings of Annual Reliability and Maintainability Symposium*, IEEE, New York, 1978, pp. 157–161.

Hilton, H. H. and Feigen, M., "Minimum Weight Analysis Based on Structural Reliability," *Journal of the Aerospace Sciences*, Vol. 27, Sept. 1960, pp. 641–652.

Hinely, J. T. Jr. and Shelley, B. F., "Method of Allocating Reliability Goals to Aircraft Sybsystems and Components," *Proceedings of Aerospace Reliability and Maintainability Conference*, AIAA, Washington, DC, May 1963, pp. 1–9.

Hock, C. D., "Some Remarks on Optimum Reliability Testing for Space Systems," *Proceedings of Eighth National Symposium on Reliability and Quality Control*, IEEE, New York, 1962, pp. 110–116.

Hoedemaker, R., Mowle, E., and Gordon, G. S., "TIROS: A Case History in Reliability," *Proceedings of Annual Symposium on Reliability*, IEEE, New York, Jan. 1968, pp. 315–322.

Holbert, C. D. and Heitert, H. E., "Reliability/Maintainability Field Test Evaluation on Flying Crane Helicopter," *Proceedings of Eighth Reliability and Maintainability Conference (1969 Annals of Assurance Sciences)*, Gordon and Breach Science Publishers, New York, July 1969, pp. 174–180.

Holmes, J., "Rocket Reliability Drive Gains Fresh Momentum," *Missiles and Rockets*, Vol. 5, 1959, p. 28.

Horwitz, D., "General Method for Incipient Failure Detection," *IEEE Transactions on Reliability*, Vol. R-25, No. 5, Dec. 1976, pp. 299–301.

Houck, S., "Maintainability (A Tutorial Session with the F-111B Aircraft)," *Proceedings of Fifth Reliability and Maintainability Conference (Annals of Reliability and Maintainability*, Vol. 5), AIAA, Washington, DC, July 1966, pp. 823–833.

Howard, T. W. III and Hutcheson, G. J., "High Payoff Product Assurance Technology Programs at the U.S. Army Missile Command," *Proceedings of Annual Reliability and Maintainability Symposium*, IEEE, New York, 1986, pp. 201–205.

Hughes, P. M. and Milkie, R. W., "Saturn Systems Automatic Checkout," *Proceedings of Tenth National Symposium on Reliability and Quality Control*, IEEE, New York, 1964, pp. 338–344.

Hull, A. M., "Mechanization of Airline Equipment Reliability Analysis," *Proceedings of Fourth National Symposium on Reliability and Quality Control*, IEEE, New York, 1958, pp. 73–77.

Hull, A. M. and Ogden, J. E., "Commercial Airplane Equipment Reliability," *Proceedings of Third National Symposium on Reliability and Quality Control*, IEEE, New York, 1957, pp. 187–192.

Hull, J. D., "US Air Force Aircraft Battle Damage Repair (ABDR) Program," *Proceedings of Annual Reliability and Maintainability Symposium*, IEEE, New York, 1987, pp. 72–74.

Hunt, G. C. and Ridley, H., "The Optimization of Earth-Orbiting Space Station Missions," *Proceedings of Fifth Reliability and Maintainability Conference (Annals of Reliability and Maintainability*, Vol. 5), AIAA, Washington, DC, July 1966, pp. 193–201.

Interian, A., Kugath, D. A., and Allen, W. H., "Satellite Maintenance Using Remote Manipulator Spacecraft," *Proceedings of Eighth Reliability and Maintainability Conference (1969 Annals of Assurance Sciences)*, Gordon and Breach Science Publishers, New York, July 1969, pp. 22–32.

Ireton, E. T., "Evolutionary Growth of Helicopter Maintainability, Reliability and Safety," *Proceedings of Seventh Reliability and Maintainability Conference (1968 Annals of Assurance Sciences)*, ASME, New York, July 1968, pp. 167–176.

Jacobson, A. J. and Nusbaum, M. R., "DIFAR Reliability Experience: 781 and Field," *Proceedings of Annual Reliability and Maintainability Symposium*, IEEE, New York, 1976, pp. 107–112.

Jacobson, R. H., and Levine, M. B., "Dynamic Environment Testing of Airborne Electronic Components," *IRE Convention Record*, Pt. 5, March 1955, pp. 95–100.

Jailer, R. W., Freilich, G., and Castellon, A. W., "Flight Vehicle Power Systems-Reliability Criteria," American Power Jet Co., Ridgefield, NJ, Tech. Rept. ASD-TR-61-736, March 1962, pp. 233–262.

Jennings, H. A. and Harlan, R. W., "Aircraft Operations and Logistics Simulation," *Proceedings of Annual Reliability and Maintainability Symposium*, IEEE, New York, 1980, pp. 320–326.

Jennings, H. A. and Trott, E. P., "Parametric Relationships of Factors Affecting Maintainability of Long Duration Manned Space Flights," *Proceedings of Sixth Reliability and Maintainability Conference* (*Annals of Reliability and Maintainability*, Vol. 6), SAE, Warrendale, PA, July 1967, pp. 516–523.

Johnson, M. D., "Reliability—Small Companies—Space Subsystems," *Proceedings of Ninth National Symposium on Reliability and Quality Control*, IEEE, New York, 1963, pp. 74–79.

Johnson, R. F. and Siegler, L., "Cost Effective Spares Provisioning Models for Airline Operations," *Proceedings of the Tenth Annual West Coast Reliability Symposium*, Western Periodicals, North Hollywood, CA, Feb. 1969, pp. 165–216.

Johnson, W. H., "Talos Missile Reliability and Quality Education Programme," *Proceedings of Fourth National Symposium on Reliability and Quality Control in Electronics*, IEEE, New York, Jan. 1958, pp. 267–271.

Johnson, W. S., Heller R. A., and Yang, J. N., "Flight Inspection Data and Crack Initiation Times," *Proceedings of Annual Reliability and Maintainability Symposium*, IEEE, New York, 1977, pp. 148–154.

Jones, L. G. and Thompson, D., "Essential Elements of Analysis from Army Test and Field Data," *Proceedings of Annual Symposium on Reliability*, IEEE, New York, 1971, pp. 85–90.

Kamins, M., "Some Effects of Scheduled Maintenance on Large Transport Aircraft," *Proceedings of Tenth Reliability and Maintainability Conference* (*Annals of Reliability and Maintainability*, Vol. 10), ASME, New York, June 1971, p. 205.

Karmiol, E. D., Weir, W. T., and Youtcheff, J. S., "The Reliability/Maintainability Relationship in Aerospace Programs," *Proceedings of Fourth Annual Reliability and Maintainability Conference* (*Annals of Reliability and Maintainability*, Vol. 4), Spartan Books, Washington, DC, July 1965, pp. 261–272.

Kasouf, G. and Weiss, D., "An Integrated Missile Reliability Growth Program," *Proceedings of Annual Reliability and Maintainability Symposium*, IEEE, New York, 1984, pp. 465–470.

Katz, L., "Pyrotechnics in Missile Design," *Ordnance*, Vol. 44, Sept.–Oct. 1959, p. 334.

Kececioglu, D., "Reliability of Aircraft Splines with or without a Wear Induction Period," *Journal of Vibration, Acoustics, Stress and Reliability in Design*, Vol. 105, April 1983, pp. 163–170.

Kececioglu, D. et al., "Aircraft Spline Reliability Predictive Technique Development and Design Improvement Methodology," Second Progress Report prepared for Naval Weapons Engineering Support Activity, Washington, DC, under Contract 0156-75-c-0944 by Engineering Experiment Station, University of Arizona, Tucson, June 30, 1975, p. 201.

Kececioglu, D., Chester, L. B., Koharcheck, A., and Shehata, M., "Aircraft Spline Reliability Predictive Technique Development and Design Improvement Methodology," Report prepared for Naval Weapons Engineering Support Activity, Washington, DC, under Contract 0156-75-c-0944 by Engineering Experiment Station, University of Arizona, Tucson, Jan. 31, 1975, p. 191.

Kececioglu, D. and Koharcheck, A., "Wear Reliability of Aircraft Splines," *Proceedings of Annual Reliability and Maintainability Symposium*, IEEE, New York, 1977, pp. 155–163.

Keel, A. G. Jr., "Combat-Aircraft Readiness and Producibility," *Proceedings of Annual Reliability and Maintainability Symposium*, IEEE, New York, 1981, p. 192.

Kelley, B., "Applying the Assurance Technologies to Helicopters and Other VTOL Aircraft," *Proceedings of Ninth Reliability and Maintainability Conference* (*Annals of Reliability and Maintainability*, Vol. 9), SAE, Warrendale, PA, July 1970, p. 112.

Kennedy, J. J., "US Naval Forces Aircraft Battle Damage Repair," *Proceedings of Annual Reliability and Maintainability Symposium*, IEEE, New York, 1987, pp. 75–77.

Kern, G. A., "Operational Influences on Avionics Reliability," *Proceedings of Annual Reliability and Maintainability Symposium*, IEEE, New York, 1978, pp. 231–242.

Kern, G. and Wong, K., "Non-Operating Failure Rates for Avionics Study," Rome Air Development Center, Griffiss Air Force Base, NY, Rept. RADC-TR-80-136, 1980 (available from NTIS as AD-A087048).

Kiefer, F. P., "Impact of Corrective Action on Flight Readiness," *Proceedings of Fourth Annual Reliability and Maintainability Conference* (*Annals of Reliability and Maintainability*, Vol. 4), Spartan Books, Washington, DC, July 1965, pp. 727–736.

Kiefer, F. P., Bishop, W., and Carter, J. W., "Man-Rating the Gemini Launch Vehicle (Crew Hazard and Mission Analysis)," *Proceedings of Annual Symposium on Reliability*, IEEE, New York, Jan. 1966, pp. 260–268.

Kimel, H. and Harnett, D., "Reliability of the Atlas Radio — Command Guidance and Re-Entry Vehicle Systems," *Proceedings of Aerospace Reliability and Maintainability Conference*, AIAA, Washington, DC, May 1963, pp. 441–449.

Kimmel, J., "Management and Organization of Space-Age Reliability Programs," *Proceedings of Annual Symposium on Reliability*, IEEE, New York, Jan. 1967, pp. 858–864.

Kirby, M. J. and Powell, H. R. "Prediction of Missile Reliability," *IRE Convention Record*, Pt. 6, March 1955, pp. 54–63 (available from IEEE).

Kirchner, W. R., "Space — A Sound Investment," *IEEE Transactions on Reliability*, Vol. R-12, No. 2, June 1963, pp. 10–14.

Kirkpatrick, C. L. and O'Brien, R. R. "F-15 Readiness—The Maintainability Contribution," *Proceedings of Annual Reliability and Maintainability Symposium*, IEEE, New York, 1984, pp. 499–504.

Klawans, B. and Thomas, E., "Flight Performance Analysis of Space Systems," *Proceedings of Annual Symposium on Reliability*, IEEE, New York, Jan. 1969, pp. 238–243.

Klein, W. A. and Lehr, S. N., "Reliability of Solar Arrays," *IRE Transactions on Reliability and Quality Control*, Vol. RQC-11, No. 3, Oct. 1962, pp. 71–80 (available from IEEE).

Klein, W. E., "Modified Aerospace R&QA Method for Wind Turbines," *Proceedings of Annual Reliability and Maintainability Symposium*, IEEE, New York, 1980, pp. 254–258.

Kleinhofer, B. A., "Components Evaluation for Airborne Systems Performance," *Proceedings of Second National Symposium on Reliability and Quality Control*, IEEE, New York, 1956, pp. 2–7.

Klivans, L. S., "An Improved Management Approach to Upgrade Avionic System Reliability," *IEEE Transactions on Reliability*, Vol. R-26, No. 1, April 1977, pp. 23–28.

Koss, E., "Developing Reliable Space-Flight Software," *Proceedings of Annual Reliability and Maintainability Symposium*, IEEE, New York, 1986, pp. 171–176.

Kovatch, D. H., "Modeling Aircraft Combat Damage Repair," *Proceedings of Annual Reliability and Maintainability Symposium*, IEEE, New York, 1984, pp. 440–445.

Krohn, C. A., "Space Shuttle Component Reuse Study," *IEEE Transactions on Reliability*, Vol. R-25, No. 4, Oct. 1976, pp. 234–238.

Kruvand, D. H., "Army Aviation Warranty Concepts," *Proceedings of Annual Reliability and Maintainability Symposium*, IEEE, New York, 1987, pp. 392–394.

Lalli, V. R. and Speck, C. E., "Travelling-Wave Tube Reliability Estimates, Life Tests, and Space Flight Experience," *Proceedings of Annual Reliability and Maintainability Symposium*, IEEE, New York, 1977, pp. 12–20.

Lambert, J. S., "Air Force Electronic Reliability Program," *IRE Transactions*, Vol. PGRQC-14, Sept. 1958, pp. 17–21 (available from IEEE).

Langwost, C. H., "Life Cycle Costing Applied in Evaluating Alternative Short Take-off and Landing (STOL) Aircraft and Tracked Air Cushion Vehicles (TACV) as Modes of Transportation," *Proceedings of Tenth Reliability and Maintainability Conference* (*Annals of Reliability and Maintainability*, Vol, 10), ASME, New York, June 1971, pp. 231–235.

Lauver, P. E., "Small Propulsion Engine LCC Considerations," *Proceedings of Annual Reliability and Maintainability Symposium*, IEEE, New York, 1985, pp. 348–356.

LaValle, R. S. and Stoller, D. S., "A Field Study of the Maintainability and Reliability of an Interceptor Squadron," *IRE Transactions on Reliability and Quality Control*, Vol. RQC-11, No. 3, Oct. 1962, pp. 28–34 (available from IEEE).

Layton, D. M., "Aviation Accidents—The Final Test of Reliability and Safety," *Proceedings of Fourth Annual Reliability and Maintainability Conference* (*Annals of Reliability and Maintainability*, Vol. 4), Spartan Books, Washington, DC, July 1965, pp. 317–324.

Lebach, J. L. and Morris, R. L. K., "Weld Reliability of a Space Structure," *Proceedings of Annual Symposium on Reliability*, IEEE, New York, Jan. 1967, pp. 766–772.

Lehr, S. N. and Tronolone, V. J., "The Space Environment and Its Effects on Materials and Component Parts," *IRE Transactions on Reliability and Quality Control*, Vol. RQC-10, No. 2, Aug. 1961, pp. 24–37 (available from IEEE).

Lennon, J. R., "A Reliability/Cost of Ownership Approach to Microavionics," *Proceedings of Eleventh National Symposium on Reliability & Quality Control*, IEEE, New York, Jan. 1965, pp. 34–40.

Lessels, J. D., "Numerical Calculation of Mean Mission Duration," *IEEE Transactions on Reliability*, Vol. R-31, No. 5, Dec. 1982, pp. 420–422.

Leventhal, A. and Bloomquist, C. E., "Spacecraft Mission Effectiveness," *Proceedings of Annual Symposium on Reliability*, IEEE, New York, Jan. 1968, pp. 615–621.

Leventhal, A., Bloomquist, C., and Joseph, J., "Spacecraft Failure Rates—Where are We?," *Proceedings of Annual Symposium on Reliability*, IEEE, New York, Jan. 1969, pp. 444–452.

Livson, B. U., Melamed, I., and Samuel, S., "Optimization of Reliability for Mini Remotely-Piloted-Vehicle (RPV)," *Proceedings of Annual Reliability and Maintainability Symposium*, IEEE, New York, 1982, pp. 30–33.

Lloyd, D. K., "Multienvironmental Life Testing of Parts and Components in Rockets and Guided Missiles by Statistical Design," *IRE Transactions, Reliability and Quality Control*, Vol. 15, 1958, pp. 34–42 (available from IEEE).

Locks, M. O. and Pauler, G. L., "Simulation of Dispatch Reliability for a Fleet of Large Commercial Aircraft," *Proceedings of Annual Reliability and Maintainability Symposium*, IEEE, New York, 1974, pp. 226–228.

Locurto, C. A., "Impact of BIT on Avionic Maintainability," *Proceedings of Annual Reliability and Maintainability Symposium*, IEEE, New York, 1983, pp. 333–338.

Lohmar, J. W. and Mandel, D. B., "Guided Missile Frigate Tiger Assessment," *Proceedings of Annual Reliability and Maintainability Symposium*, IEEE, New York, 1981, pp. 459–464.

Love, J. E. and Young, W. R., "Component Performance and Flight Operations of the X-15 Research Airplane Program," *Proceedings of Annual Symposium on Reliability*, IEEE, New York, Jan. 1966, pp. 125–132.

Lusser, R. K., "Twenty-Seven Rules for Guided Missile Design Engineers," *Tele-Tech*, Vol. 14, No. 8, 1955, pp. 86–87.

MacCrehan, W. A. Jr., "Commercial Airborne Reliability," *Proceedings of Third National Symposium on Reliability and Quality Control*, IEEE, New York, 1957, pp. 177–179.

Madden, J. L., "Cost-Effective Maximum Times for Aircraft Engines," *Proceedings of Annual Symposium on Reliability*, IEEE, New York, Jan. 1966, pp. 118–124.

Majesty, M. S., "Personnel Subsystem Reliability for Aerospace Systems," *Proceedings of the IAS Aerospace Systems Reliability Symposium*, April 1962, pp. 199–204 (available from AIAA).

Mallios, W. S. and Reuben, R. R., "Structural Predictions of Mean Times Between Failures Through Maintenance Activity for Aircraft Power Systems," *Proceedings of Annual Symposium on Reliability*, IEEE, New York, Jan. 10–12, 1967, pp. 91–103.

Mancino, V. J., Monshaw, V. R., and Slusark, W. J. Jr., "Reliability Considerations for Communication Satellites," *Proceedings of Annual Reliability and Maintainability Symposium*, IEEE, New York, 1986, pp. 389–396.

Mani, V. and Sarma, V. V. S., "Queuing Network Models for Aircraft Availability and Spares Management," *IEEE Transactions on Reliability*, Vol. R-33, No. 3, Aug. 1984, pp. 257–262.

Maniwa, T., Fukushima, Y., and Nakamura, T., "Reliability Prediction of ETA (Explosive Transfer Assembly) for the H-I Launch Vehicle," *Proceedings of Annual Reliability and Maintainability Symposium*, IEEE, New York, 1987, pp. 347–351.

Manning, S. D. and Lemon, G. H., "Plan for Developing Structural Criteria for Composite Airframes," *Proceedings of Annual Reliability and Maintainability Symposium*, IEEE, New York, 1974, pp. 155–162.

Mariani, G., "The Role of Reliability in Commercial and Military Telecommunication Satellite System Planning," *Operations Research and Reliability*, edited by D. Grouchko, Gordon and Breach Science Publishers, New York, 1971, pp. 5–46.

Markowitz, O., "Aviation Supply Office FFW/RIW Case History #2, Abex Pump," *Proceedings of Annual Reliability and Maintainability Symposium*, IEEE, New York, 1976, pp. 357–362.

Masterson, R. J. and Miller, R. N., "Testing of Spacecraft in Long-Term Storage," *Proceedings of Annual Reliability and Maintainability Symposium*, IEEE, New York, 1973, pp. 149–155.

Mastran, D. V., "A Bayesian Approach for Assessing the Reliability of Air Force Re-entry Systems," *Proceedings of Seventh Reliability and Maintainability Conference* (*1968 Annals of Assurance Sciences*), ASME, New York, July 1968, pp. 380–383.

Matthews, A. R., "Reliability and Longevity for Space Technology," *IRE Convention Record*, Vol. 6, Pt. 6, 1958, pp. 123–136.

Maxam, F. A. and Plewes, K. C., "The Teenage Dilemma! (Or, The Prematurity Years of a Jet Transport Are the Hardest)," *Proceedings of Tenth Reliability and Maintainability Conference* (*Annals of Reliability and Maintainability*, Vol. 10), ASME, New York, June 1971, pp. 2–10.

Maybell, J. L., "System Effectiveness Task Status Report," *Proceedings of Annual Reliability and Maintainability Symposium*, IEEE, New York, 1974, pp. 509–527.

McCarty, B. J., "Space Shuttle Flight Results," *Proceedings of Annual Reliability and Maintainability Symposium*, IEEE, New York, 1982, pp. 466–467.

McCreary, T. R., "Early-Identification of High-Maintainance Helicopters," *Proceedings of Annual Reliability and Maintainability Symposium*, IEEE, New York, 1978, pp. 486–491.

McGee, R. L., "Reliability & Maintainability Contribution to Hornet Mission-Success," *Proceedings of Annual Reliability and Maintainability Symposium*, IEEE, New York, 1979, pp. 338–341.

McIntire, W. L. and Lingren, M. F., "Design Reliability of Aircraft Reduction Gears," *Proceedings of the Seventh Reliability and Maintainability Conference* (*1968 Annals of Assurance Sciences*), ASME, New York, July 1968, p. 483.

Medford, J. F., "The Challenge of Meeting Manned Mission Reliability and Maintainability Requirements in Future Systems," *Proceedings of Sixth Reliability and Maintainability Conference* (*Annals of Reliability and Maintainability*, Vol. 6), SAE, Warrendale, PA, July 1967, pp. 505–509.

Mentzer, W., "What Improvements in R&M Will be Most Cost Effective in Future Transport Aircraft?," *Proceedings of Tenth Reliability and Maintainability Conference* (*Annals of Reliability and Maintainability*, Vol. 10), ASME, New York, June 1971, pp. 1–2.

Mercer, J. C., "Reliability and Maintainability of the FAA's ARTS-III Terminal Air Traffic Control System," *Proceedings of Eighth Reliability and Maintainability Conference (1969 Annals of Assurance Sciences)*, Gordon and Breach Science Publishers, New York, July 1969, pp. 12–21.

"Metallic Materials and Elements for Aerospace Vehicle Structures," MIL-HDBK-5D, U.S. Government Printing Office, Washington, DC, June 1983.

"Metallic Materials and Elements for Flight Vehicle Structures," Military Handbook 5 (superseding MIL-HDBK-5, March 1961), U.S. Department of Defense Doc. FSC 1500, U.S. Government Printing Office, Washington, D.C., Aug. 1962.

Meyer, R. C., "Variables Analysis Applied to Solid Propellant Rocket Motors," *Proceedings of Fifth Reliability and Maintainability Conference (Annals of Reliability and Maintainability*, Vol. 5), AIAA, Washington, DC, July 1966, pp. 555–562.

Mickel, L. S. and Newman, E. E., "Sparing to Optimize Naval Airship Availability," *Proceedings of Annual Reliability and Maintainability Symposium*, IEEE, New York, 1987, pp. 264–267.

Miller, C. O., "Legal Ramifications of Aircraft Accident/Malfunction Data," *Proceedings of IAS Aerospace Systems Reliability Symposium*, April 1962, pp. 174–178 (available from AIAA).

Minner, E. S. and Romero, H. A., "Reliability Testing of F-111A Avionics Systems," *Proceedings of Annual Symposium on Reliability*, IEEE, New York, Jan. 1968, pp. 567–575.

Mintz, F. and Levine, M. B., "Aircraft Electronics—Environment, Specifications, and Survival," *IRE Convention Record*, Pt. 5, March 1955, p. 3.

Moe, T. A., "ALE—A Carrier Aircraft Availability Model," *IEEE Transactions on Reliability*, Vol. R-25, No. 4, Oct. 1976, pp. 243–247.

Mona, A. G., "Maintenance Vs. Redundancy for Manned Space Stations," *Proceedings of Annual Symposium on Reliability*, IEEE, New York, Feb. 1970, pp. 404–409.

Moreno, F. J., "Reliability Estimate of a Space Deployable Antenna," *Proceedings of Annual Reliability and Maintainability Symposium*, IEEE, New York, 1973, pp. 182–185.

Morris, B. T., "Spacecraft Life for Deep Space Exploration," *Proceedings of Annual Symposium on Reliability*, IEEE, New York, Jan. 1968, pp. 298–302.

Morris, R. S. and Bartels, B. E., "Determining Operating and Support Costs with USAF D056 Maintenance Data Collection System," *Proceedings of Annual Reliability and Maintainability Symposium*, IEEE, New York, 1984, pp. 197–206.

Morrison, S. C., "Requirements for Achieving Reliability of the Minuteman Missileborne Electronic Equipment," *Proceedings of Fifth National Symposium on Reliability and Quality Control*, IEEE, New York, 1959, pp. 73–78.

Morton, J. A., "The Role of Components in Satellite Reliability," *IRE Transactions on Reliability and Quality Control*, Vol. RQC-11, No. 2, July 1962, pp. 104–106 (available from IEEE).

Moxley, F. H., "Effects of Pan Am's Reliability Program on Maintenance," *Proceedings of Annual Symposium on Reliability*, IEEE, New York, Jan. 1971, pp. 163–166.

Mumma, G. B. and O'Halloran, W. R., "Hazard Analysis—Space Applications to Mass Transit," *Proceedings of Annual Reliability and Maintainability Symposium*, IEEE, New York, 1976, pp. 251–256.

Murata, T., "Reliability Case History of an Airborne Air Data Computer," *IEEE Transactions on Reliability*, Vol. R-24, No. 2, June 1975, pp. 98–102.

Murata, T., "Reliability Case History of an Airborne Air Turbine Starter," *IEEE Transactions on Reliability*, Vol. R-25, No. 5, Dec. 1976, pp. 302–303.

Murphy, A., "Maintenance on the Moon," *Proceedings of Seventh Reliability and Maintainability Conference* (*1968 Annals of Assurance Sciences*), ASME, New York, July 1968, pp. 61–66.

Murphy, J. L., "Microelectronics and Space Vehicle Reliability," *Proceedings of Annual Symposium on Reliability*, IEEE, New York, Jan. 1966, pp. 437–449.

Murphy, M. R., "Analysis of Eighty-four Commercial Aviation Incidents: Implications for a Resource Management Approach to Crew Training," *Proceedings of Annual Reliability and Maintainability Symposium*, IEEE, New York, 1980, pp. 298–306.

Murray, W. A., "What Parts Lessons from Lunar Orbiter?," *Proceedings of Annual Symposium on Reliability*, IEEE, New York, Jan. 1969, pp. 453–464.

Meyers, J. L. and Moon, E. L., "Service Life Prediction Program for the Minuteman LGM 30 Propulsion System," *Proceedings of Seventh Reliability and Maintainability Conference* (*1968 Annals of Assurance Sciences*), ASME, New York, July 1968, pp. 194–202.

Myers, R. H., "Which Road to Satellite Reliability?," *IRE Transactions on Reliability and Quality Control*, Vol. RQC–11, No. 1, May 1962, pp. 24–32 (available from IEEE).

Meyers, R. H., Dewitt, C. M. III, and Feduccia, A. J., "Airborne Electronic Equipment Guarantees," *Proceedings of Ninth Reliability and Maintainability Conference* (*Annals of Reliability and Maintainability*, Vol. 9), SAE, Warrendale, PA, July 1970, pp. 480–505.

Naresky, J. J., "Reliability and Maintainability Research in the USAF," *Proceedings of Fifth Reliability and Maintainability Conference* (*Annals of Reliability and Maintainability*, Vol. 5), AIAA, Washington, DC, July 1966, pp. 769–787.

Naresky, J. J., "Air Force Ground Electronic Equipment Reliability Improvement Program," *IRE Convention Record*, Vol. 5, Pt. 10, 1957, pp. 21–31.

Neogy, R. and McGrath, R., "Technical Monitoring in a Communication Satellite Program," *Proceedings of Annual Reliability and Maintainability Symposium*, IEEE, New York, 1987, pp. 355–358.

Nerber, P. O., "Warranty Cost Estimates for Avionic Subsystems," *Proceedings of Annual Symposium on Reliability*, IEEE, New York, Jan. 1969, pp. 270–275.

Newirth, D. M. and Muller, H. G., "Engine Reliability with Electronic Controllers," *Proceedings of Annual Reliability and Maintainability Symposium*, IEEE, New York, 1975, pp. 102–104.

Newman, D. G. and Nesbitt, L. D., "USAF Experience with RIW," *Proceedings of Annual Reliability and Maintainability Symposium*, IEEE, New York, 1978, pp. 55–61.

Nicely, H. P. Jr., "An Automatic Reliability Assessment and Analysis System for Spacecraft Automatic Checkout Equipment," *Proceedings of Sixth Reliability and Maintainability Conference* (*Annals of Reliability and Maintainability*, Vol. 6), SAE, Warrendale, PA, July 1967, pp. 628–641.

Nicely, H. P. and Mayer, C. B., "Design Tools for the Optimization of Redundancy for a Planetary Vehicle," *Proceedings of Seventh Reliability and Maintainability Conference* (*1968 Annals of Assurance Sciences*), ASME, New York, July 1968, pp. 50–60.

Noriega, J. R., "Electrical, Electronic, Electromechanical Part Control Program for the Space Shuttle Orbiter Program," *Proceedings of Annual Reliability and Maintainability Symposium*, IEEE, New York, 1982, pp. 384–389.

Norris, H. P. and Timmins, A. R., "Failure-Rate Analysis of Goddard Space Flight Center Spacecraft Performance during Orbital Life," *Proceedings of Annual Reliability and Maintainability Symposium*, IEEE, New York, 1976, pp. 120–125.

Nowak, T. J., "Reliability of Integrated Circuits for Minuteman," *Proceedings of Annual Symposium on Reliability*, IEEE, New York, Jan. 1966, pp. 469–482.

Nucci, E. J., "Concepts and Research Needs of Reliability in Military Systems," *IEEE Transactions on Reliability*, Vol. R-12, No. 3, Sept. 1963, pp. 1–4.

Ogden, J. E. and Cavanaugh, F., "Operational Readiness and Maintenance Testing of the B–1 Strategic Bomber," *Proceedings of Annual Reliability and Maintainability Symposium*, IEEE, New York, 1973, pp. 293–297.

Ostrander, V. P., "Saturn S-II Confidence Development Plan," *Proceedings of Tenth National Symposium on Reliability and Quality Control*, IEEE, New York, 1964, pp. 494–503.

Ostrander, V. P., "Space Shuttle Orbiter—A Reliability Challenge and Achievement," *Proceedings of Annual Reliability and Maintainability Symposium*, IEEE, New York, 1982, pp. 468–473.

Pace, F. D., "Reliability for Minuteman Operational Ground Equipment," *Proceedings of Eleventh National Symposium on Reliability and Quality Control*, IEEE, New York, Jan. 1965, pp. 324–330.

Packard, A. C. and Weller, R. K., "The Reliability Problems in Guided Missile Development," *IRE Transactions on Reliability and Quality Control*, Vol. PGQC-1, Aug. 1952, pp. 34–40 (available from IEEE).

Packard, K. S. and Goldstein, M. H., "Effectivity: Its Application to a Long-Lived Space System," *IEEE Transactions on Reliability*, Vol. R-12, No. 1, March 1963, pp. 23–31.

Parrish, P., "Assessing Missile System Reliability," *Proceedings of Annual Reliability and Maintainability Symposium*, IEEE, New York, 1972, pp. 520–525.

Parsons, D. W., "Reliability Demonstration of Shillelagh Missile Subsystem," *Proceedings of Fifth Reliability and Maintainability Conference* (*Annals of Reliability and Maintainability*, Vol. 5), AIAA, Washington, DC, July 1966, pp. 710–721.

Parsons, J. H. and Rubel, J. H., "A Case History—Developing an Airborne Digital Computer," *Proceedings of Second National Symposium on Reliability and Quality Control*, IEEE, New York, 1956, pp. 84–98.

Patterson, M., "Design Techniques for Upgrading the Reliability of Weapon Systems during Flight-Readiness Check-out," *IRE Convention Record*, Vol. 6, Pt. 6, Aug. 1958, pp. 3–9.

Pedar, A. and Sarma, V. V. S., "Phased-Mission Analysis for Evaluating the Effectiveness of Aerospace Computing-Systems," *IEEE Transactions on Reliability*, Vol. R-30, No. 5, Dec. 1981, pp. 429–437.

Peratino, G. S., "Air Force Approach to Life Cycle Costing," *Proceedings of Annual Symposium on Reliability*, IEEE, New York, Jan. 1968, pp. 184–187.

Peterson, E. L., "Maintainability Design Requirements for Future Space Systems," *IEEE Transactions on Reliability*, Vol. R-15, No. 1, May 1966, pp. 17–21.

Peterson, K. and Babin, R. S., "Integrated Reliability and Safety Analysis of the DC-10 All-Weather Landing System," *Proceedings of Annual Reliability and Maintainability Symposium*, IEEE, New York, 1973, pp. 403–409.

Philipson, L. L., "Operational Reliability Model for Reconnaissance System," *IRE Convention Record*, Vol. 7, Pt. 6, 1959, pp. 79–88.

Pieruschka, E. G., "Failure Categories of Guided Missiles," *Proceedings of Sixth National Symposium on Reliability and Quality Control*, IEEE, New York, 1960, pp. 497–509.

Pieruschka, E. G., "The Measurability of the Reliability of an Expensive Guided Missile Type," *Proceedings of Seventh National Symposium on Reliability and Quality Control*, IEEE, New York, 1961, pp. 394–406.

Pieruschka, E., "Optimum Allocation of Funds for Reliability Programs of Guided Missiles," Reliability Co-ordination Office, U.S. Army Redstone Arsenal, Huntsville, AL, Jan. 1955.

Pilny, M. J., "Viking Lander Reliability Program," *Proceedings of Annual Reliability and Maintainability Symposium*, IEEE, New York, 1978, pp. 136–142.

Pizzano, F., "Reliability and Maintenance Simulation of the Hubble Space Telescope," *Proceedings of Annual Reliability and Maintainability Symposium*, IEEE, New York, 1986, pp. 83–87.

Plouff, W. E., "Avionic Reliability Experience: AR-104 and 781B," *Proceedings of Annual Reliability and Maintainability Symposium*, IEEE, New York, 1977, pp. 206–212.

Pulliam, T. L., "Analysis of Failures in Spacecraft and Aircraft Components," *Proceedings of Annual Symposium on Reliability*, IEEE, New York, Jan. 1969, pp. 40–47.

Punches, K., "Airplane Reliability in a Nutshell," *IEEE Transactions on Reliability*, Vol. R-32, No. 2, June 1983, pp. 130–133.

Raabe, P. H., "A Study of Rocket Engine System Reliability," *Proceedings of Aerospace Reliability and Maintainability Conference*, AIAA, Washington, DC, May 1963, pp. 59–71.

Radio Technical Commission for Aeronautics, "Airborne Electronics and Electrical Equipment Reliability," RTCA Secretariat, Washington, DC, 1958.

Rado, L. G., "The Effect of the NOL Corona and Battelle Data Interchange Programs on Librascope Reliability Test Efforts," *IRE Transactions on Reliability and Quality Control*, Vol. RQC-10, No. 1, March 1961, pp. 26–28 (available from IEEE).

Ralph, J. A. and Cicchetti, C., "System Effectiveness Can Be Achieved in a Manned Space Program through Correct Management Decisions," *Proceedings of Fifth Reliability and Maintainability Conference* (*Annals of Reliability and Maintainability*, Vol. 5), AIAA, Washington, DC, July 1966, pp. 951–964.

Ramquist, A. H., "Manned Space Stations and Mission Effectiveness," *Proceedings of Eleventh National Symposium on Reliability and Quality Control*, IEEE, New York, Jan. 1965, pp. 533–541.

Rappel, R. F. and Johnson, E. G., "Adaptation of Digital Controls to Rocket Engines," *Proceedings of Annual Reliability and Maintainability Symposium*, IEEE, New York, 1975, pp. 92–97.

Ratynski, M. V., "Selling Your Reliability Program to the Air Force Electronic Systems Division," *Proceedings of Eleventh National Symposium on Reliability and Quality Control*, IEEE, New York, Jan. 1965, pp. 542–556.

Redler, W. M., "Parts Reliability Problems in Aerospace Systems," *Proceedings of Annual Symposium on Reliability*, IEEE, New York, Jan. 1966, pp. 212–220.

Reesing, H. A., "DC-9 Reliability Program," *Proceedings of Fourth Annual Reliability and Maintainability Conference* (*Annals of Reliability and Maintainability*, Vol. 4), Spartan Books, Washington, DC, July 1965, pp. 447–456.

Reesing, H. A., "DC-10 Reliability Program," *Proceedings of Eighth Reliability and Maintainability Conference* (*1969 Annals of Assurance Sciences*), Gordon and Breach Science Publishers, New York, July 1969, pp. 309–314.

Remond, F. X. and Dee, P. A., "An Approach to the Launch Availability of a Typical Space System," *Proceedings of Seventh Annual West Coast Reliability Symposium*, Western Periodicals, North Hollywood, CA, May 1966, pp. 77–92.

Ricketts, M. P., "F/A-18 Hornet Reliability Challenge—Status Report," *Proceedings of Annual Reliability and Maintainability Symposium*, IEEE, New York, 1982, pp. 491–496.

Roelands, D. L., "Reliability Testing in the Saturn S-II Stage Project," *Proceedings of Fifth Reliability and Maintainability Conference* (*Annals of Reliability and Maintainability*, Vol. 5), AIAA, Washington, DC, July 1966, pp. 672–677.

Rogger, W. R., "F/A-18 Hornet: Reliability Development Testing," *Proceedings of Annual Reliability and Maintainability Symposium*, IEEE, New York, 1985, pp. 373–378.

Rose, J. and Phelps, E., "Cost of Ownership Application to Airplane Design," *Proceedings of Annual Reliability and Maintainability Symposium*, IEEE, New York, 1979, pp. 47–50.

Rue, H. D. and Thomas, J. M., "Productivity in the Avionic Field Environment," *Proceedings of Annual Reliability and Maintainability Symposium*, IEEE, New York, 1983, p. 521

Ruther, F., "Ballistic Missile & Aircraft Operational Effectiveness," *Proceedings of Annual Symposium on Reliability*, IEEE, New York, Jan. 1969, pp. 235–237.

Ryerson, C. M., "Correlation of Ground, Air and Space Failure Rates," *Proceedings of Eleventh National Symposium on Reliability and Quality Control*, IEEE, New York, Jan. 1965, pp. 273–278.

Ryerson, C. M., "Spacecraft Component Reliability," *Proceedings of Annual Symposium on Reliability*, IEEE, New York, Jan. 1966, pp. 173–179.

Rygwalski, E. E., "Estimating Reliability for Future Space Systems," *Proceedings of Annual Symposium on Reliability*, IEEE, New York, Jan. 1967, pp. 816–823.

Salt, T. L., "Evaluation of Flight Mission Severity in Cumulative Damage," *Proceedings of Tenth Reliability and Maintainability Conference* (*Annals of Reliability and Maintainability*, Vol. 10), ASME, New York, June 1971, pp. 104–113.

Samuel, S. and Tamir, Y., "Reliability Program Development and Implementation for a Remote Piloted Vehicle," *IEEE Transactions on Reliability*, Vol. R-32, No. 3, Aug. 1983, pp. 230–235.

Sandler, G. H., "Reliability Models in Space Systems Planning and Decision Making," *Proceedings of Ninth National Symposium on Reliability and Quality Control*, IEEE, New York, 1963, pp. 493–498.

Sandler, G. H., "Application of Reliability Models to the Choice of a Booster for Lunar Missions," *Proceedings of Aerospace Reliability and Maintainability Conference*, AIAA, Washington, DC, May 1963, pp. 450–456.

Sarma, V. V. S., Ramchand, K., and Rao, A. K., "Queuing Models for Estimating Aircraft Fleet Availability," *IEEE Transactions on Reliability*, Vol. R-26, No. 4, Oct. 1977, pp. 253–256.

Scheer, G. H., "Reliability in Complex Airborne Equipment," *Tele-Tech*, Vol. 41, No. 7, 1955, pp. 56–57, 86–88.

Scheer, G. H., "Maintenance Problems in New Airborne Radio Equipment," *Tele-Tech*, Vol. 39, May 1953, pp. 88–89, 148.

Scheller, J. A., "Space Transportation System Payload Safety Policy," *Proceedings of Annual Reliability and Maintainability Symposium*, IEEE, New York, 1977, pp. 130–133.

Schmidt, D. L., "Systems Testing—Final Check on Titan Reliability," *Proceedings of Fourth Annual Reliability and Maintainability Conference* (*Annals of Reliability and Maintainability*, Vol. 4), Spartan Books, Washington, DC, July 1965, pp. 585–590.

Schneider, L. L., "Reliability Contributions to Minuteman Logistics," *Proceedings of Fourth Annual Reliability and Maintainability Conference* (*Annals of Reliability and Maintainability*, Vol. 4), Spartan Books, Washington, DC, July 1965, pp. 105–110.

Schneider, L. L., "Reliability Contribution to Minuteman Logistics," *IEEE Transactions on Reliability*, Vol. R-15, No. 1, May 1966, pp. 6–10.

Scoggins, J. R. and Vaughan, W. W., "Problem of Atmospheric Wind Inputs for Missile and Space Vehicle Design," AIAA Paper 63–xxx, July 1963.

Sellers, W. H., "Missile and Spacecraft Countdown Probabilities by Markov Chain Techniques," *Proceedings of Fourth Annual Reliability and Maintainability Conference* (*Annals of Reliability and Maintainability*, Vol. 4), Spartan Books, Washington, DC, July 1965, pp. 437–446.

Senator, F. E., Calderon, S. S., and Buchanan, R. L., "Earth Orbital Space Station Effectiveness—Reliability, Safety and Maintainability," *Proceedings of Seventh Reliability and Maintainability Conference (1968 Annals of Assurance Sciences)*, ASME, New York, July 1968, pp. 11–28.

Sepmeyer, L. W., "Prediction of Command Control System Dependability and Performance," *IRE Transactions on Reliability and Quality Control*, Vol. RQC-11, No. 3, Oct. 1962, pp. 35–42.

Shawyer, W. R., Sloter, L. E. II, Stracener, J. T., and White, D. J., "Determination of Aircraft Structural Inspection Intervals," *Proceedings of Annual Reliability and Maintainability Symposium*, IEEE, New York, 1981, pp. 452–458.

Shelley, B. F., "Support Equipment Optimization for Aircraft," *Proceedings of Annual Reliability and Maintainability Symposium*, IEEE, New York, 1983, pp. 383–387.

Shelley, B. F. and Hamilton, D. O., "A Mechanized Aircraft Reliability Analysis Model," *Proceedings of Tenth National Symposium on Reliability and Quality Control*, IEEE, New York, 1964, pp. 560–566.

Sheppard, R. R., "Space Shuttle EEE (electrical, electronic and electromechanical) Parts Control," *Proceedings of Annual Reliability and Maintainability Symposium*, IEEE, New York, 1980, pp. 149–151.

Shooman, M. L. and Tenenbaum, S., "Hazard Function Monitoring of Airline Components," *Proceedings of Annual Reliability and Maintainability Symposium*, IEEE, New York, 1974, pp. 383–390.

Shumaker, M. J. and Cary, R. C., "A Mission Success Model for Unmanned Space Exploration," *Proceedings of Ninth Reliability and Maintainability Conference (Annals of Reliability and Maintainability*, Vol. 9), SAE, Warrendale, PA, July 1970, pp. 385–396.

Shurman, M. B., "Time-Dependent Failure Rates for Jet Aircraft," *Proceedings of Annual Reliability and Maintainability Symposium*, IEEE, New York, 1978, pp. 198–203.

Shurman, M. B., "Development of In-Flight Steady-State Failure Rates," *Proceedings of Annual Reliability and Maintainability Symposium*, IEEE, New York, 1979, pp. 323–325.

Simiu, E., "Aerodynamic Coefficients and Risk-Consistent Design," *ASCE Journal of Structural Engineering*, Vol. 109, May 1983, pp. 1278–1289.

Simpson, N. H., "Accelerated Life Test In Airframe Manufacture," *IRE Transactions*, Vol. PGRQC-14, Sept. 1958, pp. 33–49.

Sineath, R. M., "Complex Aircrew Escape System—R&M Requirements," *Proceedings of Annual Reliability and Maintainability Symposium*, IEEE, New York, 1979, pp. 301–307.

Singletary, J. B. and Rittenhouse, J. B., "The Effects of Space Environment on Spacecraft Reliability," *Proceedings of Annual Symposium on Reliability*, IEEE, New York, Jan. 1967, pp. 305–311.

Sinnott, J. J., "F/A-18 Supportability Assurance Readiness Program," *Proceedings of Annual Reliability and Maintainability Symposium*, IEEE, New York, 1984, pp. 513–519.

Skeie, K., "Right Road to Rocket Reliability," *Missile and Rockets*, Vol. 7, 1960, p. 16.

Sleight, R. B., "Human Factors in the Use and Acceptance of the Aircar," *Proceedings of IAS Aerospace Systems Reliability Symposium*, April 1962, pp. 79–81 (available from AIAA).

Slocum, R. W., Roberts, E. I., and Smith, R. J., "Future Needs and Trade Offs for Long Life Spacecraft," *Proceedings of Annual Symposium on Reliability*, IEEE, New York, Jan. 1968, pp. 303–314.

Smiley, R. W., "Military Management of Missile Quality Control/Reliability Programs," *Proceedings of Ninth National Symposium on Reliability and Quality Control*, IEEE, New York, 1963, pp. 66–68.

Smith, A. M., "Risk Assessment in Complex Unattended Aerospace Systems," *Proceedings of Sixth Reliability and Maintainability Conference* (*Annals of Reliability and Maintainability*, Vol. 6), SAE, Warrendale, PA, July 1967, pp. 478–486.

Smith, A. M. and Matteson, T. D., "The Role of Testing in Achieving Aerospace Systems Effectiveness," *Proceedings of Annual Reliability and Maintainability Symposium*, IEEE, New York, 1973, pp. 36–41.

Smith, A. M. and Waltz, W. R., "Testing for Spacecraft Reliability—A Management Overview," *Proceedings of Seventh Reliability and Maintainability Conference* (*1968 Annals of Assurance Sciences*), ASME, New York, July 1968, pp. 214–220.

Smith, M. T., "Reliability Initiatives: US Air Force," *Proceedings of Annual Reliability and Maintainability Symposium*, IEEE, New York, 1987, pp. 3–4.

Smith, R. L. and Smith, S. E., "Guidelines for Space-Station Maintainability," *Proceedings of Annual Reliability and Maintainability Symposium*, IEEE, New York, 1986, pp. 69–76.

Smith, T. E., "The Impact of Maintainability Engineering Techniques on the Design of the F-111," *Proceedings of Sixth Reliability and Maintainability Conference* (*Annals of Reliability and Maintainability*, Vol. 6), SAE, Warrendale, PA, July 1967, pp. 216–235.

Smith, W. J., "Minuteman Guidance System—Evaluation of Achieved Reliability," *Proceedings of Aerospace Reliability and Maintainability Conference*, AIAA, Washington, DC, May 1963, pp. 434–440.

Snyder, J. E., "Assessment of Ground System Performance and Integration of Reliability and Maintainability with Simulated System Operations," *Proceedings of Sixth Reliability and Maintainability Conference* (*Annals of Reliability and Maintainability*, Vol. 6), SAE, Warrendale, PA, July 1967, pp. 555–561.

Society of Automotive Engineers, *Reliability Control in Aerospace Equipment Development*, Tech. Prog. Series, Vol. 4, prepared under auspices of Subcommittee on Reliability, SAE, Warrendale, PA, 1963.

Sohn, H. M., "C-5A Quantitative Maintainability Program," *Proceedings of Fifth Reliability and Maintainability Conference* (*Annals of Reliability and Maintainability*, Vol. 5), AIAA, Washington, DC, July 1966, pp. 104–111.

Soucy, C. I., "Reliability Control of Electronic Equipment in Aircraft and Weapon Systems: General and Management Aspects," *Canadian Aeronautics Journal*, Vol. 3, Sept. 1957, pp. 222–231.

Sparling, R. H., "Testing in the Guided Missile Industry," *ASTM Bulletin*, No. 218, Dec. 1956, pp. 52–56.

Spatzer, H. M., "The Effect of Programming Factors on the Initial Schedule Reliability of the 707 Jet Transport," *Proceedings of Aerospace Reliability and Maintainability Conference*, AIAA, Washington, DC, May 1963, pp. 395–397.

Spergel, J. and Godwin, E. F., "Reliable Interconnections for Army Avionics," *Proceedings of Annual Reliability and Maintainability Symposium*, IEEE, New York, 1972, pp. 499–509.

Spiegel, J. and Bennett, E. M., "Military System Reliability: Some Department of the Air Force Contributions," *IRE Transactions on Reliability and Quality Control*, Vol. RQC-10, No. 2, Aug. 1961, pp. 53–63 (available from IEEE).

Spraker, W. A., Loomis, J. P., and Bagby, F. L., "Trend Analysis as Applied to Gas-Turbine Engines and Aircraft Mechanical Equipment," *Proceedings of Aerospace Reliability and Maintainability Conference*, AIAA, Washington, DC, May 1963, pp. 375–380.

Stahle, C. V., Gongloff, H. R., Young, J. P., and Keegan, W. B., "Shuttle-Payload Minimum-Cost Vibroacoustic Tests," *Proceedings of Annual Reliability and Maintainability Symposium*, IEEE, New York, 1977, pp. 290–296.

Stahr, S., "Commercial Supersonic Propulsion—Maintainability by Design," *Proceedings of Aerospace Reliability and Maintainability Conference*, AIAA, Washington, DC, May 1963, p. 535.

Steinberg, A., "Estimating Guided Missile System Reliability From R&D Systems Firing Tests," *Proceedings of Eighth National Symposium on Reliability and Quality Control*, IEEE, New York, 1962, pp. 129–134.

Steinberg, A., "Reliability Assessment of Liquid Rocket Engines," *Proceedings of Tenth National Symposium on Reliability and Quality Control*, IEEE, New York, 1964, pp. 478–484.

Steiner, R. F., "A Simple Method to Reduce Avionic Repair Costs," *Proceedings of Annual Reliability and Maintainability Symposium*, IEEE, New York, 1983, pp. 432–434.

Stember, L. H. Jr., Mark, D. G., Mesloh, R. E., and Putnam, A. A., "Research and Study of Analytical Techniques for Predicting Reliability of Flight Control Systems," Battelle Memorial Institute, Columbus, OH, Tech. Doc. Rept. ASD-TDR-63-287, Jan. 1963.

Stevenson, H. L., "Apollo CSM Parts Management: A Legacy for Future Space Programs," *Proceedings of Fifth Reliability and Maintainability Conference* (*Annals of Reliability and Maintainability*, Vol. 5), AIAA, Washington, DC, July 1966, pp. 19–30.

Stoller, D. S., "Measuring Missile Reliability in Prelaunch Environments," *IRE Transactions on Reliability and Quality Control*, Vol. RQC-10, No. 1, March 1961, pp. 19–25 (available from IEEE).

Stout, C. L., "DC-10 Dispatch Reliability—A New Challenge for Flight Test," *Proceedings of Tenth Reliability and Maintainability Conference* (*Annals of Reliability and Maintainability*, Vol. 10), ASME, New York, June 1971, pp. 100–103.

Stovall, F. A., "C-5A Reliability Program," *Proceedings of Eighth Reliability and Maintainability Conference* (*1969 Annals of Assurance Sciences*), Gordon and Breach Science Publishers, July 1969, pp. 115–122.

Swett, B. H., "Avionics Testing," *Proceedings of Annual Reliability and Maintainability Symposium*, IEEE, New York, 1975, pp. 86–87.

Symonds, J., "Reliability Requirements of Electronic Equipment in Civil Aviation," *Symposium on Electronic Equipment Reliability*, Institution of Electrical Engineers, London, May 1960, pp. 16–17.

Tamsen, J. J., "Reliability Estimation and Demonstration of Gas Turbine and Rocket Propulsion Systems," *Proceedings of 7th Reliability and Maintainability Conference* (*1968 Annals of Assurance Sciences*), ASME, New York, July 1968, pp. 108–109.

Teal, T. W., "Aviation Maintenance Data Collection in the U.S. Navy," *Proceedings of Fourth Annual Reliability and Maintainability Conference* (*Annals of Reliability and Maintainability*, Vol. 4), Spartan Books, Washington, DC, July 1965, pp. 839–862.

Tendeman, L. and Kohlheyer, R. C., "Interaction of Reliability and Safety on the Spacelab Programme," *Proceedings of Annual Reliability and Maintainability Symposium*, IEEE, New York, 1982, pp. 474–475.

Thomas, F. C., "DC-9 Maintainability Design Features," *Proceedings of Fourth Annual Reliability and Maintainability Conference* (*Annals of Reliability and Maintainability*, Vol. 4), Spartan Books, Washington, DC, July 1965, pp. 493–500.

Thomas, J. M., Derbalian, G., and Bischel, K., "Parametric Sensitivity of Rocket Motor Reliability," *Proceedings of Annual Reliability and Maintainability Symposium*, IEEE, New York, 1980, pp. 121–128.

Timmons, A. R. and Heuser, R. E., "A Study of First Day Space Malfunctions," *Proceedings of Tenth Reliability and Maintainability Conference* (*Annals of Reliability and Maintainability*, Vol. 10), ASME, New York, June 1971, pp. 90–99.

Toohey, E. F. and Calvo, A. B., "Cost Analysis for Avionics Acquisition," *Proceedings of Annual Reliability and Maintainability Symposium*, IEEE, New York, 1980, pp. 85–90.

Tornatore, A. S., "The AVSCOM Flight-Safety Part Program," *Proceedings of Annual Reliability and Maintainability Symposium*, IEEE, New York, 1987, pp. 395–398.

Trott, E. P., "Case Study—MARCEP Applied to Space Systems," *Proceedings of Sixth Reliability and Maintainability Conference* (*Annals of Reliability and Maintainability*, Vol. 6), SAE, Warrendale, PA, July 1967, pp. 524–528.

Tyszkiewicz, A. M., "Comparative Cost Model for Aircraft Investment Spares," *Proceedings of Annual Reliability and Maintainability Symposium*, IEEE, New York, 1983, pp. 376–382.

Unger, R. F., "The 'Astral' Parts Program: A New Dimension for Testing High Reliability Parts for the Saturn V Inertial Guidance System," *Proceedings of Fifth Reliability and Maintainability Conference* (*Annals of Reliability and Maintainability*, Vol. 5), AIAA, Washington, DC, July 1966, pp. 515–527.

Veca, A. C., "S-65-200 Commercial Compound Aircraft Design for Maintenance," *Proceedings of Ninth Reliability and Maintainability Conference* (*Annals of Reliability and Maintainability*, Vol. 9), SAE, Warrendale, PA, July 1970, pp. 113–122.

Venditti, R. A. and Sineath, R. M., "Saturn V System Reliability Analysis," *Proceedings of Annual Reliability and Maintainability Symposium*, IEEE, New York, Jan. 1969, pp. 567–571.

Verlander, J. M., "Cost Considerations in Launch Operations," *Proceedings of Fourth Annual Reliability and Maintainability Conference* (*Annals of Reliability and Maintainability*, Vol. 4), Spartan Books, Washington, DC, July 1965, pp. 661–668.

Viehmeyer, G. F. Jr., "An Approach to Implementation of AFSCM 375-5 in the Aircraft Industry," *Proceedings of Fifth Reliability and Maintainability Conference* (*Annals of Reliability and Maintainability*, Vol. 5), AIAA, Washington, DC, July 1966, pp. 472–480.

Voegtlen, H. D. and Retterer, B. L., "Advanced Maintainability Techniques for Aircraft Systems," *Proceedings of Aerospace Reliability and Maintainability Conference*, AIAA, Washington, DC, May 1963, pp. 429–433.

Vogt, S., "Possibilities of Checking the Reliability Assumptions Made during the Design of a Weapon System Using the Results of Flight Tests," *Operations Research and Reliability*, edited by D. Grouchko, Gordon and Breach Science Publishers, New York, 1971, pp. 251–268.

VonGuerard, H. W., "Analyzing the Interface of Reliability and Economics of Unmanned Satellites," *IEEE Transactions on Reliability*, Vol. R-23, No. 3, Aug. 1974, pp. 208–213.

Warsher, A., "A New Approach to Attainment of Reliability of the Production of Airborne Electronic Systems," *IRE Convention Record*, Pt. II, March 1954, pp. 13–22.

Webster, L. R., "Field Vs Predicted Reliability for Commercial Satcom Terminals," *Proceedings of Annual Reliability and Maintainability Symposium*, IEEE, New York, 1986, pp. 88–91.

Webster, L. R. and Mader, J. M., "VLSI Impact on RAMS Strategies in Avionics Design," *Proceedings of Annual Reliability and Maintainability Symposium*, IEEE, New York, 1986, pp. 303–306.

Weisberg, S. A., "Reliability Design Analysis for Space Systems," *Proceedings of Sixth Reliability and Maintainability Conference* (*Annals of Reliability and Maintainability*, Vol. 6), SAE, Warrendale, PA, July 1967, pp. 594–599.

Welsh, E. C., "Our National Space Program," *IEEE Transactions on Reliability*, Vol. R-13, No. 1, March 1964, pp. 5–7.

Werner, H. C. and Gross, T. W., "Some Aspects of Satellite and Space Probe Reliability," *IRE Transactions on Reliability and Quality Control*, Vol. 10, No. 3, 1961, pp. 7–14 (available from IEEE).

Westerheid, R. J., "The Apollo Spacecraft Qualification Program," *Proceedings of Eighth Reliability and Maintainability Conference* (*1969 Annals of Assurance Sciences*), Gordon and Breach Science Publishers, New York, July 1969, pp. 153–159.

White, G. C. Jr. and Hock, C. D., "Reliability Prediction Activities in the Apollo Program," *Proceedings of Sixth Reliability and Maintainability Conference* (*Annals of Reliability and Maintainability*, Vol. 6), SAE, Warrendale, PA, July 1967, pp. 587–593.

White, R. N., "Designing Airborne Electronic Equipment for Improved Reliability," *Proceedings of First National Symposium on Reliability & Quality Control*, 1954, IEEE, New York, pp. 114–119.

Whiteman, I. R., "A Mission Reliability Reporting System," *Proceedings of Eighth National Symposium on Reliability and Quality Control*, IEEE, New York, 1962, pp. 268–271.

Wiggins, J. W., "Rocket Readiness," *Ordnance*, Vol. 43, 1959, p. 101.

Willstadter, R., "Dormat Missile System Checkout Effectiveness," *Proceedings of Tenth Nation Symposium on Reliability and Quality Control*, IEEE, New York, 1964, pp. 612–623.

Wilson, R. B. and Hug, N. L., "A Prime Contractor's Reliability Program for Components/Parts for the Douglas S-IV B Stage Project," *Proceedings of Fifth Reliability and Maintainability Conference* (*Annals of Reliability and Maintainability*, Vol. 5), AIAA, Washington, DC, July 1966, pp. 3–11.

Winokur, H. S. Jr. and Goldstein, L. J., "Analysis of Mission-Oriented Systems," *IEEE Transactions on Reliability*, Vol. R-18, No. 4, Nov. 1969, pp. 144–148.

Wissel, J. W., "Human Error Potential in Today's Ballistic Missile Systems," *Proceedings of IAS Aerospace Systems Reliability Symposium*, April 1962, pp. 213–214 (available from AIAA).

Wolman, W., "The Status of Reliability Assessment for Unmanned Satellites," *Proceedings of Eleventh National Symposium on Reliability and Quality Control*, IEEE, New York, Jan. 1965, pp. 313–319.

Wolman, W., "A Model for the Reliability Estimation of Space Systems," *IEEE Transactions on Reliability*, Vol. R-12, No. 1, March 1963, pp. 32–39.

Wong, H. P. and Meyer, J. B., "Skylab Test Program-Management and Requirements," *Proceedings of Ninth Reliability and Maintainability Conference* (*Annals of Reliability and Maintainability*, Vol. 9), SAE, Warrendale, PA, July 1970, pp. 63–74.

Wong, K. L., Quart, I., Kallis, J. M., and Burkhard, A. H., "Culprits of Avionic Equipment Failures," *Proceedings of Annual Reliability and Maintainability Symposium*, IEEE, New York, 1987, pp. 416–421.

Wood, B. B., "Development of the Reliability Program for the Advanced Medium Range Air-to-Air Missile (AMRAAM)," *Proceedings of Annual Reliability and Maintainability Symposium*, IEEE, New York, 1982, pp. 510–514.

Wood, E. H., "Navy Operational Problems with Modern Aircraft," *Proceedings of Fifth Reliability and Maintainability Conference* (*Annals of Reliability and Maintainability*, Vol. 5), AIAA, Washington, DC, July 1966, pp. 818–822.

Wood, J. K. R., "A Method of Determining the Cost Effectiveness of Different Levels of Reliability and Maintainability for a Training Aircraft," *Operations Research and Reliability*, edited by D. Grouchko, Gordon and Breach Science Publishers, New York, 1971, pp. 145–170.

Wyatt, M. W., "RAM Growth of the Patriot Missile System," *Proceedings of Annual Reliability and Maintainability Symposium*, IEEE, New York, 1987, pp. 381–385.

Wynholds, H. W. and Esq, L. B., "Economics of Commercial Aviation Safety," *Proceedings of Annual Reliability and Maintainability Symposium*, IEEE, New York, 1978, pp. 162–166.

Young, P. J., "Inspection Intervals for Fail-Safe Structure," *IEEE Transactions on Reliability*, Vol. R-33, No. 2, June 1984, pp. 165–170.

Yurkowsky, W., "Non-electronic Reliability Note Book," Rome Air Development Center, Air Force Systems Command, Griffiss Air Force Base, NY, Rept. RADC-TR-69-458, March 1970.

A4 STRUCTURAL RELIABILITY

Allen, D. E., "Limit States Design — A Probabilistic Study," *Canadian Journal of Civil Engineering*, Vol. 2, 1975, pp. 36–49.

Anderson, T. and Misund, A., "Pipeline Reliability: An Investigation of Pipeline Failure Characteristics and Analysis of Pipeline Failure Rates for Submarine and Cross-Country Pipelines," *Journal of Petroleum Technology*, Vol. 35, April 1983, pp. 709–717.

Anderson, W. D., Silbert, M. N., and Lloyd, J. R., "Reliability Procedure for Fixed Offshore Platforms," *ASCE Journal of the Structural Division*, Vol. 108. No. ST11, Nov. 1982, pp. 2517–2538.

Ang, A. H. S., "Extended Reliability Basis of Structural Design Under Uncertainties," *Proceedings of Ninth Reliability and Maintainability Conference* (*Annals of Reliability and Maintainability*, Vol. 9), SAE, Warrendale, PA, July 1970, pp. 642–649.

Ang, A. H. S., "Structural Risk Analysis and Reliability-Based Design," *ASCE Journal of the Structural Division*, Vol. 99, No. ST9, Sept, 1973, pp. 1891–1910.

Ang, A. H. S. and Amin, M., "Reliability of Structures and Structural Systems," *ASCE Journal of the Engineering Mechanics Division*, Vol. 94, No. EM2, April 1968, pp. 671–691.

Ang, A. H. S. and Amin, M., "Safety Factors and Probability in Structural Design," *ASCE Journal of the Structural Division*, Vol. 95, No. ST7, July 1969, pp. 1389–1405.

Ang, A. H. S. and Cornell, C. A., "Reliability Bases of Structural Safety and Design," *ASCE Journal of the Structural Division*, Vol. 100, No. ST9, Sept. 1974, pp. 1755–1769.

Angelides, D. C., Vereziano D., and Sunder, S. S., "Random Sea and Reliability of Offshore Foundations," *ASCE Journal of the Engineering Mechanics Division*, Vol. 107, No. EM1, Feb. 1981, pp. 131–148.

ASCE 4th Speciality Conference on Probabilistic Mechanics and Structural Reliability, ASCE, New York, 1984.

Asplund, S. O., "The Risk of Failures," *Structural Engineer*, Vol. 36, No. 8, April 1958, pp. 268–270.

Augusti, G., "Probabilistic Methods in Plastic Structural Analysis," *Nuclear Engineering and Design*, Vol. 57, 1980, pp. 403–415.

Augusti, G., "Some Considerations on the Assessment of Structural Reliability," *Nuclear Engineering and Design*, Vol. 60, 1980, pp. 139–144.

Augustin, W. et al., "A Complex Study on the Reliability Assessment of the Containment of a PWR, Parts I–III," *Nuclear Engineering and Design*, Vol. 48, 1978, pp. 563–564.

Ayyub, B. M. and Halder, A., "Practical Structural Reliability Techniques," *ASCE Journal of Structural Engineering*, Vol. 110, Aug. 1984, pp. 1707–1724.

Baba, S., Nakagawa, K., and Naruoka, M., "A Model for Structural Safety Using the Extreme Value Approach," *Applied Mathematical Modelling*, Vol. 4, Aug. 1980, pp. 275–282.

Baker, H. C., "Reliability Analysis and Test of Aluminum Honeycomb Structures," *Proceedings of Fourth Reliability and Maintainability Conference* (*Annals of Reliability and Maintainability*, Vol. 4), Spartan Books, Washington, DC, July 1965, pp. 431–436.

Bangash, Y., "Safety and Reliability of Prestressed Concrete Reactor Vessels," *Nuclear Engineering and Design*, Vol. 51, 1979, pp. 473–486.

Barrett, J. D. and Foschi, R. O., "Duration of Load and Probability of Failure in Wood, Part I: Modelling Creep Rupture," *Canadian Journal of Civil Engineering*, Vol. 5, 1978, pp. 505–514.

Barrett, J. D. and Foschi, R. O., "Duration of Load and Probability of Failure in Wood, Part II: Constant, Ramp, and Cycle Loadings," *Canadian Journal of Civil Engineering*, Vol. 5, 1978, pp. 515–532.

Barry, D. M. and Garty, R. J. F., "A Motorway Lighting System Maintenance Model Based on Simulation," *Reliability Engineering*, Vol. 6, 1983, pp. 63–81.

Bauer, J. and Choi, H. S., "Application of the Spectral Method to Estimate the Structural Reliability of a Reactor Containment," *Computers and Structures*, Vol. 12, No. 4, 1980, pp. 421–426.

Bauer, J. and Schueller, G. I., "Structural Reliability Assessment under Internal and External Load Conditions, Part III," *Nuclear Engineering and Design*, Vol. 48, 1978, pp. 585–593.

Bea, R. G., "Reliability Considerations in Offshore Platform Criteria," *ASCE Journal of the Structural Division*, Vol. 106, No. ST9, Sept, 1980, pp. 1835–1853.

Bea, R. G., "Selection of Environmental Criteria for Offshore Platform Design," *Journal of Petroleum Technology*, Vol. 26, Nov. 1974, pp. 1206–1214.

Beason, W. L. and Morgan, J. R., "Glass Failure Prediction Model," *ASCE Journal of Structural Engineering*, Vol. 110, No. 2, Feb. 1984, pp. 197–212.

Beaulieu, D., Kelker, D., and Adams, P. F., "Probability Considerations in Assessing the Effects of Structural Out-of-Plumbs," *Canadian Journal of Civil Engineering*, Vol. 6, 1979, pp. 610–616.

Becher, P. E. and Pedersen, A., "Application of Statistical Linear Elastic Fracture Mechanics to Pressure Vessel Reliability Analysis," *Nuclear Engineering and Design*, Vol. 27, 1974, pp. 413–425.

Benjamin, J. R., "Probabilistic Structural Analysis and Design," *ASCE Journal of the Structural Division*, Vol. 94, No. ST7, July 1968, pp. 1665–1679.

Benjamin, J. R., "Risk and Decision Analyses Applied to Dams and Levees," *Structural Safety*, Vol. 1, Sept. 1983, pp. 257–268.

Benjamin, J. R., "The Basics of Structural Reliability," *Nuclear Engineering and Design*, Vol. 60, 1980, pp. 3–9.

Benjamin, J. R. and Lind, N. C., "A Probabilistic Basis for a Deterministic Code," *Journal of the American Concrete Institute*, Vol. 66, Nov. 1969, pp. 857–865.

Benjamin, J. R., Schueller, G. I., and Witt, F. J., "A Critical Review of the Results of the Second International Seminar on Structural Reliability of Mechanical Components and Subassemblies of Nuclear Power Plants," *Reliability Engineering*, Vol. 2, 1981, pp. 125–134.

Besuner, P. M. and Tetelman, A. S., "Probabilistic Fracture Mechanics," *Nuclear Engineering and Design*, Vol. 43, 1977, pp. 99–114.

Beynet, P. A., Berman, M. Y., and VonAshwege, J. T., "Motion, Fatigue and the Reliability Characteristics of a Vertically Moored Platform," *Journal of Petroleum Technology*, Vol. 31, March 1979, pp. 267–274.

Birdsall, C. R. and Frank, H. J., "Space Station Reliability," *Proceedings of Annual Reliability and Maintainability Symposium*, IEEE, New York, 1985, pp. 197–202.

Blake, R. E., "Predicting Structural Reliability for Design Decisions," *Journal of Spacecraft and Rockets*, Vol. 4, March 1967, pp. 392–399.

Bljuger, F., "Models of Reinforced Concrete Structures for Reliability Analysis," *Structural Safety*, Vol. 2, Jan. 1985, pp. 215–224.

Bljuger, F., "Safety and Serviceability of Thin-Section Structures," *Journal of the American Concrete Institute*, Vol. 79, March–April 1982, pp. 94–99.

Blockley, D. J., *The Nature of Structural Design and Safety*, Wiley, New York, 1980.

Blume, J. A. and Kiremidjian, A. S., "Probabilistic Procedures for Peak Ground Motions," *ASCE Journal of the Structural Division*, Vol. 105, No. ST11, Nov. 1979, pp. 2293–2311.

Bonnicksen, L. W. and Suddarth, S. K., "Structural Reliability Analysis of a Wood Load-Sharing System," *ASTM Journal of Materials*, Vol. 1, No. 3, Sept. 1966, pp. 491–508.

Bourne, A. J., "System Reliability Analysis Techniques and their Relationship to Mechanical/Structural Reliability Problems," *Nuclear Engineering and Design*, Vol. 60, 1980, pp. 133–136.

Bouton, I., Trent, D. J., and Chenoweth, H. B., "Design Factors for Structural Reliability," *Proceedings of Fifth Reliability and Maintainability Conference* (*Annals of Reliability and Maintainability*, Vol. 5), AIAA, Washington, DC, July 1966, pp. 229–235.

Bresler, B. and Hanson, J. M., "Damageability and Reliability of Existing Structures," *Proceedings of Ninth U.S. National Congress of Applied Mechanics*, ASME, New York, July 1982, pp. 309–313.

Brewer, J. W., "Interim Scaler Reliability Statements Consistent with Conventional Materials Strength Criteria," *Proceedings of Tenth Reliability and Maintainability Conference* (*Annals of Reliability and Maintainability*, Vol. 10), ASME, New York, 1971, pp. 63–71.

Broding, W. C., Drederich F. W., and Parker, P. S., "Structural Optimization and Design Based on a Reliability Design Criterion," *Journal of Spacecraft and Rockets*, Vol. 1, Jan. 1964, pp. 56–61.

Brown, C. B., "Concepts of Structural Safety," *ASCE Journal of the Structural Division*, Vol. 86, No. ST12, Dec. 1960, pp. 39–57.

Casciati, F., "A Probabilistic Approach to the Deformation Analysis of Elastic-Plastic Frames," *Journal of Structural Mechanics*, Vol. 6, No. 1, 1978, pp. 45–60.

Casciati, F. and Faravelli, L., "Load Combination By Partial Safety Factors," *Nuclear Engineering and Design*, Vol. 75, 1982, pp. 439–452.

Casciati, F. and Faravelli, L., "Safety Analysis for Random Elastic-Plastic Frames in the Presence of Second-Order Geometrical Effects," *Applied Mathematical Modelling*, Vol. 4, Feb, 1980, pp. 16–22.

Chalk, P. L. and Corotis, R. B., "Probability Model for Design Live Loads," *ASCE Journal of the Structural Division*, Vol. 106, No. ST10, Oct. 1980, pp. 2017–2033.

Chang, P. Y., "The Application of Collapse Theorems for Reliability Analysis of Structures Subjected to the Impact of Other Structural Bodies," *Proceeding of Annual Reliability and Maintainability Symposium*, IEEE, New York, 1981, pp. 1–8.

Chapuis, J. and Galambos, T. V., "Reliability of Aluminum Beam-Columns," *ASCE Journal of the Structural Division*, Vol. 108, No. ST4, April 1982, pp. 709–727.

Chenoweth, H. B., "A Theory of Reliability Growth in Structural Systems," *Proceedings of Annual Reliability and Maintainability Symposium*, IEEE, New York, 1980, pp. 106–111.

Chou, K. C., McIntosh, C., and Corotis, R. B., "Observations on Structural System Reliability and the Role of Modal Correlations," *Structural Safety*, Vol. 1, April 1983, pp. 189–198.

Cohn, M. Z. and Parimi, S. R., "Multi-Criteria Probabilistic Structural Design," *Journal of Structural Mechanics*, Vol. 1, No. 4, 1973, pp. 479–496.

Committee on Fatigue and Fracture Reliability, "Fatigue Reliability: Development of Criteria For Design," *ASCE Journal of the Structural Division*, Vol. 108, No. ST1, Jan. 1982, pp. 71–88.

Committee on Fatigue and Fracture Reliability, "Fatigue Reliability: Introduction," *ASCE Journal of the Structural Division*, Vol. 108, No. ST1, Jan. 1982, pp. 3–23.

Committee on Fatigue and Fracture Reliability, "Fatigue Reliability: Quality Assurance and Maintainability," *ASCE Journal of the Structural Division*, Vol. 108, No. ST1, Jan 1982, pp. 25–46.

Committee on Fatigue and Fracture Reliability, "Fatigue Reliability: Variable Amplitude Loading," *ASCE Journal of the Structural Division*, Vol. 108, No. ST1, Jan. 1982, pp. 47–69.

Committee on Reliability of Offshore Structures, "Application of Reliability Methods in Design and Analysis of Offshore Platforms," *Journal of Structural Engineering*, Vol. 109, Oct. 1983, pp. 2265–2291.

Committee on Structural Safety, "Structural Safety—A Literature Review," *ASCE Journal of the Structural Division*, Vol. 98, No. ST4, April 1972, pp. 845–884.

Cornell, C. A., "A Reliability-Based Structural Code," *Journal of the American Concrete Institute*, Vol. 66, Dec. 1969, pp. 974–985.

Cornell, C. A., "Bounds on the Reliability of Structural Systems," *ASCE Journal of the Structural Division*, Vol. 93, ST1, Feb. 1967, pp. 171–200.

Cornell, C. A., "Some Thoughts on Systems and Structural Reliability," *Nuclear Engineering and Design*, Vol. 60, 1980, pp. 115–116.

Cornell, C. A., "Structural Safety Specifications Based on Second-Moment Reliability Analysis," Paper presented at IABSE Symposium on Concepts of Safety of Structures and Methods of Design, International Association for Bridge and Structural Engineering, Zurich, 1969.

Cornell, C. A., "Utilization of Present Knowledge of Probabilistic Structural Reliability in Analysis of Nuclear Power Plants," *Nuclear Engineering and Design*, Vol. 60, 1980, pp. 33–36.

Corotis, R. B., "Statistical Analysis of Live Load in Column Design," *ASCE Journal of the Structural Division*, Vol. 98, No. ST8, Aug. 1972, pp. 1803–1815.

Corotis, R. B., "Probabilistic Load Duration Model for Live Loads," *ASCE Journal of Structural Engineering*, Vol. 109, April 1983, pp. 859–874.

Costello, J. F. and Chu, K., "Failure Probabilities of Reinforced Concrete Beams," *ASCE Journal of the Structural Division*, Vol. 95, No. ST10, Oct. 1969, pp. 2281–2304.

Das, P. K., Frieze, P. A., and Faulkner, D., "Structural Reliability Modelling of Stiffened Components of Floating Structures," *Structural Safety*, Vol. 2, May 1984, pp. 3–16.

Davidson, J. R., "Reliability, After Inspection, of Unique Structures," *Proceedings of Annual Reliability and Maintainability Symposium*, IEEE, New York, 1983, pp. 361–366.

Davidson, J. W., Felton, L. P., and Hart, G. C., "On Reliability-Based Structural Optimization for Earthquakes," *Computers and Structures*, Vol. 12, 1980, pp. 99–105.

Davidson, J. W., Felton, L. P., and Hart, G. C., "Reliability-Based Optimization for Dynamic Loads," *ASCE Journal of the Structural Division*, Vol. 103, No. ST10, Oct. 1977, pp. 2021–2035.

DeBonis, A. L., "Stochastic Simulation of Wood Load-Sharing Systems," *ASCE Journal of the Structural Division*, Vol. 106, No. ST2, Feb. 1980, pp. 393–410.

Der Kiureghian, A., "Reliability Analysis under Stochastic Loads," *ASCE Journal of the Structural Division*, Vol. 106, No. ST2, Feb. 1980, pp. 411–429.

Der Kiureghian, A., "Second-Moment Combination of Stochastic Loads," *ASCE Journal of the Structural Division*, Vol. 104, No. ST10, Oct. 1978, pp. 1551–1567.

Dicks, C. and Wilson, S., "Structural Reliability—The General Engineering Design Approach," *Proceedings of 11th National Syposium on Reliability and Quality Control*, IEEE, New York, Jan. 1965, pp. 169–181.

Dicks, C. and Wilson, S., "Structural System Design Technique for Reliability and Weight Tradeoffs," *Proceedings of Reliability and Maintainability Conference* (*Annals of Reliability and Maintainability*, Vol. 4), Spartan Books, Washington, DC, July 1965, pp. 421–430.

Ditlevsen, O., "First Outcrossing Probability Bounds," *Journal of Engineering Mechanics*, Vol. 110, Feb. 1984, pp. 282–292.

Ditlevsen, O., "Fundamental Postulate in Structural Safety," *Journal of Engineering Mechanics*, Vol. 109, No. 4, Aug. 1983, pp. 1096–1102.

Ditlevsen, O., "Generalized Second Moment Reliability Index," *Journal of Structural Mechanics*, Vol. 7, No. 4, 1979, pp. 435–451.

Ditlevsen, O., "Model Uncertainty in Structural Reliability," *Structural Safety*, Vol. 1, Sept. 1982, pp. 73–86.

Ditlevsen, O., "Narrow Reliability Bounds for Structural Systems," *Journal of Structural Mechanics*, Vol. 7, No. 4, 1979, pp. 453–472.

Ditlevsen, O., "Reliability Against Defect Generated Fracture," *Journal of Structural Mechanics*, Vol. 9, No. 2, 1981, pp. 115–137.

Ditlevsen, O., "System Reliability Bounding by Conditioning," *ASCE Journal of the Engineering Mechanics Division*, Vol. 108, No. EM5, Oct. 1982, pp. 708–718.

Ditlevsen, O., "Taylor Expansion of Series System Reliability," *Journal of Engineering Mechanics*, Vol. 110, No. 2, Feb. 1984, pp. 293–307.

Ditlevsen, O. and Bjerager, P., "Reliability of Highly Redundant Plastic Structures," *Journal of Engineering Mechanics*, Vol. 110, May 1984, pp. 671–693.

Dolinski, K., "First-Order Second-Moment Approximation in Reliability of Structural Systems: Critical Review and Alternative Approach," *Structural Safety*, Vol. 1, April 1983, pp. 211–231.

Dolinski, K., "Kinematic Approach to Reliability Estimate of Stochastically Loaded Plastic Structures," *Journal of Structural Mechanics*, Vol. 6, No. 3, 1978, pp. 247–266.

Downey, E. and MacLeod, I. A., "Serviceability Failures Due to Structural Movements," *Journal of the Technical Councils of ASCE*, Vol. 108, No. TC1, May 1982, pp. 1–9.

Dumonteil, P., "On the Role of Uncertainty in the Limit States Design of Steel Structures," *Canadian Journal of Civil Engineering*, Vol. 7, 1980, pp. 588–596.

Edwards, G., "A Bayesian Procedure for Drawing Inferences from Random Data," *Reliability Engineering*, Vol. 9, 1984, pp. 1–17.

Eisenberg, I. M., "Safety of Structural Systems with Reserve Elements for Earthquake Protection," Earthquake Engineering and Structural Dynamics, Vol. 5, 1977, pp. 305–318.

Ellingwood, B., "Analysis of Reliability for Masonry Structures," *ASCE Journal of the Structural Division*, Vol. 107, No. ST5, May 1981, pp. 757–773.

Ellingwood, B., "Reliability Based Criteria for Reinforced Concrete Design," *ASCE Journal of the Structural Division*, Vol. 105, No. ST4, April 1979, pp. 713–727.

Ellingwood, B., "Reliability of Current Reinforced Concrete Designs," *ASCE Journal of the Structural Division*, Vol. 105, No. ST4, April 1979, pp. 699–712.

Ellingwood, B., "Reliability of Wood Structural Elements," *ASCE Journal of the Structural Division*, Vol. 107, No. ST1, Jan. 1981, pp. 73–87.

Ellingwood, B., "Safety Checking Formats For Limit States Design," *ASCE Journal of the Structural Division*, Vol. 108, No. ST7, July 1982, pp. 1481–1493.

Ellingwood, B. and Ang, A. H. S., "Risk-Based Evaluation of Design Criteria," *ASCE Journal of the Structural Division*, Vol. 100, No. ST9, Sept. 1974, pp. 1771–1788.

Ellingwood, B. and Galambos, T. V., "Probability-Based Criteria for Structural Design," *Structural Safety*, Vol. 1, Sept. 1982, pp. 15–26.

Ellingwood, B., Leyendecker, E. V., and Yao, J. T. P., "Probability of Failure from Abnormal Load," *ASCE Journal of Structural Engineering*, Vol. 109, No. 4, April 1983, pp. 875–890.

Ellingwood, B., MacGregor, J. G., Galambos, T. V., and Cornell, C. A., "Probability Based Load Criteria: Load Factors and Load Combinations," *ASCE Journal of the Structural Division*, Vol. 108, No. ST5, May 1982, pp. 978–997.

Ellingwood, B. and Shaver, J. R., "Reliability of RC Beams Subjected to Fire," *ASCE Journal of the Structural Division*, Vol. 103, No. ST5, May 1977, pp. 1047–1059.

Ellyin, F. and Ghannoum, E., "An Assessment of Reliability of Structures," *Engineering Journal*, Vol. 55, No. A-1, 1972, pp. i–vi.

Fardis, M. N. and Cornell, C. A., "Containment Linear Seismic Reliability under Statistical Uncertainty," *Nuclear Engineering and Design*, Vol. 49, 1978, pp. 279–294.

Ferry Borges, J. and Castanheta, M., *Structural Safety*, 2nd ed., National Laboratory of Civil Engineering, Lisbon 1971.

Fiessler, B., Neumann, H. J., and Rackwitz, R., "Quadratic Limit States in Structural Reliability," *ASCE Journal of the Engineering Mechanics Division*, Vol. 105, No. EM4, Aug. 1979, pp. 661–676.

Fjeld, S., "Reliability of Offshore Structures," *Journal of Petroleum Technology*, Vol. 30, Oct. 1978, pp. 1486–1496.

Foschi, R. O., "A Discussion on the Application of the Safety Index Concept to Wood Structures," *Canadian Journal of Civil Engineering*, Vol. 6, 1979, pp. 51–58.

Foschi, R. O., "Reliability of Wood Structural Systems," *Journal of Structural Engineering*, Vol. 110, Dec. 1984, pp. 2995–3013.

Foschi, R. O., and Barrett, J. D., "Load-Duration Effects in Western Hemlock Lumber," *ASCE Journal of the Structural Division*, Vol. 108, No. ST7, July 1982, pp. 1494–1510.

Freudenthal, A. M. (ed.), *International Conference on Structural Safety and Reliability*, Pergamon Press, New York, 1969.

Freudenthal, A. M. (ed.), *Reliability Approach in Structural Engineering*, Maruzen Co., Tokyo, 1975.

Freudenthal, A. M. (ed.), *Structural Safety and Reliability*, Pergamon Press, New York, 1972.

Freudenthal, A. M., "Reliability of Reactor Components and Systems Subject to Fatigue and Creep," *Nuclear Engineering and Design*, Vol. 28, 1974, pp. 196–217.

Freudenthal, A. M., "Safety and the Probability of Structural Failure," *Transactions of the American Society of Civil Engineers*, Vol. 121, 1956, pp. 1337–1397.

Freudenthal, A. M., "Safety, Reliability and Structural Design," *ASCE Journal of the Structural Division*, Vol. 87, No. ST3, March 1961, pp. 1–16.

Freudenthal, A. M., Gerrelts, J. M., and Shinozuka, M., "The Analysis of Structural Safety," *ASCE Journal of the Structural Division*, Vol. 92, No. ST1, Feb. 1966, pp. 267–325.

Fujino, Y. and Lind, N. C., "Proof-Load Factors and Reliability," *ASCE Journal of the Structural Division*, Vol. 103, No. ST4, April 1977, pp. 853–870.

Galambos, T. V., "Reliability of Axially Loaded Columns," *Engineering Structures*, Vol. 5, Jan. 1983, pp. 73–78.

Galambos, T. V., Ellingwood, B., MacGregor, J. G., and Cornell, C. A., "Probability Based Load Criteria: Assessment of Current Design Practice," *ASCE Journal of the Structural Division*, Vol. 108, No. ST5, May 1982, pp. 959–977.

Galambos, T. V., Reinhold, T. A., and Ellingwood, B., "Serviceability Limit States: Connection Slip," *ASCE Journal of the Structural Division*, Vol. 108, No. ST12, Dec. 1982, pp. 2668–2680.

Garson, R. C., "Failure Mode Correlation in Weakest-Link Systems," *ASCE Journal of the Structural Division*, Vol. 106, No. ST8, Aug. 1980, pp. 1797–1810.

Georgiadis, C., "Probability of Failure Models in Finite Element Analysis of Brittle Materials," *Computers and Structures*, Vol. 18, No. 3, 1984, pp. 537–549.

Goodman, J. R., Vanderbilt, M. D., and Criswell, M. E., "Reliability-Based Design of Wood Transmission Line Structures," *Journal of Structural Engineering*, Vol. 109, March 1983, pp. 690–704.

Green, A. E., "Systems Reliability/Structural Reliability—Interaction and Differences in Approach," *Nuclear Engineering and Design*, Vol. 60, 1980, pp. 107–113.

Grigoriu, M., "Approximate Analysis of Complex Reliability Problems," *Structural Safety*, Vol. 1, Sept. 1983, pp. 277–288.

Grigoriu, M., "Methods for Approximate Reliability Analysis," *Structural Safety*, Vol. 1, Dec. 1982, pp. 155–165.

Grigoriu, M., "Reliability of Chain and Ductile-Parallel Systems," *Journal of Engineering Mechanics*, Vol. 109, No. 5, Oct. 1983, pp. 1175–1188.

Grigoriu, M. and Lind, N. C., "Probabilistic Models For Prototype Testing," *ASCE Journal of the Structural Division*, Vol. 108, No. ST7, July 1982, pp. 1511–1525.

Grigoriu, M. and Turkstra, C., "Safety of Structural Systems with Correlated Resistances," *Applied Mathematical Modelling*, Vol. 3, April 1979, pp. 130–136.

Grigoriu, M. and Turkstra, C., "Structural Safety Indices for Repeated Loads," *ASCE Journal of the Engineering Mechanics Division*, Vol. 104, No. EM4, Aug. 1978, pp. 829–844.

Grigoriu, M., Veneziano, D., and Cornell, C. A., "Probabilistic Modelling as Decision Making," *ASCE Journal of the Engineering Mechanics Division*, Vol. 105, No. EM4, Aug. 1979, pp. 585–596.

Grimmelt, M. J. and Schueller, G. I., "Benchmark Study on Methods to Determine Collapse Failure Probabilities of Redundant Structures," *Structural Safety*, Vol. 1, Dec. 1982, pp. 93–106.

Gromala, D. S., "Calculating Apparent Reliability of Wood Scaffold Planks," *Structural Safety*, Vol. 2, May 1984, pp. 47–57.

Hashimoto, T., Stedinger, J. R., and Loucks, D. P., "Reliability, Resiliency, and Vulnerability Criteria for Water Resource System Performance Evaluation," *Water Resources Research*, Vol. 18, No. 1, Feb. 1982, pp. 14–20.

Hasofer, A. M., "Exact Second-Moment Reliability for Sum of Products Failure Equation," *Journal of Structural Mechanics*, Vol. 3, No. 1, 1974, pp. 97–106.

Hasofer, A. M., "Reliability Index and Failure Probability," *Journal of Structural Mechanics*, Vol. 3, No. 1, 1974, pp. 25–27.

Haugen, E. B., "Implementing a Structural Reliability Program," *Proceedings of 11th National Symposium on Reliability and Quality Control*, IEEE, New York, Jan. 1965, pp. 158–168.

Haugen, E. B., "Statistical Methods for Structural Reliability Analysis," *Proceedings of 10th National Symposium on Reliability and Quality Control*, IEEE, New York, Jan. 1964, pp. 97–121.

Heller, R. A. and Heller, A. S., "Fatigue Life and Reliability of a Redundant Structure," *Proceedings of Fourth Reliability and Maintainablity Conference* (*Annals of Reliability and Maintainability*, Vol. 4), Spartan Books, Washington, DC, July 1965, pp. 881–896.

Heller, R. A. and Shinozuka, M., "Reliability Techniques in Materials and Structures," *Proceedings of Ninth Reliability and Maintainability Conference* (*Annals of Reliability and Maintainability*, Vol. 9), SAE, Warrendale, PA, July 1970, pp. 635–636.

Hilton, H. H. and Freigen, M., "Minimum Weight Analysis Based on Structural Reliability," *Journal of the Aerospace Sciences*, Vol. 29, March 1962, pp. 355–356.

Hohenbichler, M. and Rackwitz, R., "First-Order Concepts in System Reliability," *Structural Safety*, Vol. 1, April 1983, pp. 177–188.

Hohenbichler, M. and Rackwitz, R., "Non-Normal Dependent Vectors in Structural Safety," *ASCE Journal of the Engineering Mechanics Division*, Vol. 107, No. EM6, Dec. 1981, pp. 1227–1238.

Hohenbichler, M. and Rackwitz, R., "On Structural Reliability of Parallel Systems," *Reliability Engineering*, Vol. 2, 1981, pp. 1–6.

Hohenbichler, M. and Rackwitz, R., "Reliability of Parallel Systems under Imposed Uniform Strain," *Journal of Engineering Mechanics*, Vol. 109, No. 3, June 1983, pp. 896–907.

Hong, S. Y. and Yeater, M. L., "Critical Parameter Study of PWR Primary Coolant Pipe Leak Failure Using Probabilistic Fracture Mechanics," *Nuclear Engineering and Design*, Vol. 78, 1984, pp. 3–22.

Howey, R. A. and Gaarder, O. H., "Application of Reliability Engineering to Offshore Production Equipment," *Journal of Petroleum Technology*, Vol. 29, April 1977, pp. 339–346.

Ichikawa, M., "A Theoretical Study of Defect Detection Probability of Nondestructive Inspection," *Reliability Engineering*, Vol. 10, 1985, pp. 175–182.

Ichikawa, M., "A Theoretical Study of Flaw Size Distributions," *Reliability Engineering*, Vol. 9, 1984, pp. 221–228.

Ichikawa, M., "Confidence Limit of Probability of Failure Based on Stress-Strength Model," *Reliability Engineering*, Vol. 8, 1984, pp. 75–83.

Ichikawa, M., "Possible Models for Quasi-Composite Distributions," *Reliability Engineering*, Vol. 8, 1984, pp. 117–128.

Inagaki, T., Inoue, K., and Akashi, H., "Optimal Reliability Design of Systems Under Changing Environment," *Proceedings—HOPE International Symposium*, Japan Society of Mechanical Engineers, Tokyo, Nov. 1977, pp. 473–480.

Ingles, O. G., "Safety in Civil Engineering—Its Perception and Promotion," *Reliability Engineering*, Vol. 1, 1980, pp. 15–27.

Ingman, D. and Reznik, L., "A Dynamic Model For Element Reliability," *Nuclear Engineering and Design*, Vol. 70, 1982, pp. 209–213.

Ingram, G. E., "A Basic Approach For Structural Reliability," *Proceedings of 11th National Symposium on Reliability & Quality Control*, IEEE, New York, Jan. 1965, pp. 154–157.

Ingram, G. E., Haugen, E. B., Ducks, C., and Wilson, S., "Panel Discussion—Structural Reliability," *National Symposium on Reliability*, IEEE, New York, 1965, pp. 154–169.

Ingram, G. E. and Herrman, C. R., "Structural Reliability Methodology—Application at the System Level," *Nuclear Engineering and Design*, Vol. 50, 1978, pp. 185–193.

International Research Seminar on Safety of Structures under Dynamic Loading, Norwegian Institute of Technology, Trondheim, June 1977.

Iwatsubo, T. and Narimatsu, T., "Dynamical Approach to the Design of Structures Containing Uncertainty," *Proceedings—HOPE International Symposium*, Japan Society of Mechanical Engineers, Tokyo, Nov. 1977, pp. 65–72.

Jacobi, S. K. W., Schneider, R. O., and Weber, G. G., "Reliability and Availability of a Safety Shutdown System," *Proceedings of Annual Reliability and Maintainability Symposium*, IEEE, New York, 1974, pp. 186–196.

Johnson, C. W. and Maxwell, R. E., "Reliability Analysis of a Structure—A New Approach," *Proceedings of Annual Reliability and Maintainability Symposium*, 1976, IEEE, New York, pp. 213–217.

Jorgenson, J. L. and Goldberg, J. E., "Probability of Plastic Collapse Failure," *ASCE Journal of the Structural Division*, Vol. 95, No. ST8, Aug. 1969, pp. 1743–1761.

Kafka, P., "Some Thoughts on the Interaction between System and Structural Reliability," *Nuclear Engineering and Design*, Vol. 60, 1980, pp. 129–132.

Kafka, P. and Augustin, W., "Magnitude and Probability of Internal Load Behavior, Pt. I," *Nuclear Engineering and Design*, Vol. 48, 1978, pp. 565–574.

Kalaba, R., "Design of Minimal-Weight Analysis Based on Structural Reliability and Cost," *Journal of the Aerospace Sciences*, Vol. 27, No. 9, 1960, pp. 641–652.

Kam, T. Y., Corotis, R. B., and Rossow, E. C., "Reliability of Nonlinear Framed Structures," *Journal of Structural Engineering*, Vol. 109, No. 7, July 1983, pp. 1585–1601.

Kameda, H. and Koike, T., "Reliability Theory of Deteriorating Structures," *ASCE Journal of the Structural Division*, Vol. 101, No. ST1, Jan. 1975, pp. 295–310.

Kamil, H., "A Computer Oriented Deterministic-Cum-Probabilistic Approach For the Extreme Load Design of Complex Structures," *Computers and Structures*, Vol. 6, 1976, pp. 375–379.

Kao, S. C. and Shooman, M. L., "Probabilistic Approaches to the Combination of Loads in Structural Design," *Proceedings of Annual Reliability and Maintainability Symposium*, IEEE, New York, 1981, pp. 28–32.

Kapur, K. C. and Lamberson, L. R., *Reliability in Engineering Design*, Wiley, New York, 1977, pp. 164–212.

Karnopp, D. and Valls, F. C., "Structural Reliability Predictions Using Finite Element Programs," *Proceedings of Tenth Reliability and Maintainability Conference* (*Annals of Reliability and Maintainability*, Vol. 10), ASME, New York, 1971, pp. 72–80.

Karshenas, S. and Ang. A. H. S., "A Structural Safety Analysis of Steel Buildings during Construction," *Structural Safety*, Vol. 1, Sept. 1983, pp. 239–255.

Kececioglu, D. and Lamarre, G., "Designing Mechanical Components to a Specified Reliability and Confidence Level," *Nuclear Engineering and Design*, Vol. 50, 1978, pp. 149–162.

Kececioglu, D., Smith, R. E., and Felsted, E. A., "Distributions of Strength in Simple Fatigue and the Associated Reliabilities," *Proceedings of Ninth Reliability and Maintainability Conference* (*Annals of Reliability and Maintainability*, Vol. 9), SAE, Warrendale, PA, July 1970, pp. 659–672.

Kellerman, O., "Some Aspects of Probabilistic Structural Reliability to Meet the Needs of Risk Analysis of Nuclear Power Plants," *Nuclear Engineering and Design*, Vol. 60, 1980, pp. 157–158.

Kinkead, A. N., "Some Aspects of Redundancy in Relationship to Structural Reliability," *Reliability Engineering*, Vol. 3, 1982, pp. 363–377.

Kiremidjian, A. A., "Reliability of Structures Subjected to Differential Fault Slip," *Earthquake Engineering and Structural Dynamics*, Vol. 12, 1984, pp. 603–618.

Kjerengtroen, L. and Wirshing, P. H., "Structural Reliability Analysis of Series Systems," *Journal of Structural Engineering*, Vol. 110, No. 7, July 1984, pp. 1495–1511.

Kluger, P., "Evaluation of the Effects of Manufacturing Processes on Structural Design Reliability," *Proceedings of the 10th National Symposium on Reliability and Quality Control*, IEEE, New York, Jan. 1964, pp. 538–544.

Lance, J. R., Mutone, G. A., and Bjoro, E. F., "Safety Analysis of an Advanced Energy System Facility," *Proceedings of Annual Reliability and Maintainability Symposium*, IEEE, New York, 1980, pp. 268–274.

Larabee, R. and Cornell, C. A. (eds.), *Probabilistic Mechanics and Structural Reliability*, ASCE, New York, 1979.

Lee, H. L. and Mays, L., "Improved Risk and Reliability Model for Hydraulic Structures," *Water Resources Research*, Vol. 19, No. 6, Dec. 1983, pp. 1415–1422.

Lee, L. T. and Collins, J. D., "Engineering Risk Management For Structures," *ASCE Journal of the Structural Division*, Vol. 103, No. ST9, Sept. 1977, pp. 1739–1756.

Leipholz, H. H., Topper, T., and Menoufy, M. E., "Lifetime Prediction for Metallic Components Subjected to Stochastic Loading," *Computers and Structures*, Vol. 16, 1983, pp. 499–507.

Lemon, G. H. and Manning, S. D., "Literature Survey on Structural Reliability," *IEEE Transactions on Reliability*, Vol. R-23, No. 4, Oct. 1974, pp. 263–266.

Lenz, J., Ravindra, M. K., and Galambos, T. V., "Reliability Based Design Rules for Column Buckling," *Computers and Structures*, Vol. 3, 1973, pp. 573–588.

Leporati, E., *The Assessment of Structural Safety*, Research Studies Press, New York, 1979.

Leve, H. L., "Reliability Framework for Structural Design," *ASCE Journal of the Structural Division*, Vol. 89, No. ST4, Aug. 1963, pp. 315–332.

Lie, T. T., "Optimum Fire Resistance of Structures," *ASCE Journal of the Structural Division*, Vol. 98, No. ST1, Jan. 1972, pp. 215–232.

Lie, T. T., "Safety Factors for Fire Loads," *Canadian Journal of Civil Engineering*, Vol. 6, 1979, pp. 617–628.

Lin, T. S. and Nowak, A. S., "Proof Loading and Structural Reliability," *Reliability Engineering*, Vol. 8, 1984, pp. 85–100.

Lin, Y. K., *Probabilistic Theory of Structural Dynamics*, McGraw-Hill, New York, 1967.

Lind, N. C., "Approximate Analysis and Economics of Structures," *ASCE Journal of the Structural Division*, Vol. 102, No. ST6, June 1976, pp. 1177–1196.

Lind, N. C., "Consistent Partial Safety Factors," *ASCE Journal of the Structural Division*, Vol. 97, No. ST6, June 1971, pp. 1651–1669.

Lind, N. C. (ed.), *Structural Reliability and Codified Design*, Solid Mechanics Div., University of Waterloo Press, Waterloo, Ontario, Canada, 1970.

Lind, N. C., "Formulation of Probabilistic Design," *ASCE Journal of the Engineering Mechanics Division*, Vol. 103, No. EM2, April 1977, pp. 273–284.

Lind, N. C., "Models of Human Error in Structural Reliability," *Structural Safety*, Vol. 1, No. 3, 1983, pp. 167–175.

Lind, N. C., "Optimal Reliability Analysis by Fast Convolution," *ASCE Journal of the Engineering Mechanics Division*, Vol. 105, No. EM3, June 1979, pp. 447–452.

Longinow, A., Chu, K. -H., and Thomopoulos, N. T., "Probability of Survival in Blast Environment," *ASCE Journal of the Engineering Mechanics Division*, Vol. 108, No. EM2, April 1982, pp. 309–330.

Lutes, L. D. and Tzuang, S. H., "First-Passage Probability for Two-Mode Systems," *Journal of Engineering Mechanics*, Vol. 109, No. 6, Dec. 1983, pp. 1358–1374.

MacGregor, J. G., "Safety and Limit States Design for Reinforced Concrete," *Canadian Journal of Civil Engineering*, Vol. 3, 1976, pp. 484–513.

Madsen, H. O., "First Order vs Second Order Reliability Analysis of Series Structures," *Structural Safety*, Vol. 2, Jan. 1985, pp. 207–214.

Madsen, P. H. and Lind, N. C., "Bayesian Approach to Prototype Testing," *ASCE Journal of the Structural Division*, Vol. 108, No. ST4, April 1982, pp. 753–770.

Mancini, G. and Volta, G., "Structural Reliability in the Continuous Feedback Process for Model Validation and Regulations Improvement," *Nuclear Engineering and Design*, Vol. 60, 1980, pp. 125–128.

Manning, S. D. and Lemon, G. H., "Plan for Developing Structural Criteria for Composite Airframes," *Proceedings of Annual Reliability and Maintainability Symposium*, IEEE, New York, 1974, pp. 155–162.

Marshall, P. W., "Risk Evaluations For Offshore Structures," *ASCE Journal of the Structural Division*, Vol. 95, No. ST12, Dec. 1969, pp. 2907–2929.

Martindale, S. G. and Wirsching, P. H., "Reliability-Based Progressive Fatigue Collapse," *Journal of Structural Engineering*, Vol. 109, No. 8, Aug. 1983, pp. 1792–1811.

Maurer, H. A., "A Note on the Application of Probabilistic Structural Reliability Methodology to Nuclear Power Plants," *Nuclear Engineering and Design*, Vol. 50, 1978, pp. 213–215.

Mazumdar, M., Marshall, J. A., and Chay, S. C., "Propagation of Uncertainties in Problems of Structural Reliability," *Nuclear Engineering and Design*, Vol. 50, 1978, pp. 163–167.

McMartin, K. C., Quaile, A. T., and Keenan, F. J., "Strength and Structural Safety of Long-Span Light Wood Roof Trusses," *Canadian Journal of Civil Engineering*, Vol. 11, 1984, pp. 978–992.

Melchers, R. E., "Human Error in Structural Reliability Assessments," *Reliability Engineering*, Vol. 7, 1984, pp. 61–75.

Melchers, R. E., "Reliability of Parallel Structural Systems," *Journal of Structural Engineering*, Vol. 109, Nov. 1983, pp. 2651–2665.

Melchers, R. E. and Tang, L. K., "Dominant Failure Modes in Stochastic Structural Systems," *Structural Safety*, Vol. 2, Oct. 1984, pp. 127–143.

Meyerhof, G. G., "Limit States Design in Geotechnical Engineering," *Structural Safety*, Vol. 1, Sept, 1982, pp. 67–71.

Milford, R. V., "Structural Reliability and Cross Wind Response of Tall Chimneys," *Engineering Structures*, Vol. 4, Oct. 1982, pp. 263–270.

Mittenbergs, A. A., "The Materials Problem in Structural Reliability," *Proceedings of Fifth Reliability and Maintainability Conference* (*Annals of Reliability and Maintainability*, Vol. 5), July 1966, pp. 148–158.

Mjelde, K. M., "Inspection-Optimization for Serial Systems of Dependent Elements," *Structural Safety*, Vol. 2, Oct. 1984, pp. 119–125.

Mjelde, K. M., "Reliability Bounds for Serial Systems," *Journal of Structural Mechanics*, Vol. 12, No. 1, 1984, pp. 79–85.

Moan, T. and Shinozuka, M. (eds.), *Structural Safety and Reliability, Proceedings of ICOSSAR '81*, Elsevier Scientific Publishing, New York, 1981.

Moghtaderizadeh, M. and Kiureghian, A. D., "Reliability Upgrading of Lifeline Networks for Post-Earthquake Serviceability," *Earthquake Engineering and Structural Dynamics*, Vol. 11, 1983, pp. 557–566.

Monnier, P. Y. and Schmalz, P., "Probability of Failure of a Box Beam Using Simulation Techniques," *Canadian Journal of Civil Engineering*, Vol. 8, 1981, pp. 22–30.

Moses, F., "Design of Reliability—Concept and Applications," *Optimum Structural Design Theory and Applications*, edited by R. H. Gallagher and O. C. Zienkiewitz, Wiley, New York, 1973, pp. 241–265.

Moses, F., "Reliability of Structural Systems," *ASCE Journal of the Structural Division*, Vol. 100, No. ST9, Sept. 1974, pp. 1813–1820.

Moses, F., "Reliability of Structural Systems," *Proceedings of International Conference on Behavior of Offshore Structures (BOSS '76)*, Norwegian Institute of Technology, Trondheim, Aug. 1976, Vol. 1, pp. 912–923.

Moses, F., "Structural System Reliability and Optimization," *Computers and Structures*, Vol. 7, 1977, pp. 283–290.

Moses, F., "System Reliability Developments in Structural Engineering," *Structural Safety*, Vol. 1, Sept. 1982, pp. 3–13.

Moses, F. and Kinser, D. E., "Analysis of Structural Reliability," *AIAA Journal*, Vol. 5, June 1967, pp. 1152–1158.

Moses, F. and Kinser, D. E., "Analysis of Structural Reliability," *ASCE Journal of the Structural Division*, Vol. 93, No. ST5, Oct. 1967, pp. 147–164.

Moses, F. and Stahl, B., "Reliability Analysis Format for Offshore Structures," *Journal of Petroleum Technology*, Vol. 31, March 1979, pp. 347–354.

Moses, F. and Stevenson, J. D., "Reliability-Based Structural Design," *ASCE Journal of the Structural Division*, Vol. 96, No. ST2, Feb. 1970, pp. 221–244.

Mozer, J. D., Peyrot, A. H., and Digioia, A. M., "Probabilistic Design of Transmission Line Structures," *Journal of Structural Engineering*, Vol. 110, Oct. 1984, pp. 2513–2528.

Murotsu, Y., Okada, H., Niwa, K., and Mura, S., "Reliability Analysis of Truss Structures by Using a Matrix Method," *Transactions of ASME, Journal of Mechanical Design*, Vol. 102, No. 4, 1980, pp. 749–756.

Murotsu, Y., Okada, H., Taguchi, K., Grimmelt, M., and Yonezawa, M., "Automatic Generation of Stochastically Dominant Failure Modes of Frame Structures," *Structural Safety*, Vol. 2, May 1984, pp. 17–25.

Murotsu, Y., Yonezawa, M., Oba, F., and Niwa, K., "Optimum Design Problems in Reliability-Based Structural Design," *Proceedings—HOPE International Symposium*, Japan Society of Mechanical Engineers, Tokyo, Nov. 1977, pp. 461–466.

Murthy, P. M. and Subramaniam, G., "Minimum Weight Analysis Based on Structural Reliability," *AIAA Journal*, Vol. 10, 1968, pp. 2037–2038.

Nowak, A. S. and Boutros, M. K., "Probabilistic Analysis of Timber Bridge Decks," *Journal of Structural Engineering*, Vol. 110, Dec. 1984, pp. 2939–2954.

Nowak, A. S. and Regupathy, P. V., "Reliability of Spot Welds in Cold-Formed Channels," *Journal of Structural Engineering*, Vol. 110, No. 6, June 1984, pp. 1265–1277.

O'Neill, R., "Future Developments of Probabilistic Structural Reliability," *Nuclear Engineering and Design*, Vol. 60, 1980, pp. 151–156.

Paez, T. L., Tang, J.-P., and Yao, J. T. P., "Reliability of Maintainable Structures," *Proceedings of Annual Reliability and Maintainability Symposium*, IEEE, New York, 1973, pp. 17–24.

Paez, T. L. and Yao, J. P., "Probabilistic Analysis of Elastoplastic Structures," *ASCE Journal of the Engineering Mechanics Division*, Vol. 102, No. EM1, Feb. 1976, pp. 105–120.

Paloheimo, E. and Hannus, M., "Structural Design Based on Weighted Fractiles," *ASCE Journal of the Structural Division*, Vol. 100, No. ST7, July 1974, pp. 1367–1378.

Parkinson, D. B., "Alternative Reliability Formats," *Engineering Structures*, Vol. 5, July 1983, pp. 207–214.

Parkinson, D. B., "Computer Solution for the Reliability Index," *Engineering Structures*, Vol. 2, Jan. 1980, pp. 57–62.

Parkinson, D. B., "First-Order Reliability Analysis Employing Translation Systems," *Engineering Structures*, Vol. 1, Oct. 1978, pp. 31–40.

Parkinson, D. B., "Solution For Second Moment Reliability Index," *ASCE Journal of the Engineering Mechanics Division*, Vol. 104, No. EM5, Oct. 1978, pp. 1267–1275.

Pearce, H. T. and Wen, Y. K., "Stochastic Combination of Load Effects," *Journal of Structural Engineering*, Vol. 110, July 1984, pp. 1613–1629.

Peres, J. E. C. S. and Rogerson, J. H., "The Use of Probabilistic Fracture Mechanics in Devising Quality Control Policies in the Fabrication Industry," *Reliability Engineering*, Vol. 8, 1984, pp. 149–164.

Peyrot, A. H., Criswell, M. E., Folse, M. D., and Aznavour, J.-P., "Reliability Analysis of Wood Transmission Poles," *ASCE Journal of the Structural Division*, Vol. 108, No. ST9, Sept. 1982, pp. 1981–1994.

Peyrot, A. H. and Dagher, H. J., "Reliability-Based Design of Transmission Lines," *Journal of Structural Engineering*, Vol. 110, Nov. 1984, pp. 2758–2777.

Pickup, A., "A Method for Evaluating the Reliability of Complex Multiple Redundant Systems," *Reliability Engineering*, Vol. 5, 1983, pp. 53–62.

Proceedings of 3rd International Conference on Applications of Statistics & Probability in Soil and Structural Engineering, Unisearch Ltd., Kensington, NSW, Australia, Jan.–Feb. 1979.

Proceedings of 4th International Conference on Applications of Statistics & Probability in Soil and Structural Engineering, Pitagora Editrice, Bologna, Italy, June 1983.

Proceedings of ASCE Speciality Conference on Probabilistic Mechanics and Structural Reliability, ASCE, New York, Jan. 1979.

Proceedings of 2nd International Conference on Structural Safety and Reliability, International Association for Structural Safety and Reliability, New York, Sept. 1977.

Prucz, Z. and Soong, T. T., "Reliability and Safety of Tension Leg Platforms," *Engineering Structures*, Vol. 6, April 1984, pp. 142–149.

Pugsley, A. G., *The Safety of Structures*, Edward Arnold, London, 1966.

Rackwitz, R., "Failure Rates for General Systems Including Structural Components," *Reliability Engineering*, Vol. 9. 1984, pp. 229–242.

Rackwitz, R. and Fiessler, B., "Structural Reliability under Combined Random Load Sequences," *Computers and Structures*, Vol. 9, 1978, pp. 489–494.

Rachwitz, R. and Schrupp, K., "Quality Control, Proof Testing and Structural Reliability," *Structural Safety*, Vol. 2, Jan. 1985, pp. 239–244.

Rao, C. V. S. K., "Safety of Glass Panels against Wind Loads," *Engineering Structures*, Vol. 6, July 1984, pp. 232–234.

Rao, S. S., "Reliability-Based Optimization under Random Vibration Environment," *Computers and Structures*, Vol. 14, No. 5-6, 1981, pp. 345–355.

Ravindra, M. K., "Probability-Based Design Criteria for Nuclear Structures," *Nuclear Engineering and Design*, Vol. 59, 1980, pp. 197–204.

Ravindra, M. K., Lind N. C., and Siu, W., "Illustrations of Reliability-Based Design," *ASCE Journal of the Structural Division*, Vol. 100, No. ST9, Sept. 1974, pp. 1789–1811.

Ravindra, M. K. and Walser, A., "Probabilistic Design of Nuclear Structures: A Summary of State of the Art and Research Needs," *Nuclear Engineering and Design*, Vol. 50, 1978, pp. 115–122.

Reemsnyder, H. S., "Procurement and Analysis of Structural Fatigue Data," *ASCE Journal of the Structural Division*, Vol. 95, No. ST7, July 1969, pp. 1533–1551.

Reese, R. C., "Probabilistic Approaches to Structural Safety," *Journal of the American Concrete Institute*, Vol. 79, Jan. 1976, pp. 37–49.

Reich, R. and Hwang, H., "Probability-Based Load Combinations for Design of Category I Structures—Overview of Research Program and Recent Results," *Nuclear Engineering and Design*, Vol. 79, 1984, pp. 129–135.

Rojiani, K. B. and Wen, Y.-K., "Effect of Estimation and Sampling Errors in Distributions of Extreme Winds on Structural Reliability," *Engineering Structures*, Vol. 2, Oct. 1980, pp. 237–243.

Rojiani, K. and Wen, Y.-K., "Reliability of Steel Buildings under Winds," *ASCE Journal of the Structural Division*, Vol. 107, No. ST1, Jan. 1981, pp. 203–221.

Rosenblueth, E., "Probabilistic Design to Resist Earthquakes," *ASCE Journal of the Engineering Mechanics Division*, Vol. 90, No. EM5, Oct. 1964, pp. 189–220.

Rosenblueth, E. and Mendoza, E., "Reliability Optimization in Isostatic Structures," *ASCE Journal of the Engineering Mechanics Division*, Vol. 97, No. EM6, Dec. 1971, pp. 1625–1642.

Rowe, R. E., "Current European Views on Structural Safety," *ASCE Journal of the Structural Division*, Vol. 96, No. ST3, March 1970, pp. 461–467.

Ruiz, S. E., "Reliability Index for Offshore Piles Subjected To Bending," *Structural Safety*, Vol. 2, Oct. 1984, pp. 83–90.

Saether, K., Sandberg, H. R., and Beck, C. F., "Reliability of Computer-Aided Design," *Journal of the Technical Councils of ASCE*, Vol. 105, No. TC2, Dec. 1979, pp. 371–384.

Schmitt, W., "Some Aspects of Systems and Structural Reliability," *Nuclear Engineering and Design*, Vol. 60, 1980, pp. 137–138.

Schnurer, H., "Future Developments of Probabilistic Structural Reliability to Meet the Needs of Risk Analysis of Nuclear Power Plants," *Nuclear Engineering and Design*, Vol. 60, 1980, pp. 145–149.

Schueller, G. I., "Impact of Probability Risk Assessment on Containment," *Nuclear Engineering and Design*, Vol. 80, 1984, pp. 203–216.

Schueller, G. I., "On the Structural Reliability of Reactor Safety Containments," *Nuclear Engineering and Design*, Vol. 27, 1974, pp. 426–433.

Schueller, G. I. and Kafka, P., "Future of Structural Reliability Methodology in Nuclear Power Plant Technology," *Nuclear Engineering and Design*, Vol. 50, 1978, pp. 201–205.

Schueller, G. I. and Schwarz, R. F., "Some Aspects of the Reliability-Based Design of Reactor Containment Structures," *Nuclear Engineering and Design*, Vol. 37, 1976, pp. 299–305.

Seguchi, Y., Iwatsubo, T., and Gen, M., "A Reliability Approach to Limit Analysis and Design of Shells," *Proceedings—HOPE International Symposium*, Japan Society of Mechanical Engineers, Tokyo, Nov. 1977, pp. 453–459.

Shah, H. C. and Tang, W. H., "Statistical Evaluation of Load Factors in Structural Design," *Proceedings of Ninth Reliability and Maintainability Conference* (*Annals of Reliability and Maintainability*, Vol. 9), SAE, Warrendale, PA, July 1970, pp. 650–658.

Shinozuka, M., "Basic Analysis of Structural Safety," *Journal of Structural Engineering*, Vol. 109, March 1983, pp. 721–740.

Shinozuka, M., "Probability of Structural Failure under Random Loading," *ASCE Journal of the Engineering Mechanics Division*, Vol. 90, No. EM5, Oct. 1964, pp. 147–170.

Shinozuka, M., "Safety Against Dynamic Forces," *ASCE Journal of the Structural Division*, Vol. 100, No. ST9, Sept. 1974, pp. 1821–1826.

Shinozuka, M., Hwang, H. and Reich, M., "Reliability Assessment of Reinforced Concrete Containment Structures," *Nuclear Engineering and Design*, Vol. 80, 1984, pp. 247–267.

Shinozuka, M. and Itagaki, H., "On the Reliability of Redundant Structures," *Proceedings of Fifth Reliability and Maintainability Conference* (*Annals of Reliability and Maintainability*, Vol. 5), AIAA, Washington, DC, July 1966, pp. 605–610.

Shinozuka, M. and Hanai, M., "Structural Reliability of a Simple Rigid Frame," *Proceedings of Sixth Reliability and Maintainability Conference* (*Annals of Reliability and Maintainability*, Vol. 6), SAE, Warrendale, PA, July 1967, pp. 63–67.

Shinozuka, M. and Shao, L. C., "Basic Probabilistic Considerations on Safety of Prestressed Concrete Pressure Vessels," *Nuclear Engineering and Design*, Vol. 29, 1974, pp. 338–345.

Shinozuka, M. and Yang, J-N., "On the Bound of First Excursion Probability," *ASCE Journal of the Engineering Mechanics Division*, Vol. 95, No. EM2, April 1969, pp. 363–377.

Shinozuka M. and Yang, J-N., "Optimum Structural Design Based on Reliability and Proof-Load Test," *Proceedings of Eighth Reliability and Maintainability Conference (1969 Annals of Assurances Sciences)*, Gordon and Breach Science Publishers, New York, July 1969, pp. 375–391.

Shiraishi, N. and Furuta, H., "Reliability Analysis Based on Fuzzy Probability," *ASCE Journal of Engineering Mechanics*, Vol. 109, No. 6, Dec. 1983, pp. 1445–1459.

Shiraishi, N., Furuta, H., and Sugimoto, M., "Effects of Gross Errors on Structural Design," *Structural Safety*, Vol. 2, Jan. 1985, pp. 245–250.

Simiu, E., "Aerodynamic Coefficients and Risk-Consistent Design," *ASCE Journal of Structural Engineering*, Vol. 109, May 1983, pp. 1278–1289.

Simiu, E., "Wind Climate and Failure Risks," *ASCE Journal of the Structural Division*, Vol. 102, No. ST9, Sept. 1976, pp. 1703–1707.

Spanos, P. T. D. and Solomos, G. P., "Markov Approximation to Transient Vibration," *Journal of Engineering Mechanics*, Vol. 109, No. 4, Aug. 1983, pp. 1134–1150.

Spidsoe, N. and Sigbjornsson, R., "On the Reliability of Standard Wave Spectra in Structural Response Analysis," *Engineering Structures*, Vol. 2, April 1980, pp. 123–135.

Springer, W. T., Lawrence, K. L., and Lawley, T. J., "Vibration Based NDE for Structural Components," *Proceedings of Annual Reliability and Maintainability Symposium*, IEEE, New York, 1982, pp. 51–55.

Stahl, B. and Geyer, J. F., "Fatigue Reliability of Parallel Member Systems," *Journal of Structural Engineering*, Vol. 110, Oct. 1984, pp. 2307–2323.

Stancampiano, P. A., "Simulation Approaches to Probabilistic Structural Design at the Component Level," *Nuclear Engineering and Design*, Vol. 50, 1978, pp. 123–129.

Stancampiano, P. A., "Some Thoughts on the Future of Probabilistic Structural Design of Nuclear Components," *Nuclear Engineering and Design*, Vol. 50, 1978, pp. 207–211.

Sterman, B. P. and Schofer, J. L., "Factors Affecting Reliability of Urban Bus Services," *Transportation of Engineering Journal*, Vol. 102, No. TE1, Feb. 1976, pp. 147–159.

Stevenson, J. and Moses, F., "Reliability Analysis of Frame Structures," *ASCE Journal of the Structural Division*, Vol. 96, No. ST11, Nov. 1970, pp. 2409–2427.

Surahman, A. and Rojiani, K. B., "Reliability Based Optimum Design of Concrete Frames," *Journal of Structural Engineering*, Vol. 109, No. 3, March 1983, pp. 741–757.

Switzky, H., "Minimum Weight Design with Structural Reliability," *Proceedings of AIAA 5th Annual Structures and Materials Conference*, AIAA, Washington, DC, 1964, pp. 316–322.

Takaoka, N. and Shiraki, W., "Reliability Analysis of Structural Members Composed of Several Random Components, Making Use of Theory of Random Processes," *Journal of Structural Mechanics*, Vol. 12, No. 2, 1984, pp. 151–180.

Tang, J-P. and Yao, J. T. P., "Fatigue Damage Factor in Structural Design," *ASCE Journal of the Structural Division*, Vol. 98, No. ST1, Jan. 1972, pp. 125–134.

Thoft-Christensen, P. (ed.), *Fundamentals of Structural Reliability, Lectures on Structural Reliability*, Aalborg University Center, Aalborg, Denmark, 1980.

Thoft-Christensen, P. (ed.), *Reliability Theory and Its Application in Structural and Soil Mechanics*, Martinus Nijhoff Publishers, The Hague, 1983.

Thoft-Christensen, P. and Baker M. J., *Structural Reliability Theory and Its Applications*, Springer-Verlag, Berlin, 1982.

Thoft-Christensen, P. and Nielsen, S. R. K., "Bounds on the Probability of Failure in Random Vibration," *Journal of Structural Mechanics*, Vol. 10, No. 1, 1982, pp. 67–91.

Thoft-Christensen, P. and Sorensen, J. D., "Reliability of Structural Systems with Correlated Elements," *Applied Mathematical Modelling*, Vol. 6, June 1982, pp. 171–178.

Thomas, J. M., Hanagud, S., and Hawk, J. D., "Decision Theory in Structural Reliability," *Proceedings of Annual Reliability and Maintainability Symposium*, IEEE, New York, 1975, pp. 255–262.

Thomopoulos, N. T. and Longinow, A., "Bivariate Lognormal Probability Distribution," *Journal of Structural Engineering*, Vol. 110, No. 12, Dec. 1984, pp. 3045–3049.

Tung, C. C., "Life Expectancy of Highway Bridges to Vehicle Loads," *ASCE Journal of the Engineering Mechanics Division*, Vol. 95, No. EM6, Dec. 1969, pp. 1417–1428.

Turkstra, C. J., "Choice of Failure Probabilities," *ASCE Journal of the Structural Division*, Vol. 93, No. ST6, Dec. 1967, pp. 189–200.

Turkstra, C. J. and Daly, M. J., "Two Moment Structural Safety Analysis," *Canadian Journal of Civil Engineering*, Vol. 5, 1978, pp. 414–426.

Twisdale, L. A., "Probabilistic Analysis of Tornado Wind Risks," *Journal of Structural Engineering*, Vol. 109, No. 2, Feb. 1983, pp. 468–488.

Vanmarcke, E. H., "Matrix Formulation of Reliability Analysis and Reliability-Based Design," *Computers and Structures*, Vol. 3, 1971, pp. 757–770.

Vanmarcke, E. H. and Angelides, D., "Risk Assessment for Offshore Structures: A Review," *Journal of Structural Engineering*, Vol. 109, No. 2, Feb. 1983, pp. 555–571.

Veneziano, D., "New Index of Reliability," *ASCE Journal of the Engineering Mechanics Division*, Vol. 105, No. EM2, April 1979, pp. 277–296.

Veneziano, D., "Proof Loading For Target Reliability," *ASCE Journal of the Structural Division*, Vol. 104, No. ST1, Jan. 1978, pp. 79–93.

Vinnem, J. E., "Quantitative Risk Analysis in the Design of Offshore Installations," *Reliability Engineering*, Vol. 6, 1983, pp. 1–12.

Warner, R. F. and Kabaila, A. P., "Monte Carlo Study of Structural Safety," *ASCE Journal of the Structural Division*, Vol. 94, No. ST12, Dec. 1968, pp. 2847–2859.

Webster, F., "Probabilistic Analysis of a Simple Portal Structure," *Journal of the American Concrete Institute*, Vol. 70, Sept. 1973, pp. 649–651.

Webster, F. A., "Reliability of Multistory Slab Structures against Progressive Collapse During Construction," *Journal of the American Concrete Institute*, Vol. 77, Nov.–Dec. 1980, pp. 449–457.

Wen, Y. K., "Methods for Reliability of Structures under Multiple Time Varying Loads," *Nuclear Engineering and Design*, Vol. 60, 1980, pp. 61–71.

Wen, Y. K., "Statistics of Extreme of Live Load on Buildings," *ASCE Journal of the Structural Division*, Vol. 105, No. ST10, Oct. 1979, pp. 1893–1900.

Wen, Y. K., "Wind Direction and Structural Reliability," *Journal of Structural Engineering*, Vol. 109, April 1983, pp. 1028–1041.

Wen, Y. K., "Wind Direction and Structural Reliability II," *Journal of Structural Engineering*, Vol. 110, June 1984, pp. 1253–1264.

Winterstein, S. R. and Cornell, C. A., "Load Combinations and Clustering Effects," *Journal of Structural Engineering*, Vol. 110, Nov. 1984, pp. 2690–2708.

Wirsching, P. H., "Fatigue Reliability for Offshore Structures," *Journal of Structural Engineering*, Vol. 110, Oct. 1984, pp. 2340–2356.

Wirsching, P. H., Stahl, B., and Nolte, K. G., "Probabilistic Fatigue Design for Ocean Structures," *ASCE Journal of the Structural Division*, Vol. 103, No. ST10, Oct. 1977, pp. 2049–2062.

Wirsching, P. H. and Haugen, E. B., "Probabilistic Design For Random Fatigue Loads," *ASCE Journal of the Engineering Mechanics Division*, Vol. 99, No. EM6, Dec. 1973, pp. 1165–1179.

Wirsching, P. H. and Yao, J. T. P., "Statistical Methods in Structural Fatigue," *ASCE Journal of the Structural Division*, Vol. 96, No. ST6, June 1970, pp. 1201–1219.

Wu, Y. T. and Wirsching, P. H., "Advanced Reliability Method for Fatigue Analysis," *Journal of Engineering Mechanics*, Vol. 110, April 1984, pp. 536–553.

Yamaguchi, A., Kondo, S., and Togo, Y., "A Case Study on Effectiveness of Structural Reliability Analysis in Nuclear Reactor Safety Assessment," *Reliability Engineering*, Vol. 5, 1983, pp. 21–36.

Yang, J. N., "Optimal Periodic Proof Tests Based on Cost-Effective and Reliability Criteria," *AIAA Journal*, Vol. 15, March 1977, pp. 402–409.

Yang, J. N., "Reliability Analysis of Structures under Periodic Proof Tests in Service," *AIAA Journal*, Vol. 14, Sept. 1976, pp. 1225–1234.

Yang, J. N. and Heer, E., "Reliability of Randomly Excited Structures," *AIAA Journal*, Vol. 9, July 1971, pp. 1262–1268.

Yao, J. T. P., "A General Note on Structural Reliability," *Nuclear Engineering and Design*, Vol. 50, 1978, pp. 145–147.

Yao, J. T. P., "Damage Assessment and Reliability Evaluation of Existing Structures," *Engineering Sturctures*, Vol. 1, Oct. 1979, pp. 245–251.

Yao, J. T. P., "Damage Evaluation for Structural Reliability Assessment," *Nuclear Engineering and Design*, Vol. 75, 1982, pp. 205–212.

Yao, J. T. P., "Fatigue Reliability and Design," *ASCE Journal of the Structural Division*, Vol. 100, No. ST9, Sept. 1974, pp. 1827–1837.

Yao, J. T. P., *Safety and Reliability of Existing Structures*, Pitman Advanced Publishing Program, Boston, 1985.

Yao, J. T. P. and Yeh, H-Y., "Formulation of Structural Reliability," *ASCE Journal of the Structural Division*, Vol. 95, No. ST12, Dec. 1969, pp. 2611–2619.

Yao, J. T. P. and Yeh, H-Y., "Safety Analysis of Statistically Indeterminate Trusses," *Proceedings of Sixth Reliability and Maintainability Conference* (*Annals of Reliability and Maintainability*, Vol. 6), SAE, Warrendale, PA, 1967, pp. 54–62.

Zahn, J. J., "Reliability-Based Design Procedures for Wood Structures," *Forest Products Journal*, Vol. 27, No. 3, 1977, pp. 21–28.

Zech, B. and Wittmann, F. H., "Probabilistic Approach to Describe the Behavior of Materials—Part II," *Nuclear Engineering and Design*, Vol. 48, 1978, pp. 575–584.

Zemanick, P. P., "Evolution of Structural Reliability Methodology as Applied in Major Nuclear Risk Assessments," *Nuclear Engineering and Design*, Vol. 60, 1980, pp. 163–168.

Zemanick, P. P. and Witt, F. J., "An Engineering Assessment of Probabilistic Structural Reliability Analyses," *Nuclear Engineering and Design*, Vol. 50, 1978, pp. 173–183.

Index